Latin America

An Interpretive History

Latin America

An Interpretive History

Tenth Edition

Julie A. Charlip
Whitman College

E. Bradford Burns
University of California, Los Angeles

PEARSON

Boston Columbus Indianapolis New York San Francisco Upper Saddle River Amsterdam
Cape Town Dubai London Madrid Milan Munich Paris Montreal Toronto Delhi
Mexico City Sao Paulo Sydney Hong Kong Seoul Singapore Taipei Tokyo

Editor in Chief: *Ashley Dodge*
Program Team Lead: *Amber Mackey*
Managing Editor: *Sutapa Mukherjee*
Program Manager: *Carly Czech*
Sponsoring Editor: *Tanimaa Mehra*
Editorial Project Manager: *Lindsay Bethoney, Lumina Datamatics, Inc.*
Editorial Assistant: *Casseia Lewis*
Director, Content Strategy and Development: *Brita Nordin*
VP, Director of Marketing: *Maggie Moylan*
Director of Field Marketing: *Jonathan Cottrell*
Senior Marketing Coordinator: *Susan Osterlitz*
Director, Project Management Services: *Lisa Iarkowski*
Print Project Team Lead: *Melissa Feimer*
Operations Manager: *Mary Fischer*
Operations Specialist: *Carol Melville / Mary Ann Gloriande*
Associate Director of Design: *Blair Brown*
Interior Design: *Kathryn Foot*
Cover Art Director: *Maria Lange*
Cover Design: *Lumina Datamatics, Inc.*
Cover Art: *fotolia*
Digital Studio Team Lead: *Peggy Bliss*
Digital Studio Project Manager: *Liz Roden Hall*
Digital Studio Project Manager: *Elissa Senra-Sargent*
Full-Service Project Management and Composition: *Murugesh Rajkumar Namasivayam, Lumina Datamatics, Inc.*
Printer/Binder: *RRD-Crawfordsville*
Cover Printer: *Phoenix Color*

Acknowledgements of third party content appear on page 364, which constitutes an extension of this copyright page.

Library of Congress Cataloging-in-Publication Data

Charlip, Julie A., 1954-
Latin America : an interpretive history / Julie A. Charlip (Whitman College), E. Bradford Burns (University of California, Los Angeles).—Tenth edition.
pages cm
Includes bibliographical references and index.
ISBN 13: 978-0-13-374582-5 (alkaline paper)—ISBN 10: 0-13-374582-1 (alkaline paper)
1. Latin America—History. 2. Latin America—Social conditions. 3. Latin America—Economic conditions. I. Burns, E. Bradford. II. Title.
F1410.B8 2017
980—dc23

2015028348

10 9 8 7 6 5 4 3 2 1

ISBN 10: 0-13-374582-1
ISBN 13: 978-0-13-374582-5

For our students

Brief Contents

Contents

Maps

Preface

I have always endeavored to substantially update each edition of *Latin America: An Interpretive History*. The tenth edition offers the greatest change to date—the addition of a new chapter. In the ninth edition, I looked at the "Pink Tide," the swing left in Latin America that began at the end of the twentieth century. Now fifteen years into the twenty-first century, it is clear that this change was not a brief moment. Latin America has changed profoundly in this century, and those changes deserve a closer look. These changes have brought a new and expanded prosperity for more of the population. Unfortunately, these changes have not diminished the central paradigm of the text: "poor people inhabit rich lands" because the elites have "tended to confuse their own well being and desires with those of the nation at large."

New to this Edition

- Chapter 11, now titled "The Limits of Liberalism," has been revised to focus only on the late twentieth century, including expanded attention to the end of the Cold War and the war in Colombia while asking whether revolution has come to an end in Latin America.
- A new Chapter 12 recasts as a question the title of the old closing chapter, "Forward into the Past?" This exciting new chapter focuses on the twenty-first century and includes an expanded section, "Latin America Swings Left," updating the political changes in the region. The new section "A Mobilized Population" addresses the new movements that struggle to push the governments further left, including groups focused on race and LGBTQ rights. "The Conservative Exceptions" looks at Mexico, as well as the coups that ended progressive governments in Honduras and Paraguay. "A New Regional Independence" shows how the region has challenged the United States on the drug war and shown more independence by welcoming investment from China and joining the BRICS movement. "Rise of the Middle Class—And the Vulnerable" shows that despite profound economic changes in the region that have brought greater prosperity to more people, Latin America continues to be the region of greatest economic inequality in the world. And finally, "Change and Continuity" expands the Latin American theoretical approaches presented to include modernity/coloniality/decoloniality.
- Chapter 12 also includes a brief profile of Pope Francis, the Argentine native who has taken the world stage with a renewed focus on poverty and social justice.
- Further attention has been given to women through new sections, "Unsung Heroes and Heroines," in Chapter 3, "Independence," and "Las Soldaderas," in Chapter 7, "The Mexican Explosion." In addition, there is also an expanded discussion of the remarkable number of women presidents in the region in Chapter 12.
- In addition to updating the table "Latin America Elects Leftists," there are two new tables: "Women Presidents in Latin America" and "Latin American Inequality Data."

Acknowledgments

I owe a debt of thanks to the many colleagues and students who have taken the time to send me their feedback on the text. I am particularly grateful to the eight anonymous reviewers who evaluated the ninth edition and offered excellent suggestions for the tenth. Special thanks are due to Jennifer Johnson at Whitman College's Penrose Library for her indispensable and always gracious assistance. My students in History 188, Modern Latin America, have asked questions that made me rethink aspects of the textbook. This book is for them and shaped by them. As always, I am most indebted to my husband, Charly Bloomquist, and my daughter, Delaney; they are my joy.

Julie A. Charlip

Walla Walla, Washington

1 Land and People

When Europeans first encountered the "New World," they found a land unlike any they had ever seen. It was a lush tropical wonder, colored by brilliant plants and animals. Amerigo Vespucci marveled, "Sometimes I was so wonder-struck by the fragrant smells of the herbs and flowers and the savor of the fruits and the roots that I fancied myself near the Terrestrial Paradise."

As Spanish colonies, the New World offered wealth that other Europeans envied. The British priest Thomas Gage commented: "The streets of Christendom must not compare with those of Mexico City in breadth and cleanliness but especially in the riches of the shops that adorn them."

But the images of an earthly paradise and colonial splendor would fade over time. By the nineteenth century, Latin America was considered "backward." In the twentieth century, the region was described as "underdeveloped," "Third World," or simply "impoverished." In the twenty-first century, Latin America is the region of greatest inequality in the world.

What happened to the Garden of Eden? In 1972, E. Bradford Burns, the original author of this textbook, called the problem the enigma: "Poor people inhabit rich lands." And although in the ensuing years those lands have been exploited and subjected to substantial environmental degradation, they are still rich—and the majority of the people are still poor.

Latin America has moved from paradise to poverty as a result of historical patterns that have developed over the years. This book explores those patterns in an attempt to understand why the Latin America of the twenty-first century is still wrestling with issues it has faced throughout its history. We argue that the most destructive pattern has been the continuing tendency of the elites of the region to confuse their nations' well-being with their own. Earlier scholars, however, placed the blame on the region's climate, on racist characterizations of the populace, and on the size of the population.

1.1: The Land

In the 1490s, Christopher Columbus tried to convince himself and his disbelieving crew that the island of Cuba was actually a peninsula of China. In reality, he had stumbled upon the unexpected: a region of such vastness and geographical variety that, even today, not all of the territory is controlled by the people who have so desperately tried to do so. It has been a land of both opportunity and disaster. Geography is destiny until one has the technology to surmount it. The geographic attributes of Latin America have contributed to the region's economic organization and created challenges for settlement and state building.

The original territory claimed by the kingdoms of the Iberian Peninsula included all of Central and South America, modern Mexico, many islands off the coasts, and much of what is now the United States. Contemporary Latin America is a huge region of a continent and a half, stretching 7,000 miles southward from the Rio Grande to Cape Horn. Geopolitically the region today encompasses eighteen Spanish-speaking republics, Portuguese-speaking Brazil, and French-speaking Haiti, a total of approximately 8 million square miles.

It is a region of geographic extremes. The Andes, the highest continuous mountain barrier on earth, spans 4,400 miles and has at least three dozen peaks that are taller than Mount McKinley. The Amazon River has the greatest discharge volume, drainage basin, and length of navigable waterways on the planet. Yet Latin America also contains the driest region on earth, the Atacama Desert. Half of Latin America is forested, comprising one quarter of the world's total forest area, which has led to its description as the "lungs of the world."

In the U.S. press, Latin America often seems a tragic victim of its climate, rocked by frequent earthquakes, volcanic eruptions, punishing hurricanes, and deadly avalanches. Indeed, Latin America has more than its share of natural disasters, a result of sitting atop five active tectonic plates—Caribbean, Cocos, Nazca, Scotia, and South American. In addition, part of South America's Pacific coast lies along the "ring of fire," the region where 80 percent of the seismic and volcanic activity of the earth takes place. That we in the United States seem to know so much about these events, however, says more about the limited media portrayal of the region than it does about the frequency of climatic violence.

But climate has long been a factor in foreign views of the region. Most of Latin America lies within the tropics, which prompted Europeans to speculate that the hot, steamy climate made people lazy. It is true that a generous nature provided natural abundance that made it possible for subsistence farmers to support themselves, with no incentive to work in European-owned enterprises. As many Latin Americans gradually lost access to the best lands and were forced to eke out a living on poor soils or work on the large landholdings of elites, it became clear that the climate was no drawback to hard work.

Latin America has only one country, Uruguay, with no territory in the tropics. South America reaches its widest point, 3,200 miles, just a few degrees south of the equator, unlike North America, which narrows rapidly as it approaches the equator. However, the cold Pacific Ocean currents refresh much of the west coast of Latin America, and the altitudes of the mountains and highlands offer a wide range of temperatures that belie the latitude. For centuries, and certainly long before the Europeans arrived, many of the region's most advanced civilizations flourished in the mountain plateaus and valleys. Today many of Latin America's largest cities are in the mountains or

Latin America's Environmental Woes

Environmental degradation is one of the most serious issues facing modern Latin America. The region encompasses some of the most endangered forest habitats on Earth, as well as the most rapid rates of deforestation. Coastal and marine areas are contaminated by land-based pollution, overexploitation of fisheries, the conversion of habitat to tourism, oil and gas extraction, refining, and transport. The region increasingly suffers from desertification, a process in which productive but dry land becomes unproductive desert. Desertification is caused by overcultivation, overgrazing, deforestation, and poor irrigation practices.

And that's just the rural areas.

Eighty percent of Latin Americans live in urban areas, making it the most urbanized area in the world. This change has occurred rapidly. In the 1950s, 60 percent of Latin Americans lived in the countryside. In 1960, for the first time more than 50 percent of the population was urban. São Paulo and Mexico City each teem with more than 20 million people. Twenty-three percent of Latin America's urban population lives in slums without access to adequate housing, clean water, and sanitation. Urban environmental problems include air and water pollution, fresh water shortages, poor disposal of waste, and industrial contamination.

Brazil, the region's largest country, is illustrative of the environmental challenges. Enormous São Paulo is home to some of the oldest and most extensive shantytowns (*favelas*). But Brazil's territory also includes 60 percent of the Amazon rainforest, which consumes significant portions of the carbon dioxide released on the planet, contains one-fifth of the world's fresh water, and has the world's greatest diversity of flora and fauna. But the rainforest is constantly under attack by loggers, oil and mineral exploration, and the construction of roads and hydroelectric dams.

Brazilian leaders have made a concerted effort to reduce deforestation, from a high of 10,723 square miles in 2004 to 2,275 miles in 2013. But at the same time, deforestation rose by 29 percent in 2013, raising fears that progress might be reversed, despite legislation and the formation of an environmental police force. The fires burning to clear land were so extensive they could be seen and were photographed from the International Space Station.

Problems are exacerbated by the region's poverty and demands for job creation and economic development. In 2003, Brazil elected Luiz Inacio "Lula" da Silva, who pledged to help the country's poor. One year later, environmentalists accused him of sacrificing the Amazon to job creation efforts for the 53 million Brazilians living on less than $1 a day. A 2009 study indicates that development accompanying deforestation does bring a short-lived increase in income, literacy, and longevity. But those improvements are transitory—as developers move on to new land, the populations of the cleared areas lose all they had gained.

Lula's successor, Dilma Rousseff, continued to promote social programs reducing poverty while pursuing economic development projects, such as the Belo Monte dam, in the Amazon region. Environmentalists criticized the president in 2012, just weeks before Brazil hosted the United Nations Conference on Sustainable Development, also known as Rio+20 because it marked the twentieth anniversary of the 1992 United Nations Conference on Environment and Development in Rio de Janeiro. The Brazil Committee in Defense of the Forests said Rousseff gave in to the agricultural lobby by partially vetoing a new Forest Code to waive fines and reduce requirements for restitution of areas illegally deforested in the past.

The tensions between protecting the environment and providing for human needs were captured in the change of the UN conference titles, from 1992's "environment and development" to Rio+20's "sustainable development." As Rousseff headed into her reelection race in 2014, her main rivals were Aecio Neves, who said Brazil needed to be more business friendly, and Marina Silva, a prominent environmentalist. Despite Silva's initial burst of popularity, she ended up trailing her developmentalist foes. Rousseff won, but her pro-business opponent polled 48 percent of the vote.

Map 1.1 Political Map of Latin America

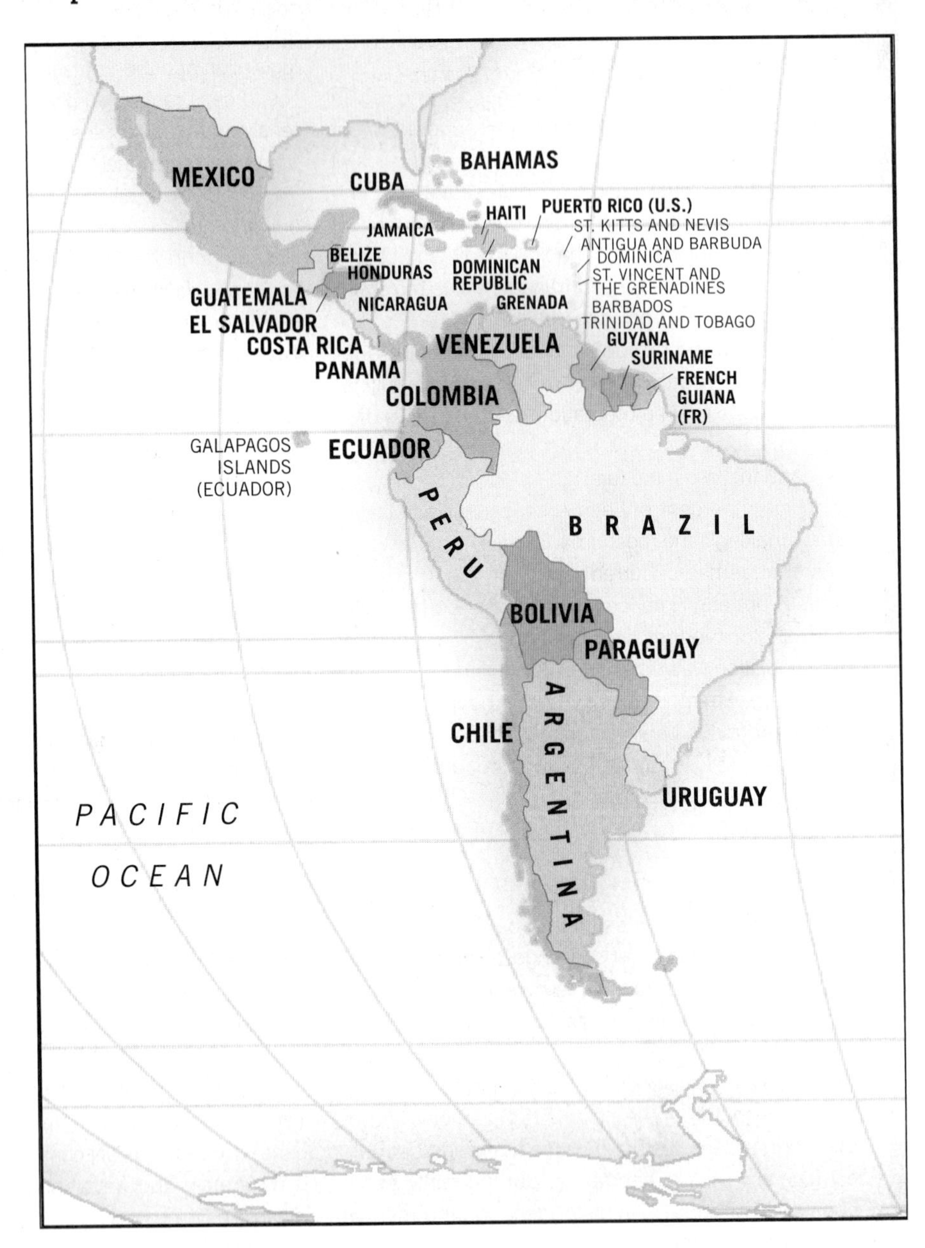

on mountain plateaus: Mexico City, Guatemala City, Bogotá, Quito, La Paz, and São Paulo, to mention only a few. Much of Latin America's population, particularly in Middle America and along the west coast of South America, concentrates in the highland areas.

In Mexico and Central America, the highlands create a rugged backbone running through the center of most of the countries, leaving coastal plains on either side.

Part of that mountain system emerges in the Greater Antilles to shape the geography of the major Caribbean islands. In South America, unlike Middle America, the mountains closely rim the Pacific coast, whereas the highlands skirt much of the Atlantic coast, making penetration into the flatter interior of the continent difficult. The Andes predominate: The world's longest continuous mountain barrier, it runs 4,400 miles down the west coast and fluctuates in width between 100 and 400 miles. Aconcagua, the highest mountain in the hemisphere, rises to a majestic 22,834 feet along the Chilean–Argentine frontier. The formidable Andes have been a severe obstacle to exploration and settlement of the South American interior from the west. Along the east coast, the older Guiana and Brazilian Highlands average 2,600 feet in altitude and rarely reach 9,000 feet. Running southward from the Caribbean and frequently fronting on the ocean, they disappear in the extreme south of Brazil. Like the Andes, they too have inhibited penetration of the interior. The largest cities on the Atlantic side are all on the coast or, like São Paulo, within a short distance of the ocean.

Four major river networks, the Magdalena, Orinoco, Amazon, and La Plata, flow into the Caribbean or Atlantic, providing access into the interior that is missing on the west coast. The Amazon ranks as one of the world's most impressive river systems. Aptly referred to in Portuguese as the "riversea," it is the largest river in volume in the world, exceeding that of the Mississippi fourteen times. In places, it is impossible to see from shore to shore, and over

This turn-of-the-century photograph captures the drama of Chile's Aconcagua Valley in the towering Andes. (Library of Congress)

much of its course, the river averages 100 feet in depth. Running eastward from its source 18,000 feet up in the Andes, it is joined from both the north and south by more than 200 tributaries. Together this imposing river and its tributaries provide 25,000 miles of navigable water. Farther to the south, the Plata network flows through some of the world's richest soil, the Pampas, a vast flat area shared by Argentina, Uruguay, and Brazil. The river system includes the Uruguay, Paraguay, and Paraná rivers, but it gets its name from the Río de la Plata, a 180-mile-long estuary separating Uruguay and the Argentine province of Buenos Aires. The system drains a basin of more than 1.5 million square miles. Shallow in depth, it still provides a vital communication and transportation link between the Atlantic coast and the southern interior of the continent.

No single country better illustrates the kaleidoscopic variety of Latin American geography than Chile, that long, lean land clinging to the Pacific shore for 2,600 miles. One of the world's most forbidding deserts in the north, the Atacama, gives way to rugged mountains with forests and alpine pastures. The Central Valley combines a Mediterranean climate with fertile plains, the heartland of Chile's agriculture and population. Moving southward, the traveler encounters dense, mixed forests; heavy rainfall; and a cold climate, warning of the glaciers and rugged coasts that lie beyond. Snow permanently covers much of Tierra del Fuego.

Yet, even with Chile's extremes from the desert to the snow, geographical differences are even greater in Bolivia, Brazil, Colombia, Mexico, and Peru, which alone encompasses 84 of the 104 ecological regions in the world and twenty-eight different climates. Latin America is the most geographically diverse area in the world. It includes seven distinct geographical zones: border, tropical highlands, lowland Pacific coast, lowland Atlantic coast, Amazon, highland and dry Southern Cone, and the temperate Southern Cone.

The United States–Mexico border is an area of arid or temperate climate and low population, and it is the only place in the world where rich and poor countries abut. Because it is home to the manufacturing assembly industry (*maquiladora*), the region has a higher gross domestic product than the rest of Latin America. To its south, the tropical highlands include the highlands of Central America and the Andean countries north of the Tropic of Capricorn. Access to the coast of this region is difficult, yet it is also an area of high population density, including most of the indigenous population of Latin America. Because of poor soil and high population, this is the poorest area in Latin America—even though the region includes the relatively high-income areas of Mexico City and Bogotá.

The lowland Pacific and Atlantic coasts are both tropical, but both have small dry areas. Although the highest population density of all Latin America is found on the Pacific coast, the Atlantic coast also has a large population. The income of the lowland coasts is about 20 percent higher than that of the tropical highlands, partly because of their advantageous position for international trade. But the lowlands are also areas that are prone to disease, and tropical soils present problems for successful agriculture.

The Amazon zone has the lowest population density of Latin America. It boasts a higher gross domestic product than neighboring areas because absentee owners earn high rents from mining and from large plantations. These economic activities are taking a toll on the delicate ecology of the area. The dry Southern Cone has only a slightly higher population than the Amazon, but there is a high population density in the temperate Southern Cone. Both are high-income areas.

Latin Americans have always been aware of the significance of their environment. Visiting the harsh, arid interior of northeastern Brazil for the first time, Euclydes da Cunha marveled in his *Rebellion in the Backlands* (*Os Sertöes*, 1902) at how the land had shaped a different people and created a civilization that contrasted sharply with that of the coast:

> Here was an absolute and radical break between the coastal cities and the clay huts of the interior, one that so disturbed the rhythm of our evolutionary development and which was so deplorable a stumbling block to national unity. They were in a strange country now, with other customs, other scenes, a different kind of people. Another language even, spoken with an original and picturesque drawl. They had, precisely, the feelings of going to war in another land. They felt that they were outside Brazil.

The variety of environment within countries has also been a trope in literature. Gabriel García Márquez grew up in Aracataca, the model for the fictional Macondo, a lush, steamy tropical zone, in his novel *One Hundred Years of Solitude* (*Cien Años de Soledad*, 1967). At fourteen, when García Márquez first went to Bogotá, he described it as "a remote and mournful city, where a cold drizzle had been falling since the beginning of the sixteenth century."

Latin American films, too, often assign nature the role of a major protagonist. Certainly in the Argentine classic *Prisoners of the Earth* (*Prisioneros de la Tierra*, 1939), the forests and rivers of the northeast overpower the outsider. Nature even forces the local people to bend before her rather than conquer her. A schoolteacher exiled by a military dictatorship to the geographically remote and rugged Chilean south in the Chilean film *The Frontier* (*La Frontera*, 1991) quickly learns that the ocean, mountains, and elements dominate and shape the lives of the inhabitants. Nature thus enforces some characteristics on the people of Latin America. The towering Andes, the vast Amazon, the unbroken Pampas, the lush rain forests provide an impressive setting for an equally powerful human drama.

The number of humans in that drama has been an issue of great concern, especially in the context of Latin American development. In the 1960s, Latin

Map 1.2 Physical Map of Latin America

SOURCE: https://en.wikipedia.org/wiki/R%C3%ADo_de_la_Plata, http://cluckfield.com/?p=320 and "Latin America's Physcial Geography" in Cathryn L. Lombardi, Latin American History: A Teaching Atlas, University of Wisconsin Press, 1983.

America's annual population growth of 2.8 percent made it the most rapidly growing area in the world. By the end of the twentieth century, the region's population growth rate had slowed to 1.5 percent, close to the world average of 1.4 percent. By 2013, Latin America and the Caribbean's 1.12 percent growth was below the world's 1.16 percent. Half of that population is either Brazilian or Mexican.

Despite concerns in the more developed world about Latin American population growth, the region is relatively underpopulated, with the exception of overcrowded El Salvador and Haiti. More than twice the size of

Europe, Latin America has a smaller population than Europe. Most countries in Latin America have a far lower population density for its agricultural land than the countries of Europe, which are able to feed its populations not just from domestic agricultural production but also by importing food from other countries.

Population growth is a serious issue for many reasons—environmental concerns, quality of life, and maternal well-being. But it is a poor explanation for Latin America's economic travails, which owe more to international and national power relations and choices. The international distribution of goods has little to do with a country's population: The 12 percent of the world's population that lives in the United States, Canada, and Western Europe accounts for 60 percent of private consumption, according to the Worldwatch Institute. The United States has 5 percent of the world's population but uses 25 percent of the world's energy.

Latin America's tropical beauty is on display in this 1911 photograph from Panama. (Library of Congress)

To many Latin Americans, northern concern about population growth in the region reflects long-term racial and cultural biases against a population that has its roots in the people who came to the region from Asia, Europe, and Africa.

1.2: The Indigenous

Some 15,000 to 25,000 years ago, when a land bridge still existed between Asia and North America, migrants crossed the Bering Strait. Moving slowly southward, they dispersed throughout North and South America. Over the millennia, at an uneven rate, some moved from hunting and fishing cultures to take up agriculture. At the same time, they fragmented into many cultural and linguistic groups, with up to 2,200 different languages, although they maintained certain general physical features in common: straight black hair, dark eyes, copper-colored skin, and short stature.

Table 1.1 Population per Hectare of Agricultural Land in Latin America and Selected European Countries

Country	Population per hectare
Netherlands	9.1
Belgium	8.4
Haiti	5.8
Germany	4.8
El Salvador	4.0
Italy	4.4
Dominican Republic	4.2
United Kingdom	3.7
Guatemala	3.5
Austria	2.7
Poland	2.6
Costa Rica	2.6
Honduras	2.5
France	2.3
Denmark	2.1
Ecuador	2.1
Hungary	1.9
Cuba	1.8
Panama	1.7
Venezuela	1.4
Peru	1.2
Nicaragua	1.2
Colombia	1.1
Chile	1.1
Mexico	1.1
Brazil	0.7
Paraguay	0.3
Argentina	0.3
Bolivia	0.3
Uruguay	0.2

SOURCE: Based on population statistics from The World Bank, 2013, http://wdi.worldbank.org. and agricultural area statistics from United Nations Food and Agriculture Organization, 2012, http://faostat.fao.org. Agricultural area comprises areas of temporary crops, permanent crops, and permanent meadows and pastures.

The indigenous people can best be understood by grouping them as nonsedentary, semisedentary, and sedentary societies. Nonsedentary societies were gathering and hunting groups that followed a seasonal cycle of moving through a delimited territory in search of food; they were mostly found in the area that now encompasses the northern Mexico frontier, the Argentine pampas, and the interior of Brazil. In semisedentary societies, hunting was still important, but they had also developed slash-and-burn agriculture, which shifted sites within their region. They populated much of Latin America and were often found on the fringes of fully sedentary peoples. Fully sedentary peoples had settled communities based on intensive agriculture, which provided enough surplus to support a hierarchical society with specialized classes. They were found in central Mexico, Guatemala, Ecuador, Peru, and Bolivia. The most advanced of these groups founded the impressive imperial societies. The social organization of the various indigenous groups was shaped by their environments, which shaped particular material cultures.

The early American cultures were varied, but a majority shared enough traits to permit a few generalizations. Family or clan units served as the basic social organization. All displayed profound faith in supernatural forces that they believed shaped, influenced, and guided their lives. For that reason, the *shamans*, those intimate with the supernatural, played important roles. They provided the contact between the mortal and the immortal, between the human and the spirit, and they served as healers. In the more complex and highly stratified societies, there was a differentiation between the more extensive landholdings of the nobility and that of the commoners. But in all sedentary indigenous societies, land was provided to everyone on the basis of membership in the community. Game roamed and ate off the land. Further, the land

A Miskito man in Nicaragua shows the continuing presence of indigenous populations. (Photograph by Julie A. Charlip)

furnished fruits, berries, nuts, and roots. Tilling the soil produced other foods, such as corn and potatoes. Many artifacts, instruments, and implements were similar from Alaska to Cape Horn. For example, spears, bows and arrows, and clubs were the common weapons of warfare or the hunt. Although these similarities are significant, the differences among the many cultures were enormous and impressive. By the end of the fifteenth century, between 9 million and 100 million people inhabited the Western Hemisphere. Scholars still heatedly debate the figures, and one can find forceful arguments favoring each extreme; a commonly used number now is 54 million.

Mistaking the New World for Asia, Christopher Columbus called the inhabitants he met "Indians." Exploration later indicated that the "Indians" belonged to myriad cultural groups, none of which had a word in their language that grouped together all the indigenous people of the hemisphere. They were as differentiated as the ancestral tribes of Europe, and their identities were locally based. The most important indigenous groups were the Mexica of the Aztec empire and the Mayas of Mexico and Central America; the Carib and the Arawak of the Caribbean area; the Chibcha of Colombia; the Inca empire of Ecuador, Peru, and Bolivia; the Araucanian of Chile; the Guaraní of Paraguay; and the Tupí of Brazil. Of these, the Aztec, Maya, and Inca exemplify the most complex cultural achievements, with fully sedentary and imperial societies.

Empire may be a misnomer for the Maya civilization, which dominated parts of what is now southern Mexico and northern Central America. The Maya had no central capital and authority, and some have likened the region to the ancient Greeks, who might be regarded as religiously and artistically united but consisting of politically sovereign states. Two distinct periods, the Classic and the Late, mark the history of Mayan civilization. During the Classic period, from the fourth to the tenth centuries, the Mayas lived in Guatemala; then they suddenly migrated to Yucatan, beginning the Late period, which lasted until the Spanish conquest. The exodus baffles anthropologists, who have suggested that exhaustion of the soil in Guatemala limited the corn harvests and forced the Mayas to move in order to survive. Corn provided the basis for Mayan civilization, and the Mayan creation account revolves around corn. The gods "began to talk about the creation and the making of our first mother and father; of yellow corn and of white corn they made their flesh; of cornmeal dough they made the arms and the legs of man," relates the *Popul Vuh*, the sacred book of the Mayas. All human activity and religion centered on the planting, growing, and harvesting of corn. The Mayas dug an extensive network of canals and water-control ditches, which made intensive agriculture possible. These efficient agricultural methods produced corn surpluses and hence the leisure time available for a large priestly class to dedicate its talents to religion and scientific study.

Extraordinary intellectual achievements resulted. The Mayas progressed from the pictograph to the ideograph and thus invented a type of writing, the only indigenous in the hemisphere to do so. Sophisticated in mathematics, they invented the zero and devised numeration by position. Astute observers of the heavens, they applied their mathematical skills to astronomy. Their careful studies of the heavens enabled them to predict eclipses, follow the path of the planet Venus, and prepare a calendar more accurate than that used in Europe. As the ruins of Copán, Tikal, Palenque, Chichén Itzá, Mayapán, and Uxmal testify, the Mayas built magnificent temples. One of the most striking features of that architecture is its extremely elaborate carving and sculpture.

To the west of the Mayas, another native empire, the Aztec, expanded and flourished in the fifteenth century. The Aztec empire originated with the Mexica, a group that migrated from the north in the early thirteenth century to the central valley of Mexico, where they conquered several prosperous and highly advanced city-states. Constant conquests gave prominence to the warriors, and, not surprisingly, among multiple divinities the gods of war and the sun predominated. Because the Mexica believed that the gods had sacrificed themselves to create the world, the propitiation of the gods in turn required human sacrifices. In 1325, the Mexica founded Tenochtitlan, their island capital, and from that religious and political center they radiated outward to absorb other cultures until they controlled all of Central Mexico. The misnomer Aztec for the people of the empire is drawn from the name of the Mexica's

The Mayan monument known as the Castillo at Chichén Itzá in the Yucatan is indicative of the sophisticated societies that predated the Spanish conquest. (Library of Congress)

original northern, and perhaps mythical, homeland, Aztlán, and refers specifically to the alliance of the three dominant city-states of Tenochtitlan, Texcoco, and Tlacopan. The people are sometimes referred to as Nahua, speakers of the Nahuatl language. Their highly productive system of agriculture included the *chinampas*, floating gardens that made effective use of their lake location. As a result, they were able to support a large population, with the population of Tenochtitlan estimated at 200,000, larger than any European city. The Mexica devised the pictograph, an accurate calendar, impressive architecture, and an elaborate and effective system of government that dominated the Central Valley of Mexico, with trade networks extending as far south as present-day Nicaragua.

Largest, oldest, and best organized of the indigenous civilizations was the so-called Incan, which flowered in the harsh environment of the Andes. The Inca was in fact the ruler of the empire, which was named Tawantinsuyu. By the early sixteenth century, the empire extended in all directions from Cuzco, regarded as the center of the universe. It stretched nearly 3,000 miles from Ecuador into Chile, and its maximum width measured 400 miles. Few empires have been more rigidly regimented or more highly centralized, an astounding feat when one realizes that it was run without the benefit—or hindrance—of written accounts or records. The only record system was the *quipu*, cords upon which knots were made to record information. Spanish chroniclers attested that the cords were used not just to record such mathematical data as censuses, inventories, and tribute records, but also for royal chronicles, records of sacred places and sacrifices, successions, postal messages, and criminal trials. The highly effective government rapidly assimilated newly conquered peoples into the empire. Entire populations were moved around the empire when security suggested the wisdom of such relocations. Every subject was required to speak Quechua, the language of the court. In weaving, pottery, medicine, and agriculture, their achievements were magnificent. Challenged by stingy soil, they developed systems of drainage, terracing, and irrigation and learned the value of fertilizing their fields. They produced impressive food surpluses, stored by the state for lean years.

Many differences separated these three high indigenous civilizations, but some impressive similarities existed. Society was highly structured. The hierarchy of nobles, priests, warriors, artisans, farmers, and slaves was ordinarily inflexible, although occasionally some mobility did occur. At the pinnacle of that hierarchy stood the omnipotent emperor, the object of the greatest respect and veneration. The sixteenth-century chronicler Pedro de Cieza de León, in his own charming style, illustrated the awe in which people held the Inca: "Thus the kings were so feared that, when they traveled over the provinces, and permitted a piece of the cloth to be raised which hung round their litter, so as to allow their vassals to behold them, there was such an outcry that the birds fell from the upper air where they were flying, insomuch that they could

Map 1.3 Pre-Columbian America

SOURCE: Adapted from Howard Spodek, *The World's History Combined*, 3rd ed. © 2006. Electronically reproduced by permission of Pearson Education, Inc., Upper Saddle River, NJ.

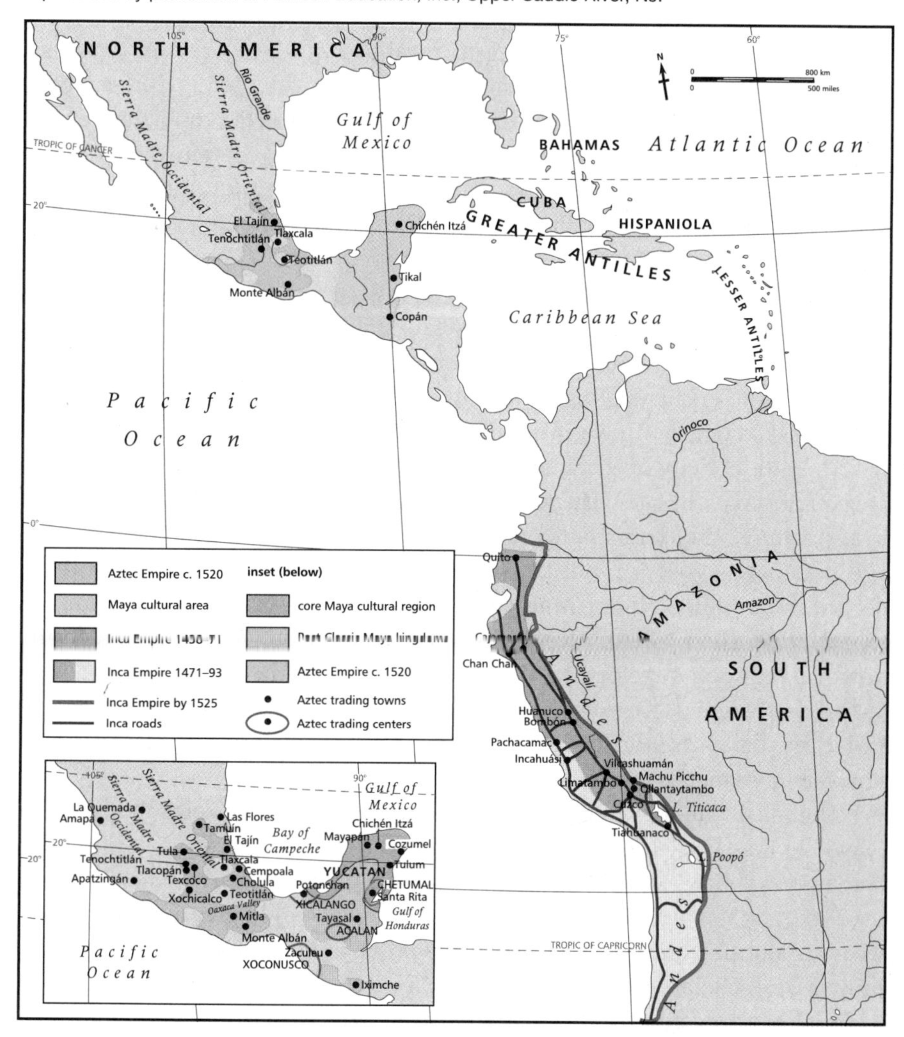

be caught in men's hands. All men so feared the king, that they did not dare to speak evil of his shadow."

Little or no distinction existed between civil and religious authority, so for all intents and purposes Church and State were one. The Incan and Aztec emperors were both regarded as representatives of the sun on earth and thus as deities, a position probably held by the rulers of the Mayan city-states as well. Royal judges impartially administered the laws of the empires and apparently enjoyed a reputation for fairness. The sixteenth-century chroniclers who saw the judicial systems functioning invariably praised them. Cieza de León, for one, noted, "It was felt to be certain that those who did evil would receive punishment without fail and that neither prayers nor bribes would avert it."

These civilizations rested on a firm rural base. Cities were rare, although a few boasted populations exceeding 100,000. They were centers of commerce, government, and religion. Eyewitness accounts as well as the ruins that remain leave no doubt that these cities were well-organized and contained impressive architecture. The sixteenth-century chronicles reveal that the cities astonished the first Spaniards who saw them. Bernal Díaz del Castillo, who accompanied Hernando Cortés into Tenochtitlan in 1519, gasped, "And when we saw all those cities and villages built in the water, and other great towns on dry land, and that straight and level causeway leading to Mexico [City], we were astounded. These great towns and cities and buildings rising from the water, all made of stone, seemed like an enchanted vision from the tale of Amadis. Indeed, some of our soldiers asked whether it was not a dream!"

The productivity of the land made possible an opulent court life and complex religious ceremonies. The vast majority of the population, however, worked in agriculture. The farmers grew corn, beans, squash, pumpkins, manioc root, and potatoes as well as other crops. Communal lands were cultivated for the benefit of the state, religion, and community. The state thoroughly organized and directed the rural labor force. Advanced as these native civilizations were, however, not one developed the use of iron or used the wheel because they lacked draft animals to pull wheeled vehicles. However, the indigenous people had learned to work gold, silver, copper, tin, and bronze. Artifacts that have survived in those metals testify to fine skills.

There is some disagreement among scholars about the roles that women played in these societies. In both the Inca and Aztec worlds, women primarily bore responsibility for domestic duties. The roles were so clearly defined among the Mexica that when a child was born, the midwife would give a girl a spindle, weaving shuttles, and a broom, whereas a boy would be given a shield and arrows. Many scholars contend that these roles were different but not necessarily unequal; they see gender complementarity, in which men and women formed equally important halves of the social order. For example, home was a sacred place, and ritual sweeping was related to religious practices. Childbirth was considered akin to battle; as one Nahuatl account described it, "Is this not a fatal time for us poor women? This is our kind of war."

In the Aztec empire, women also sold goods in the markets, where they often had supervisory positions; they were cloth makers and embroiderers, and they served as midwives. In Inca society, women took care of the home, but they also worked in agriculture. Their work tended to be seen not as private service for husbands but as a continuation of household and community. Here there was complementarity as well: men plowed, women sowed, and both harvested.

Although Mexica women in Central Mexico could not hold high political office, this was not the case at the fringes of the empire. Mixtec women inherited dynastic titles and frequently served as rulers. The same was true on the fringes of the Inca empire, for example, in the highlands of Ecuador. However, the complementary social positions and occasional positions of power do not imply political equality. In both empires, men held the highest positions of power.

The spectacular achievements of these sedentary farming cultures contrast sharply with the more elementary evolution of the gathering, hunting, and fishing cultures and semisedentary farming societies. The Tupí tribes, the single most important native element contributing to the early formation of Brazil, illustrate the status of the many intermediate farming cultures found throughout Latin America.

The Tupí tribes were loosely organized. The small, temporary villages, often surrounded by a crude wooden stockade, were, when possible, located along a riverbank. The indigenous lived communally in large thatched huts in which they strung their hammocks in extended family or lineage groups of as many as 100 people. Most of the groups had at least a nominal chief, although some seemed to recognize a leader only in time of war and a few seemed to have no concept of a leader. More often than not, the *shaman* was the most important figure. He communed with the spirits, proffered advice, and prescribed medicines. The religions abounded with good and evil spirits.

The men spent considerable time preparing for and participating in wars. They hunted monkeys, tapirs, armadillos, and birds. They also fished, trapping the fish with funnel-shaped baskets, poisoning the water and collecting the fish, or shooting the fish with arrows. They cleared the forest to plant crops. Nearly every year during the dry season, the men cut down trees, bushes, and vines; waited until they had dried; and then burned them, a method used throughout Latin America, then as well as now. The burning destroyed the thin humus, and the soil was quickly exhausted. Hence, it was constantly necessary to clear new land, and eventually the village moved in order to be near virgin soil. In general, although not exclusively, the women took charge of planting and harvesting crops and of collecting and preparing the food. Manioc was the principal cultivated crop. Maize, beans, yams, peppers, squash, sweet potatoes, tobacco, pineapples, and occasionally cotton were the other cultivated crops. Forest fruits were collected.

To the first Europeans who observed them, these people seemed to live an idyllic life. The tropics required little or no clothing. Generally nude, the Tupí developed the art of body ornamentation and painted elaborate and ornate geometric designs on themselves. Into their noses, lips, and ears they inserted stone and wooden artifacts. Feathers from the colorful forest birds provided an additional decorative touch. Their appearance prompted the Europeans to think of them as innocent children of nature. The first chronicler of Brazil, Pero Vaz de Caminha, marveled to the king of Portugal, "Sire, the innocence of Adam himself was not greater than these people's." As competition for land and resources increased, chroniclers would later tell quite a different tale, one in which the indigenous emerged as wicked villains, brutes who desperately needed the civilizing hand of Europe.

The European romantics who thought they saw utopia in native life obviously exaggerated. The indigenous by no means led the perfect life. Misunderstanding, if not outright ignorance, has always characterized outsiders' perceptions of them. Far too often since the conquest, images of native peoples have erred at either extreme—the violent savage or the noble savage—rather than showing their humanity. It is through their own words that the indigenous can be seen as more fully human, as in this Nahuatl lament over the conquest: "Broken spears lie in the roads; we have torn our hair in our grief. . . . We have pounded our hands in despair against the adobe walls, for our inheritance, our city, is lost and dead."

1.3: The European

The Europeans who came to dominate Latin America came primarily from the Iberian Peninsula, a land of as much contrast as the New World. Almost an island, the peninsula is bounded by the Bay of Biscay, Atlantic Ocean, Gulf of Cádiz, and the Mediterranean Sea. Half of its territory comprises arid tableland. But this *meseta* is bisected by one imposing mountain system—the Sierra da Estrela, Sierra de Gredos, and Sierra de Guadarrama—and circled by another—the Cantabrian Cordillera, Ibera mountains and Sierra Morena, and Cordillera Bética. Spain is divided by the Pyrenees from the rest of Europe, and separated from Africa only by the Strait of Gibraltar. The varied climate ranges from the cold winters of the north to the subtropical sunshine of the south.

The region was also the crossroads of many peoples—Iberians, Celts, Phoenicians, Greeks, Carthaginians, Romans, Visigoths, and Muslims—and these cultures blended together. The most stable periods in this varied history were under Roman rule, from about 19 B.C.E. to the late fifth century, which was followed by rule of the Visigoths, who continued many Roman customs, during the sixth and seventh centuries. The Visigoths fell to the Muslims in the invasion of 711–720, which prompted similar laments to those of the Mexica: "Who can bear to relate such perils? Who can count such terrible disasters? Even if every limb were transformed into a tongue, it would be beyond human capability to describe the ruin of Spain and its many and great evils."

Muslim control of the peninsula was never complete, and even its dominance of the south waxed and waned, due in part to divisions between Berbers and Arabs within the Muslim world. Furthermore, although many Christians converted to Islam, the majority of the rural population was still Christian as late as 948, when Arab geographer Mohammed Abul-Kassem Ibn Hawqal visited. In addition, Christian groups in the north continued to resist Muslim forces. The Umayyad caliphate was successful at slowly conquering and ruling much of the south of Spain, raising Córdoba to a cultural center and ruling from 929 to 1031. Two subsequent caliphates, the Almoravid, a Berber group from North Africa (1086–1147), and the Almohad (1146–1220s), were able to maintain unity in al-Andalus, as the Muslims called their Iberian territories. But by the thirteenth century, the Christian groups from the north had gained strength. The crusade to retake the peninsula, the *reconquista*, began in 732 at the Battle of Tours, and it would take until 1492, when Granada fell, to expel the Muslims from the peninsula.

Throughout the years of Muslim rule, Iberia was a land of three cultures: Muslim, Jewish, and Christian. There were conflicts, but the eleventh century was considered a high point of cooperation between the Jewish and Muslim communities, resulting in a flourishing of culture. Caliph Al-Hakam II (961–976) is said to have founded a library of hundreds of thousands of volumes, something impossible to imagine in the rest of Europe at that time. One of the great contributions of Muslim Spain was the preservation and translation of classical philosophy.

Both Jews and Muslims, however, would suffer with the reunification of the peninsula under the Catholic monarchs, Isabella I of Castile and Ferdinand II of Aragón, who were married in 1479. In 1492, the monarchs ordered the expulsion of all Jews and Muslims unless they converted to Christianity. Catholicism became the official religion and served as a proto-nationalism: To be Spanish was to be Catholic.

But the Catholic Church in Spain was far from a monolithic entity. One split was between the secular and the regular clergy. The secular clergy were loosely organized and charged with administering to Christian populations. These were the worldly religious, concerned with the day-to-day lives of their flocks, and as a result they needed to find economic activities in that world to sustain them. The regular clergy were tightly organized into orders—Franciscan, Dominican, Augustinian—each a world of its own, with separate rules and concerns. These clerics had largely withdrawn from the world to the solitude of monasteries, and they were accorded higher status because they were seen as devoting their lives to God rather than to man.

Another significant split within the Church was between official and folk religion. The official church dogma emphasized the one true God, the Trinity, Jesus as redeemer, and the importance of the Church and its sacraments, particularly baptism and confession. Popular religion emphasized the humanity and suffering of Jesus and made cults of Mary and the various saints, which were the patrons of guilds and towns and represented particular maladies and problems. While priests taught the official rites of the church, Spanish immigrants to the Americas were more likely to follow folk practices.

Map 1.4 Iberian Peninsula

SOURCE: Based on http://www.fordham.edu/halsall/maps/1492spain.jpg.

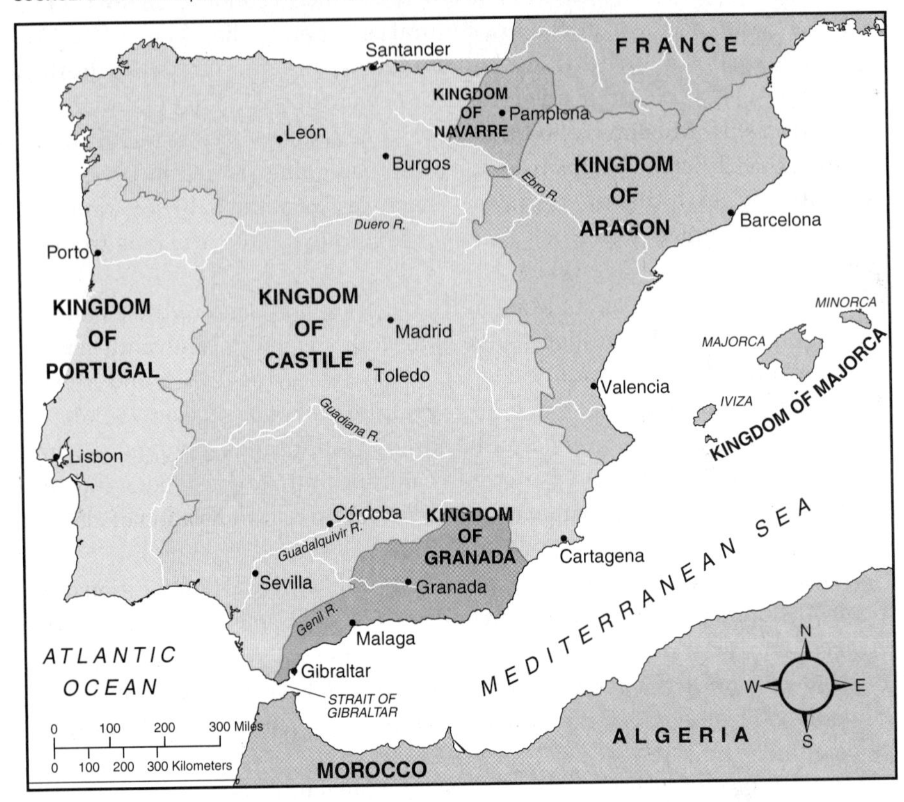

Members of the clergy served in government ministries, but the monarchs had the power to appoint bishops. The Spanish Crown had more control over the Church than any other monarchy in Europe. The two institutions were mutually dependent, equal pillars of society.

Although 1492 is popularly referred to as the reunification, there is little accuracy in the term. The peninsula had always been a splintered entity, and the royal marriage did not create a territorial or administrative merger. Further, each kingdom was a loose confederation: Isabella's "Aragón" included Aragón, Catalonia, Valencia, Majorca, Sardinia, and Sicily, each autonomous in legal, administrative, and economic matters. There were still customs barriers between Aragón and Castile.

There were no Spaniards at this point, although the region was called Hispania starting in the Roman era. People identified with their region, such as Aragón or Catalonia, and more specifically with their city. This urban focus was so widespread that, even in rural areas, people tended to live in nucleated towns and went out to their fields.

The social structure was divided most importantly along the lines of nobles and commoners, with the landed nobleman at the top. This hierarchy could be further subdivided by occupation, with professionals—trained for the Church, law, or medicine—ranking at the top. Next were merchants, who despite lower rank had the advantage of access to liquid assets and were worldly through their ties to long-distance trade. These households were staffed by a variety of servants and retainers. More plebian than merchants were the artisans, though many gathered substantial assets in large shops employing staffs of journeymen, apprentices, and slaves. And at the bottom of society were the farmers and herders, but even among them there was a division based on the size and success of agricultural enterprises.

At the center of Iberian life was the extended family, with cousins as closely tied as brothers. The head of the family was the patriarch, whose status was partly based on gender and partly on age. Women's positions, in turn, were dictated by the standing of their fathers and husbands. It was common for men to have sexual relationships outside of marriage, and the offspring were usually recognized and given help, although rarely were they included in the official family. The family might be viewed as a corporation, and it was desirable for nobles to have a son at court, a son in the clergy, and daughters who married into other noble or wealthy merchant families.

Like most peoples, the Iberians believed their ways of life, customs, language, and religion were superior to all others. But they also lived in a region of great diversity, exposed to many different ethnic groups and beliefs. When they arrived in the New World, they brought with them both their prejudices and their familiarity with diversity.

1.4: The African

From the very beginning, some free Africans from the Iberian Peninsula participated in exploration and conquest of the Americas. The majority, however, came as slaves, with the first sent from Iberia as early as 1502. The slave trade brought people directly from Africa starting in Cuba in 1512 and in Brazil in 1538, continuing until the trade ended in Brazil in 1850 and in Cuba in 1866. During the course of three centuries, nearly 2.5 million slaves were sold into Spanish America and 4.8 million in Brazil.

Slaves came from West and Central Africa, a region encompassing the Sahara Desert and its oases; the savanna immediately to the south, which is semiarid grassland; and the tropical rain forest. The region's economies were based on agriculture—as in Eurasia and the Americas, Africans began plant domestication around 5000 B.C.E.—and featured both domestic and imported crops brought in through trade networks. Iron technology, begun circa 500 B.C.E. and spread by Bantu-language speaker expansion, allowed the extension of agriculture into formerly unavailable land. Early states (200–700 C.E.)

This woman in Bluefields, Nicaragua, is part of a substantial population that is of African descent. (Photograph by Julie A. Charlip)

in West and Central Africa, such as Djenne, developed through trade within sub-Saharan Africa. From 700 to 1600 C.E., trans-Saharan trade with Arabs and Muslims led to the growth of Western and Central African states with stronger governments, more pronounced class stratification, and larger urban systems.

Around 700, the first Muslim traders established commerce between the northern savanna regions and their home bases north of the Sahara. By 900 C.E., trade between sub-Saharan Africa and the Muslim world was substantial and regular. The Muslim traders brought cloth, salt, steel swords, glass, and luxury goods in exchange for gold, slaves, ostrich feathers, fine leathers, decorative woods, and kola nuts. Gold had been mined in West Africa since 800 C.E. But with the advent of the Muslim trade, there was a larger market for the mineral, which led to greater production, the development of larger cities, and a more powerful elite, with greater class stratification and stronger governments. There was also some conversion to Islam, especially among merchants, because Islam provided a code of ethics leading to the trust necessary for long-distance trade. Although rulers and commoners sometimes followed suit, conversion was often only nominal, and traditional practices continued. A similar pattern would be seen with Christianity in the New World.

The more organized states of 700–1600 C.E. often included an opulent life at court, furnished by talented artisans, who provided bronze castings, carved wooden sculptures, ivory carving, cast gold, feather work, and painted leather. Commoners were employed in public works projects, including royal tombs, walled palaces, mosques, irrigation and drainage systems, and great walls around cities.

Although most independent polities were small (comparable to city-states), there were three imperial societies: Ghana, Mali, and Songhai. Ghana was an inland empire centered on the western portion of present-day Mali. It consisted of a number of chiefdoms joined together. Oral tradition says Ghana had twenty kings before the time of Muhammad circa 600 C.E. A powerful empire from 700–1100 C.E., it was destroyed by the Almoravids, and its fall led to the dispersal of the Soninke people. The kingdom of Mali, founded around 1200 C.E., was even larger and richer than Ghana. Centered on the city of Niani in the old Ghana empire, it incorporated all of Ghana and extended west to the Atlantic Ocean. In the process, Mali overextended its military, and in the 1400s, its capital of Timbuktu was taken over by Tuareg nomads. Songhai was once a part of the eastern Mali Empire. Founded in 1350, the empire grew by 1515 to become

Map 1.5 African Kingdoms

SOURCE: Adapted from Howard Spodek, *The World's History Combined*, 3rd ed. © 2006. Electronically reproduced by permission of Pearson Education, Inc., Upper Saddle River, NJ.

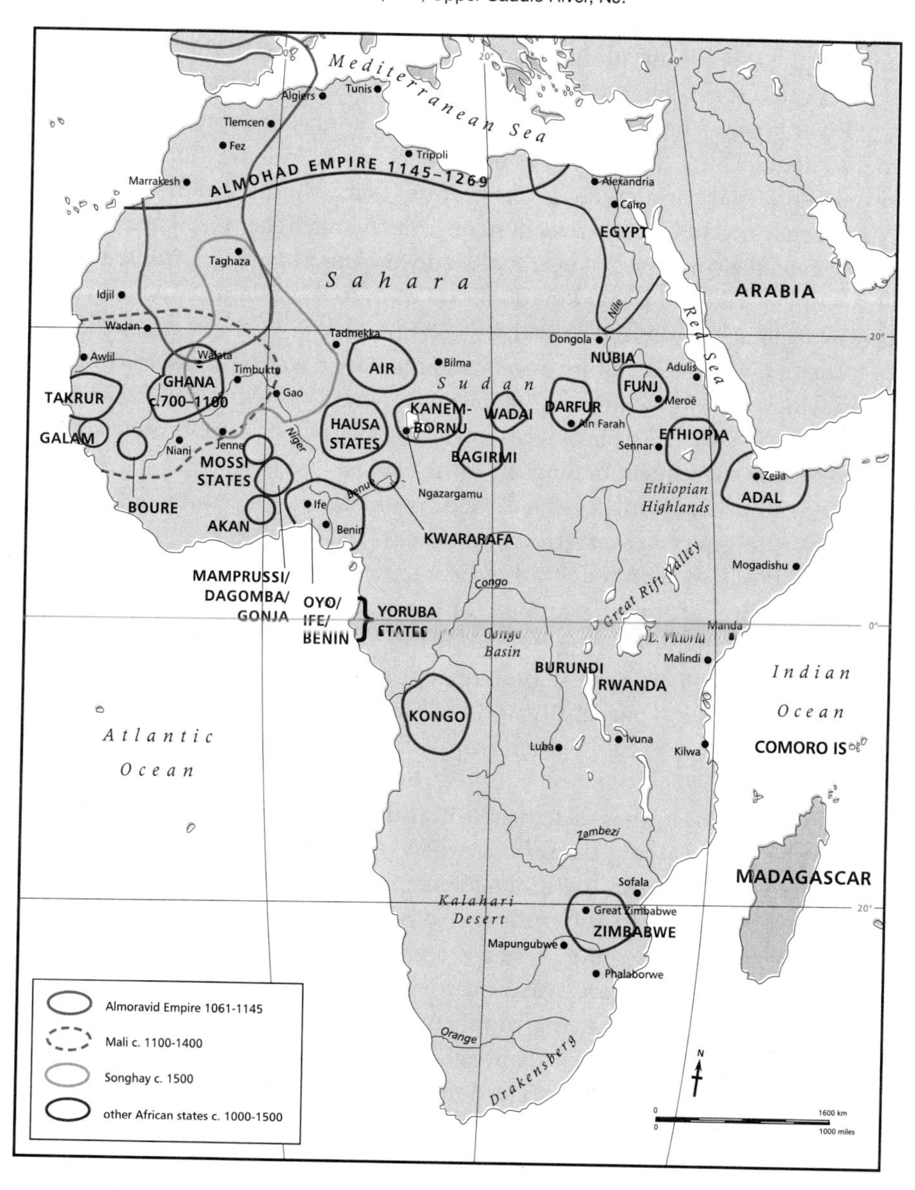

the largest sub-Saharan empire, with more than 1 million people. Eventually weakened by internal dissension—the rich province of Hausaland was lost after a local ethnic group staged a revolt—the empire collapsed in 1590, conquered by Moroccan forces while the Songhai fought over the royal succession.

The development of large and complex polities came later to the forest than savanna. The rise of complexity parallels the rise of large-scale trade networks with the savanna kingdoms to the north. They probably traded with Ghana, leading to the rise of a wealthy merchant class, which gained power and led the forest communities. By 1200, the Ife and Benin states developed sophisticated royal courts with an extravagant bronze sculptural tradition.

River systems linked these African regions, connecting the western Sudan to the Atlantic. Mali was the center of political power from the thirteenth to the seventeenth centuries largely because of its location at the headwaters of the Niger, Senegal, and Gambia river systems, which linked the West African region and provided a corridor that eventually added Hausa kingdoms; Yoruba states; and Nupe, Igala, and Benin kingdoms via river to the Atlantic. This maritime culture facilitated trade and coastal protection.

The first advanced culture in sub-Saharan Africa was Nok, a society of iron users which developed around 1000 BCE in what is now central Nigeria and ended circa 300 BCE. Archaeologists have found terracotta sculptures from Nok that resemble subsequent findings in Yoruba areas.

Repeated invasions by the Phoenicians, Greeks, Romans, and Arabs brought foreigners to western Africa as early as 100 B.C.E., but the fall of the seaport of Ceuta on the Strait of Gibraltar to the Portuguese in 1415 C.E. heralded new European incursions. Europeans were attracted by Africa's commercial potential—gold, ivory, cotton, and spices. However, African naval power protected the region against raids, with the result that trade had to be carried on peacefully and on African terms. For example, Afonso I, king of Kongo, seized a French ship and crew for trading illegally on the coast in 1525.

The Portuguese soon discovered that the Africans themselves were the continent's most valuable export. Between 1441 and 1443, the Portuguese began to transport Africans to Europe for sale. The majority of these slaves were purchased from African authorities, who responded to the growing demand. Before the Europeans came, Africans were collecting slaves as prisoners of war. In societies where land was held communally, slaves were the only form of private, revenue-producing property. Land was made available to whoever could work it, and Africans would purchase slaves to fill that purpose. In practicality, the slaves functioned much like tenants and hired workers in Europe. In 1659, Giacinto Brugiotti da Vetralla commented that Central African slaves were "slaves in name only." Such comments have led to the supposition that slavery in Africa did not incorporate the brutality of slavery in the Americas.

African elites gladly participated in trade with Europeans, mostly for prestige and luxury items because European trade offered nothing that Africans did not produce. Africans manufactured sufficient steel and cloth, including beautiful varieties that Europeans said rivaled Italian manufacture and in volume that rivaled Dutch production. Furthermore, although Africans did import European arms, they were of use primarily against fortifications, of which there were few. Warfare in Africa continued primarily with African weapons and for African political reasons. The majority of Africans who came to the Americas came from three large cultural zones. The region that provided the majority of the first two waves of slaves was Senegambia. The first slaves came from Upper Guinea, ranging from the Senegal River to modern Liberia, where the Mande language family dominated. Next was Lower Guinea, ranging from the Ivory Coast to Cameroon, where Akan and Aja languages were predominant. The third primary area was West Central Africa. The region included the Angola coast, Kongo, the Costa de Mina, present-day Benin, and stretching inland as far as modern Zaire, where Bantu was the dominant family of languages. Among the people who were enslaved, and ethnicities that appeared most frequently in Latin American documents, were the Fulani, Mandingo, Asante, Yoruba, Hausa, and Ibo.

Although there was great diversity within these regions and language families—with dozens of ethnic identities in Upper Guinea and Central Africa—there was also enough commonality to help build a new culture in the Americas. Though the designs differed, the various cultures had traditions of making cloth, which the people draped and wrapped around their bodies. They all were accomplished at pottery making and basket weaving and used rhythmic drumming as part of their musical tradition. Most cultures used pounded or steamed starchy flours—made from millet, cassava, yam, and in Upper Guinea, rice—as the central component of meals.

Most important, the Africans also brought their religious beliefs to the New World. Many of the slaves from Upper Guinea were Muslims, but the majority of Africans who came to the New World had spiritist beliefs. Although many groups believed in a creator god, all of the groups' beliefs included a variety of gods or spirits responsible for various areas of life, and the spirits of ancestors continued to play a role in people's lives. In this belief system, there is a material and an otherworld, where people go when they die. Some gifted people, known as diviners, could pass between the two worlds, receiving revelations from the other world and communicating them to people in the material world. The importance of those beliefs was attested to by the Nigerian slave Olaudah Equiano, who recounted the horrors of the middle passage, noting, "I was now persuaded that I had gotten into a world of bad spirits."

1.5: Mestizaje and the Creation of New People

The three peoples of the Americas came together in sometimes violent and sometimes consensual ways. Through their interactions, they created new groups of people: *mestizos*, a mixture of indigenous and European; *mulattos*, mixing African and European; and *zambos*, the joining of indigenous and African. The Iberians developed an extensive vocabulary in an attempt to describe the exact mixture of races. Iberians were already concerned with purity of blood, shunning any trace of Muslim or Jewish heritage. These concerns would be magnified with colonial race mixture, or *mestizaje*. Although originally the term referred specifically to indigenous—European mixtures, it came to be used more broadly to refer to all "race" mixtures, which produced a variety of darker-skinned people, known generally as *castas*.

Race is a problematic term, freighted with the baggage of nineteenth- and twentieth-century usages constructing human categories by phenotype and assigning characteristics to them. Certainly, once those categories were created and social status assigned accordingly, members of "races" frequently shared experiences based on their position in one of those categories. But people do not inherently share traits based on these groupings, and the Europeans, Africans, and indigenous of the fifteenth through eighteenth centuries certainly did not think so.

On the Iberian Peninsula in the fifteenth century, privilege was allocated according to legal status—noble, commoner, and slave. Legal status was typically linked with kinship, which meant status was linked to bloodlines. As a result, elites became concerned that bloodlines be pure, uncontaminated by Jewish and Muslim links. However, Spaniards were puzzled by the indigenous; intellectuals debated their status, wondering whether they had souls and where they fit into the hierarchical social system. When the Spanish decided that the indigenous were inherently inferior, a new ideology of race began to take form. That assignment of race would turn out to be fluid rather than fixed; nonetheless, Spaniards made an issue of whiteness and ranked darker people at the lower end of the social scale.

Labels for Miscegenation in Eighteenth-Century New Spain

1. Spaniard and Indian beget mestizo
2. Mestizo and Spanish woman beget castizo
3. Castizo woman and Spaniard beget Spaniard
4. Spanish woman and Negro beget mulatto
5. Spaniard and mulatto woman beget morisco
6. Morisco woman and Spaniard beget albino
7. Spaniard and albino woman beget torna atrás
8. Indian and torna atrás woman beget lobo
9. Lobo and Indian woman beget zambaigo
10. Zambaigo and Indian woman beget cambujo
11. Cambujo and mulatto woman beget albarazado
12. Albarazado and mulatto woman beget barcino
13. Barcino and mulatto woman beget coyote
14. Coyote woman and Indian beget chamison
15. Chamiso woman and mestizo beget coyote mestizo
16. Coyote mestizo and mulatto woman beget ahí te estás

SOURCE: Magnus Mörner, *Race Mixture in the History of Latin America* (Boston, MA: Little, Brown, 1967), 58.

Labels for Miscegenation in Eighteenth-Century Peru

1. Spaniard and Indian woman beget mestizo
2. Spaniard and mestizo woman beget cuarterón de mestizo
3. Spaniard and cuarteróna de mestizo beget quinterón
4. Spaniard and quinterona de mestizo beget Spaniard or requinterón de mestizo
5. Spaniard and Negress beget mulatto
6. Spaniard and mulatto woman beget quarterón de mulato
7. Spaniard and cuarterόna de mulato beget quinterón
8. Spaniard and quinterona de mulato beget requinterón

SOURCE: Magnus Mörner, *Race Mixture in the History of Latin America* (Boston, MA: Little, Brown, 1967), 58–59.

Social scientists have tended to use the term *race* to refer to populations grouped by phenotype, particularly skin color, giving race a fixed, biological status, and use ethnicity to refer to practices that define and separate people on the basis of more fluid cultural signifiers. As a result, blacks in Latin America are frequently studied in terms of race, while the indigenous are most often studied in terms of culture—clothing, food, music, art, ways of life, religion, and perception. But the boundaries of race and ethnicity have been porous in Latin America, and the bridges across those borders have created new groups that defy easy identification.

Mestizaje would prove to be an extraordinarily complex issue. To Iberians, mixture was generally viewed as contamination. Yet, at other times, miscegenation was actually encouraged in order to facilitate economic relations with indigenous people or to hasten assimilation. Indigenous peoples would sometimes reject *mestizos* for abandoning the indigenous side of the equation. Independence leader Simón Bolívar used the concept of mestizaje to argue that Spain's colonial subjects were not Spanish and should be independent, whereas Cuba's José Martí argued that the new nations should be faithful to their own creative mestizaje rather than adopt foreign ideals.

Many nineteenth-century elites hoped mestizaje would whiten their populations, leading to greater progress. However, elites also echoed the theories of Count Joseph Arthur Gobineau and Gustave Le Bon, French intellectuals, who argued that people of mixed race "always inherited the most negative characteristics of the blended races." In practice, the Latin American elites who organized nineteenth-century nation-building efforts often used mestizaje to deny the continued presence of indigenous people, claiming they had all blended into the new mestizo citizens.

Twentieth-century Latin Americans would ponder the meaning of mestizaje, what it included from each group and what resulted. In Mexico in 1925, José Vasconcelos lauded the *mestizo* as "the cosmic race," supposedly combining the best of all other races, although by 1944, he dismissed the concept as "one of my silly notions." In Andean countries, the *mestizo* was often disparagingly referred to as a *cholo*, a marginalized figure rejected by both whites and the indigenous. Brazilians would sometimes point to race mixture (*mestiçagem* in Portuguese) as supposed proof that racism did not exist in their country. Brazilian sociologist and anthropologist Gilberto Freyre heralded miscegenation as the essence of national identity and a successful adaptation to the tropics in his seminal *Casa Grande e Senzala* (*The Mansions and the Shanties*), which was wildly popular when it appeared in 1933 and frankly addressed issues of sexuality.

In fact, sexuality was at the heart of this view, exemplified by Brazilian novelist Jorge Amado's contention that racial problems could only be solved by "the mixture of blood." He contended, "No other solution exists, only this one which is born from love." Nonetheless, by the late-twentieth century, the cult of mestizaje had been transformed to cynicism, summed up deftly by poet Juan Felipe Herrera in *187 Reasons Why Mexicanos Can't Cross the Border* (2007)—"because we're still waiting to be cosmic."

In the twenty-first century, mestizaje has frequently been recast as *hybridity*, a notion more conceptual than racial, a postmodern pastiche of cultures past and present. Recently, the term hybridity has been challenged as well. The concern is usually that any of these terms suggests a purity of each element in the mixture before contact—an idea that the history of the pre-Columbian Americas, Iberia, and Africa shows not to be the case. Of even more concern is the idea of assimilation or acculturation, often portrayed as a one-way street of Iberian dominant culture overpowering indigenous and African elements. Mestizaje frequently has been employed by Latin American governments as a way of effectively eliminating black and indigenous communities, identities that were actively reclaimed in the late twentieth and twenty-first centuries.

Historian David Buisseret has suggested the use of the term *creolization*. Creole refers to people of European descent (or African descent in the U.S. colonies) born in the Americas. Creolization, then, describes "something that is born or developed in the New World." It was in this way, Buisseret notes, that "a truly new world came into being in those regions that the sixteenth-century Europeans prematurely called their 'New World'."

Hybridity or creolization may be new as an analytical concept, but the reality on which it is based is as old as the conquest that brought these people together. Each contributed to the formation of new societies on the basis of mixture and conflict among the three. Overlaying these societies were powerful institutions that were imported from the Iberian Peninsula but had to be adapted to local circumstances.

¿Latin? America

Latin America did not exist as a region until the nineteenth century. Before then, it was known by Iberians as *the New World, the Indies*, and *the colonies*. Only the last term was literally true. The region was a new world only to the Europeans, and obviously, not to those who lived there. It was the Indies only because of Columbus's misplaced hopes. And it was the Americas after another Italian navigator, Amerigo Vespucci.

Certainly, before the nineteenth century, no one conceived of a Latin America. The Aztec empire saw Anáhuac as home, but it would become the Viceroyalty of New Spain and then Mexico—a nation whose boundaries would change dramatically over time. Inca leaders called their home Tawantinsuyu, but it would become the Viceroyalty of Peru, and eventually it would become several countries: Ecuador, Peru, Bolivia, and Chile.

The idea of a Latin American people and region originated with French author Michel Chevalier, who theorized in the 1830s that the world was increasingly dominated by people whose roots were Anglo-Saxon, in contrast to the sensibilities of people, like the French, whose roots, at least linguistically, were Latin. In 1833, Chevalier was sent by the French government to the United States and Mexico, where he developed the idea that the Spanish- and Portuguese-speaking Americas shared the culture of Latin Europe and perhaps even formed a Latin race.

Chevalier's idea was transformed into *América Latina* by two South Americans living in Europe in the mid-nineteenth century—Chilean Francisco Bilbao and Colombian José María Caicedo. Bilbao used the term *latinoamericano* in writings in the 1850s and most notably at a conference in Paris in 1856. Both writers used the term to differentiate the former Iberian colonies from North America, particularly the expanding United States. In his 1857 poem "Las Dos Américas," Caicedo contends that the "Latin American race" shared origins and a mission—to confront the Saxon race, an enemy threatening Latin American liberty. By then, the United States had already acquired half of Mexico, and a U.S. filibuster had tried to conquer Central America.

Bilbao and Caicedo—both steeped in French intellectual thought, as were most Latin American elites—sought Latin American unity to offset U.S. ambitions. But Chevalier continued to see the region as linked to Latin Europe, particularly to France. This contention would give France a claim in the hemisphere. As adviser to Napoleon III, Chevalier's ideas bolstered the 1861 French invasion of Mexico.

Use of the term continued throughout the nineteenth century and became engrained in the twentieth. Indeed, *Latin America* is one of the oldest world regional designations. It is a problematic designation, however. Geographer Harm de Blij suggested to the American Association of Geographers in 2008 that if *Anglo America* is no longer routinely used in geography curriculum, as it was fifty years earlier, perhaps *Latin America* as a designation should be retired as well. *Latin America*, however, remains ensconced on the association's Web page, which includes links to Latin American geography organizations.

One reason that *Latin America* lives on, while *Anglo America* does not, is that the label has been embraced by the region itself in the years since independence, giving the term historical resonance. The term was reinforced in the post–World War II years by U.S. emphasis on area studies and by the formation in 1966 of the Latin American Studies Association, encompassing scholars from around the world and across disciplines who study Latin America. But it remains a complicated term.

In terms of physical geography, *Latin America* does not simply follow continental divisions, akin to *North* and *South America*—and these designations are fraught with problems as well. Until World War II, the Americas were frequently described as one continent. By the 1950s, most geographers concluded that there were two continents—but they did not agree on the dividing point. The southern border of North America over time has been drawn at the Isthmus of Tehuantepec, the Río San Juan, and the Costa Rica–Panama border. The latter was particularly the case when Panama was still part of Colombia. After Panama seceded in 1903, the dominant marker became Darien, making

Panama's southern border the edge of North America. Maps and geographical definitions are representations of power, and it can be argued that the designation of Darien increases the size of North America, increasing the sphere of the dominant North American power.

Within that North American area is Central America, which is geographically the area from Guatemala to Panama, though geopolitically it frequently excludes Belize and Panama. The region has variously been thought of as part of the continent of North America, or as a transition zone between the two American continents.

That Latin America is not merely a category of physical geography can be seen by the region's exclusions: Belize, formerly British Honduras, is rarely considered part of Latin America. On the South American continent, Latin America excludes Guyana, which was settled by the British; Suriname, settled by the Dutch; and French Guiana, which is still part of France. The exclusion of French Guiana (as well as Quebec) would seem to also favor excluding Haiti from Latin America. The main reason that Haiti frequently gets included in Latin America is because of the impact of Haiti's revolution on the independence movements of the region.

It has been suggested that a more accurate designation for Latin America might be *Ibero-America*, indicating the predominantly Spanish- and Portuguese-speaking countries that were colonies of Spain and Portugal. However, that designation seems, even more than *Latin America*, to privilege the role of the colonizing powers. *Hispanic* or *Hispano-America* does the same and leaves out Brazil.

Critics of the name *Latin America* argue that it leaves out the indigenous, African, and mestizo aspects of Latin America. But the critics have been short of suggestions for a replacement term that would encompass all of those elements of the region in a succinct name.

Some indigenous people in the region have adopted the name *Abya-Yala*, a word from the Kuna of Panama and Colombia meaning "place of life." Aymara leader Takir Mamani suggested the adoption of the term to mean "Continent of Life," in place of the term *Americas*. The name has been embraced by some indigenous people from Canada to Chile but has made little inroads outside the activist indigenous community.

In the 1990s, scholars began to rethink the *Latin America* designation. The Ford Foundation launched a project called "Rethinking Area Studies." Many institutes and centers chose new names: *Latin American and Caribbean Studies, Hemispheric or Global Studies, Centers for the Studies of the Americas*. These new approaches were not simply reactions to the dubious nature of the title *Latin America*, but were also ways of considering the complexity of the world in the late twentieth century and new ways of knowing about the world.

Anthropologist Arturo Escobar neatly lays out six elements to consider: "Latin America is literally the world over," with immigration and the growth of Latin American populations and cultures abroad; global connections outside national polities and economies put the understanding of Latin America in a larger context; subregional groupings—such as the Caribbean, including the Anglophone and Francophone countries—stress similarities among the islands; the limited categories of the past have been superseded by new emphasis on formerly ignored groups, such as African-descended communities; new crossdisciplinary frameworks move academics to think beyond simply economics or politics and the areas those disciplines define; and new sites of knowledge, outside the university, make it possible to listen to more people involved in nongovernmental organizations and social movements, organized in sub- and supranational groupings.

Nonetheless, Escobar was writing in the *Latin American Studies Association Forum*. *Latin America* is a world regional designation that is unlikely to disappear soon.

Whatever *Latin America* originally meant, it is a designation that we should try to understand now in all its richness and complexity as a region colonized by Iberian powers and populated by indigenous, African, and European people. From its origins, it was drawn into global networks, which have become increasingly complex over time. It is a region that since the moment of conquest has been characterized by change and continuity, conflict and cooperation, multiculturalism and hybridity.

Questions for Discussion

1. Why is the naming of the region as Latin America a subject of debate? What arguments do you find convincing and why?
2. How has the poverty of the Latin American region been explained?
3. How has Latin American geography impacted the organization of human societies in the region?
4. How does an understanding of the historical backgrounds of the indigenous, Spanish, and African regions contribute to understanding Latin America? What similarities and differences did these contributing regions share?
5. What is the significance of race and ethnicity in Latin America?

2 From Conquest to Empire

The exploits of dashing *conquistadores* roving in search of "God, gold, and glory" stir the imagination. Childhood stories portray clever Christopher Columbus as the great adventurer, or fearless Hernando Cortés, who supposedly defeated an entire empire with but a handful of men. But the "true story," to borrow from the account by Bernál Díaz de Castillo, is more complicated and far more intriguing. It is a story about human beings, not gods or primitives, who fought to gain or maintain position in a changing world.

The story unfolded over 300 years of imperial rule, when patterns of empire were established that would resonate throughout Latin America's history: the creation of a hierarchical society, structured by race and ethnicity, with power and wealth in the hands of a few, and dominated from abroad.

2.1: European Exploration

The encounter of New and Old Worlds was prompted by profound changes occurring in Europe in the late 1400s. The population had finally rebounded after one-third of the populace succumbed to the Black Death, a plague that spread throughout Europe beginning in 1347. The increasing, and increasingly urbanized, population provided a larger market and greater incentives for trade. Resolution of struggles among nobles, including the Hundred Years' War (1336–1453) and War of the Roses (1455–1487), led to consolidation of larger kingdoms, financed by merchants, who hoped the kings would create stability for local markets and relaunch foreign trade.

Their goals were furthered by new shipbuilding techniques, which allowed ships to travel longer distances, and new navigation equipment that made travel more accurate. In addition, the invention of movable type led to an increase in knowledge. The second book ever printed, after the Bible, was *Marco Polo's Travels*. Published in 1477, the book told travel tales from 200 years earlier and piqued a renewed interest in Asian goods. Merchants dreamed of breaking the Arab and Italian monopolies of trade with Asia, thereby sharing the lucrative profits from spices, precious stones, pearls, dyes, silks, tapestries, porcelains, and rugs coveted by wealthy Europeans. The products were available via overland routes, but these were controlled by the Ottoman Empire, which conquered Constantinople in 1453. Iberian merchants sought a route that would cut out the middleman, a water route that would take them to the East.

Portugal led the quest for new trade routes and became Europe's foremost sea power, due in part to its fortunate location on the westernmost tip of continental Europe. Most of the sparse population, less than 1 million in the fifteenth century, inhabited the coastal area, facing the great, gray, open sea and nearby Africa. The Portuguese initiated their overseas expansion in Africa in 1415 with the conquest of strategic Ceuta, guardian of the opening to the Mediterranean.

The first to appreciate fully that the ocean was not a barrier but a vast highway of commerce was Prince Henry (1394–1460), known as "the Navigator" to English writers, although he was a confirmed landlubber. In 1488, Bartolomeu Dias rounded the Cape of Good Hope and pointed the way to a water route to India. With Dias's success, the Portuguese were not interested in suggestions made by Christopher Columbus that a quicker route to the East could be obtained by sailing west. His timing was better when he approached Spain in 1492, just as the attention previously focused on the *reconquista* was turned to overseas expansion.

Cristobal Colón, as the Spanish called him, embodied many interests of his time. One of many Genoese living in Seville, he first went to sea as a young man, serving as a trader or clerk. A Genoese mercantile firm sent him to Portugal, where he learned maritime mapmaking. He went on to live on Madeira as the agent of a Genoese firm trading sugar and traveled the coast of Africa, learning the gold trade and observing the slave trade.

The story of Columbus is filled with myths, such as his supposedly novel idea that the world was round and that Queen Isabel pawned her jewels to finance the mission. In reality, Columbus did not need to convince any educated person that the world was round—Artistotle observed circa 300 B.C.E. that Earth cast a round shadow on the moon. The queen did not offer her jewels—in fact, the crown offered little more than recognition, 10 percent of the bounty encountered, and the title Admiral of the Ocean Seas. The crown did arrange for two ships, provided by the city of Palos as repayment of its debt to the crown. Columbus was financed primarily by Genoese merchants in Seville. He rounded up a crew and headed off in the belief that Asia was but 2,400 miles away. In actuality, the distance is 16,000 miles, with an unexpected, intervening landmass.

The return of Columbus from his first voyage intensified rivalry between Spain and Portugal. War threatened until diplomacy triumphed. At Tordesillas in 1494, representatives of the two monarchs agreed to divide the world. An imaginary line running pole to pole 370 leagues west of the Cape Verde Islands gave Portugal everything discovered for 180 degrees east and Spain everything for 180 degrees west. Within the half of the world reserved for Portugal, Vasco da Gama discovered the long sought water route to India. His voyage in 1497–1499 joined East and West by sea for the first time. Although subsequent voyages by Columbus (1493–1496, 1498–1500, 1502–1504) suggested the extent of the lands he had found, they also proved he had not reached India.

Portugal for the moment monopolized the only sea lanes to India, and that monopoly promised to enrich the realm. The cargo that Vasco da Gama brought back to Lisbon repaid sixty times over the original cost of the expedition. But the Portuguese had not lost interest in the New World, despite the limits of the Treaty of Tordesillas. The coast of Brazil was likely explored in a clandestine voyage by Duarte Pacheco Pereira in 1498. Pereira joined the expedition of Pedro Alvares Cabral in 1500, which supposedly was following da Gama's route east but was blown off course to Brazil. Some scholars believe that Cabral intentionally steered farther west to lay claim to the region, which later was determined to fall within Portugal's half of the world. Along the coasts of South America, Africa, and Asia, the Portuguese eagerly established their commercial empire, whereas the Spanish concentrated more on colonization.

2.2: Patterns of Conquest

The explorers who arrived after Columbus were not soldiers of the crown. Spain had no standing army and instead had recruited men to fight throughout the reconquista. The conquest of the New World was primarily a private venture. Companies of men outfitted themselves, borrowing money or buying supplies on credit, and would be paid with a share of the wealth they found. They were primarily literate commoners, lower-ranking professionals, and marginal *hidalgos* (noblemen), taking a chance to acquire wealth and prestige. Many had never before handled a sword. For example, among those accompanying Francisco Pizarro to conquer the Inca empire were twelve notaries and twenty-four artisans, including smiths, tailors, and carpenters.

These private undertakings were formalized by a contract, known as a *capitulación*, signed between the monarch and the aspiring conquistador, who was given the title of *adelantado*. Those contracts introduced European capitalism, as it had taken shape by the early sixteenth century, into the Americas. Royal officials accompanied all the private expeditions to ensure respect for the crown's interests and fulfillment of the capitulación.

The Spanish conquest proceeded as a relay system: After establishing a base in one area, the adventurers would explore farther and create new settlements, from Hispaniola (Haiti and the Dominican Republic) on to Cuba, Mexico, and Guatemala. The Spanish learned that there were many separate indigenous peoples; although some would rush to fight the Spanish, others would seek to aid them. The conquistadors would seize the *cacique*, or chief, and induce him to force his people to submit. The first Spaniards to arrive would divide the goods, leaving little for the later arrivals who followed to seek their fortune. Latecomers would then move on in search of new conquests, encouraged by Spaniards and indigenous who spoke of El Dorado, the mythical golden city that lay beyond.

Hernando Cortés became perhaps the most famous conquistador, but he started out as a marginal nobleman, a notary who fled to the New World after a narrow escape from a jealous husband. He was only nineteen years old when he arrived in Hispaniola in 1504, where, through kinship connections from his native Estremadura, he acquired an *encomienda*. The encomienda was an institution borrowed from the reconquista, whose forces had been paid by giving them control over a group of laborers. Cortés was given a second encomienda after serving as Diego Velasquez's clerk during an expedition to Cuba. Velasquez became governor of Cuba, and in 1519, he chose Cortés to lead an expedition to search for the Aztec empire. He quickly reconsidered, fearing that the ambitious Cortés would not be loyal. Velasquez's fears were well-founded: Hearing he would be replaced, Cortés took off early with 600 men, sixteen horses, eleven ships, and some artillery and headed north.

It took Cortés two years to defeat the Aztec empire. After first trying to convince him to leave, the central city–states of the empire fought valiantly. But Cortés was aided by thousands of indigenous who were subjected to the Aztec imperial yoke. Similarly, the illegitimate and modestly prepared Francisco Pizarro, after three attempts, finally conquered the Inca empire in 1535, aided by conflict between brothers Huáscar and Atahualpa. Their rivalry for the Inca crown divided the empire and provided the Spanish with thousands of indigenous allies. Many also accompanied the Spanish in other conquests.

The Spanish were also aided by diseases that they unwittingly brought. Because the Americas had been isolated from contact with Europe and Africa, the inhabitants had no immunities to diseases such as plague, smallpox, and typhus; even influenza and measles claimed lives in the colonies. Multiple and overlapping epidemics killed millions. Spanish swords were mighty weapons, and horses served as the tanks of the invasion, moving people and supplies long distances, but these technologies were not as significant as disease and indigenous divisions. Spanish technology, in fact, had its limitations. Heavy armor was impressive glinting in the sunlight at the conquistadors' first approach, but it turned out to be cumbersome and hot in Tenochtitlan. The Spanish adopted the indigenous protective clothing of thick cotton padding, which was lighter, cooler, and effective against the obsidian blade technology of the indigenous.

The Spanish pattern of exploration and settlement changed after 1521, the year not only of Cortés's triumph but also of the circumnavigation of the globe, begun by Ferdinand Magellan in 1519 and successfully concluded by Juan Sebastián del Cano. He proved that Asia could be reached by sailing west, but his expedition around South America also proved that the westward passage was longer and more difficult than the African route used by the Portuguese. In time, Spain realized it did not need the route to India. Conquered Mexico

Table 2.1 Years when Major New World Cities were founded

Founding of Major New World Cities	Year
Havana	1519
Mexico City	1521
Quito	1534
Lima	1535
Buenos Aires	1536, refounded 1580
Asunción	1537
Bogotá	1538
Santiago	1541

revealed that the New World held far more wealth in the form of coveted gold and silver than the Spaniards could hope to reap from trade with Asia. Instead of viewing the New World as a mere way station en route to Asia, the Spanish made America their center of attention.

The rest of the New World civilizations fell rapidly to the Spanish: Within half a century after Columbus's discovery, Spanish adelantados had explored and claimed the territory from approximately 40 degrees north—Oregon, Colorado, and the Carolinas—to 40 degrees south—mid-Chile and Argentina—with the exception of the Brazilian coast. Reflecting the Spanish preference for urban living, by 1550 the settlers had already founded all of Latin America's major cities. Meanwhile, the Portuguese largely ignored Brazil from its "discovery" in 1500 until 1532. Feeling pressured by Spanish and French interests, the crown authorized Martim Afonso de Sousa to found the first town, São Vicente, near present-day Santos. But the Portuguese followed patterns they established in Africa, focusing more on trading forts than colonization.

Legend has given the conquistadores glory for the conquest, but in reality they ended their lives with far less prestige. Columbus was returned to Spain in chains, accused of misgovernment. He spent the last months before his death trying to gain money and honors that he claimed he had been promised. Cortés, too, died in Spain complaining that his power had been preempted by the viceroy and arguing with local officials over the outrageous size of his encomienda. Pizarro was assassinated by followers of his former partner, Diego de Almagro, who felt that Almagro had been cheated out of his share of the wealth. Despite their achievements and outsized reputations, they were not significant in the establishment of stable, long-lasting colonies. In fact, their supposed feats of derring-do would not have been possible without the much more prosaic role of the merchants, who advanced the money and supplies to finance expeditions and support colonies. One merchant, Hernando de Castro, complained in 1520 that because of the power struggle between "this Cortés" and Cuban Governor Velasquez, "it is clearly no time to do business." And business was the point of conquest and colonization.

2.3: Colonial Economy

Expeditions to the New World were economic enterprises for both the crown and the conquistadores. Spain operated under an economic system known as mercantilism, which held that national wealth and power were built by controlling trade and gathering bullion. The goal was for the empire to sell more than it bought, keeping bullion in the kingdom's coffers. The adventurers who headed to the New World were equally motivated by wealth. As Bernál Diaz, chronicler of the Aztec conquest, commented: "We came here to serve God and also to get rich." Economic interest involved finding not only a marketable product, particularly mineral wealth, but also agricultural products. Control over labor and land were crucial to these economic enterprises. And finally, as a colonial enterprise, the crown would try to strictly control trade through monopolies.

The Americas at first frustrated the Iberians. The Taino/Arawak people of the Caribbean and the Tupí of the Brazilian coast showed no interest in transoceanic trade. The Portuguese contented themselves for three decades with exporting the brazilwood found growing close to the coast. The Spaniards lacked even that product to stimulate commerce in the Caribbean.

The Iberians focused on the search for a single item: gold. It offered three attractive advantages: easy shipment, imperishability, and high value. Columbus wrote, "O, most excellent gold! Who has gold has a treasure with which he gets what he wants, imposes his will on the world and even helps souls to paradise." The Portuguese encountered little of it in Brazil, but the Spaniards came into limited but tantalizing contact with the prized metal almost at once, as the Taino displayed some golden ornaments. The Taino also spoke of gold on various Caribbean islands, causing the Spaniards to begin a feverish search for deposits. The Spanish forced the indigenous to bring in quotas of gold; if they failed, the Spaniards cut off their hands. Nonetheless, only modest amounts of gold were found. Between 1501 and 1519, the Caribbean produced approximately 8 million pesos of gold.

The principal source of gold in Spanish America was New Granada (Colombia), which by 1600 had exported more than 4 million ounces of gold. Most of that gold came from placer deposits worked by slave labor. Production grew steadily, with eighteenth-century production nearly tripling that of the sixteenth, for a total supply of about 30 million ounces of gold.

In Brazil, the hardy *bandeirantes* (explorers) found gold for the first time in 1695 in the interior of Minas Gerais, followed by Mato Grosso in 1721 and Goiás in 1726. Such discoveries were incentives to open the vast southern interior of Brazil to settlement. Each discovery precipitated a wild rush of people eager to seek their fortunes. The boom caused notable population growth in eighteenth-century Brazil and a population shift from the older sugar-producing region of the Northeast to the newly opened regions of the Southeast. During the eighteenth century, Brazil produced 32 million ounces of gold, a majority of the world's supply.

Most of Spain's wealth, however, came not from gold but from silver. Spaniards found silver in New Spain at Taxco (1534), Zacatecas (1546), Guanajuato (1550), and San Luis Potosí (1592). In the sixteenth century, New Spain shipped more than 35 million pesos worth of precious metals to Spain, becoming Spanish America's leading producer toward the end of the seventeenth century. The single richest mine was discovered in 1545 at Potosí, a remote area in mountainous Upper Peru (later Bolivia). Collectively, these mines produced some 100,000 tons of silver during the colonial period. The crown carefully collected their *quinto*, or one-fifth, of the precious metals, employing large bureaucracies to oversee their interests.

Mining exemplified the region's dependence on exports. The crowns were hungry for gold and silver, but more than that, they were hungry for wealth. In areas without gold and silver, Iberians and their descendants searched for other exportable goods. They sent to Iberia—and on to Europe—the rich purplish-blue dye made from the indigo plant and the deep red dye, *cochineal*, made from the insects feeding on nopal cactuses. The hides of New World cattle made fine leather goods. Cacao beans fed the new desire for chocolate, sweetened by New World sugar. Success depended on the resources provided by geography and by perhaps the greatest wealth of the colonies—the labor force to produce the goods.

In search of a workable labor system, Spaniards looked to their own traditions. Seeing the conquest of the New World as a continuation of the reconquista, they transferred the *encomienda*, or entrustment, once used to control and exploit the Moors. The institution required the Spanish *encomendero* to instruct the indigenous entrusted to him in Christianity and to protect them. In return, he could demand tribute and labor.

The crown hesitated to approve the encomienda, strengthening a class of encomenderos who could impose their will between the monarchs and their new indigenous subjects. Isabel in 1501 ordered the governor of the Indies to free the indigenous, but when that experiment resulted in indigenous refusal to work for the Spaniards, the queen changed her mind. The monarchs tried repeatedly to control the encomienda. A 1503 royal edict institutionalized the encomienda but also expressed concern about indigenous welfare. In 1513, King Ferdinand promulgated the Laws of Burgos, calling for humane treatment, a policy difficult to enforce. Charles I tried to abolish the encomienda in 1519, but the edict was ignored. The New Laws of 1542 forbade indigenous enslavement and compulsory personal service, stopped inheritance of encomiendas, and proscribed new assignments. But the colonists protested vehemently: Rebellion threatened New Spain, and in Peru, encomenderos rose up to defy the law. The monarch acquiesced, modifying some laws and revoking others.

The encomienda system was most effective among sedentary indigenous populations, accustomed to participating in draft rotary labor systems in the Aztec empire (*coatequitl*) and the Inca (*mita*). Where no such indigenous base existed, Spaniards simply could not establish the encomienda. This was apparent in Zacatecas, when the Spanish tried to induce northern indigenous groups, such as the nonsedentary Chichimec, to work in the silver mines.

Because nothing could be done without labor, the encomiendas were highly prized. But while Cortés's encomiendas numbered 100,000, latecomers and lower-ranking conquistadores were less fortunate. For example, Bartolomé García wrote plaintively to the crown that in Paraguay he was given sixteen Guaraní workers, located some eighty leagues away, "from where one can get no service." Where there were semisedentary people—Tucumán, Paraguay, and Chile north of the Bío Bío River—the Spaniards devised a modified encomienda system by integrating with the indigenous people, forming a kinship group and becoming their leaders.

The encomienda was weakened by two trends—moral opprobrium and demographic pressure. The crown received reports from the Church that encomenderos abused their indigenous charges. One of the most vocal defenders of the indigenous was Bartolomé de las Casas, a Dominican missionary and later bishop. His persuasive arguments contributed to Pope Paul III's 1537 bull declaring that the indigenous were fully capable of receiving the faith of Christ—that is, they possessed souls and should not be deprived of their liberty and property.

But moral concerns did not weaken the encomienda as much as the demographic pressure of continued immigration and demand for labor. Perhaps one in fifty Spaniards who came to the colonies was awarded encomiendas. But immigrants continued to come, inspired by letters encomenderos wrote home about their success. Melchor Verdugo wrote from Peru in 1536, "[I have] a very good encomienda of Indians, with about eight or ten thousand vassals; I think there's never a year that they don't give me 5,000 or 6,000 pesos in income."

Replacing the encomienda was the *repartimiento*, the temporary allotment of indigenous workers for a given task. The Spanish colonist in need of laborers applied to a royal official explaining the work to be done, the time it would take to complete, and the specific number of indigenous it would require. In theory, the crown officials ensured fair payment and satisfactory working conditions; in practice, abuse abounded. Planters and miners constantly badgered or bribed royal officials to bend the system, which flourished from about 1550 to 1650.

In addition to furnishing an agricultural labor force, the repartimiento system also provided the major share of workers for the mines in South America. In the Viceroyalty of Peru, the *mita* at Potosí became particularly burdensome. All adult male Quechua and Aymara were subject to serve in the mita one year out of every seven. Far from home, the miner worked under dangerous conditions and earned a wage insufficient for half of his own and his family's expenses.

As in Spanish America, landowners in Brazil relied in part on the indigenous as a source of labor. Some employed indigenous labor from the *aldeias*, the villages created by the crown and religious orders to concentrate the nomadic populations. Protected within the village, they were introduced to Christianity and European civilization. In return, they gave a portion of their labor to the Church and State. This part of the aldeia system resembled the encomienda. In addition, planters could apply to the aldeia administrators for paid indigenous workers to perform a specific task for a specified period of time, similar to the repartimiento. But the aldeia system included only a small percentage of Brazil's mostly nonsedentary indigenous people.

The existence of a sedentary indigenous workforce was a key difference between labor organization in Brazil and in Spanish America. Because the Tupí died off or fled to remote areas, the Portuguese turned to the source they used on the island of Madeira—African slaves. Africans offered the added benefit of immunity to European diseases because of the many years of contact. By the end of the sixteenth century, Africa furnished most of the productive labor in Brazil and the West Indies. Brazil and Spanish America received nearly 60 percent of the Atlantic slave trade.

Although most slaves worked in sugar production first in Brazil and eventually in Cuba and Haiti, slaves were not limited to these areas. African slaves were the backbone of Latin America's skilled labor force: artisans, muleteers, herdsmen, and skilled labor at the mines. Africans also were important as cowboys, particularly in Venezuela, Brazil, and Argentina. Urban slaves also worked as domestic servants, peddlers, mechanics, and artisans, particularly for their skilled ironwork.

Justifications for slavery included the idea that slaves were captured in just wars, or that slavery was a small price to pay for Christianization. Even Las Casas initially welcomed African slavery as preferable to enslaving the indigenous, although by the 1520s, Las Casas renounced his earlier advocacy. There were some moral qualms about African enslavement; as Alonso Montúfar, archbishop of Mexico City wrote to King Philip II in 1560: "We do not know of any cause why the Negroes should be captives any more than the Indians, since we are told that they receive the Gospel in good will and do not make war on Christians." The Spanish wrestled with the issue for decades, but in the end, economic and political interest won over morality. The Council of the Indies assured Charles II, after a review of slave-trade policy in 1685, "There cannot be any doubt as to the necessity of these slaves for the support of the kingdoms of the Indies, and . . . to the importance of the public welfare of continuing and maintaining the procedure without change."

Male slaves outnumbered females by a ratio of almost two to one. One explanation has been that large landowners preferred men because they considered them better field hands and hence more profitable. However, many African slave-owning societies preferred to own women, who had primary responsibility for agriculture and could be integrated into kin groups as concubines or wives. As a result, prices for women were higher than for men. Nonetheless, women in the New World often worked in the fields, in addition to performing domestic service. Their owners discouraged large-scale reproduction as uneconomical. Thus, the Latin American slave system was seldom self-sustaining and required constant replacement through the slave trade.

Demand for labor did not always lead to enslavement. As early as 1546, indigenous workers were paid wages to lure them to the relatively remote mines of Zacatecas. As the repartimiento system became more cumbersome, wage labor systems spread. Some employers tried to keep wages low and labor nearby via debt peonage, making deceptively friendly loans to be repaid with labor. Wages often did not suffice to liquidate the debt, which fathers passed on to sons. But the very labor scarcity that prompted debt peonage also gave workers maneuverability. Workers gravitated toward employers promising better salaries or working conditions, sometimes accepting loans from one employer, then moving on to another. Without a strong military or police force, fleeing workers were difficult to catch.

Under all the labor systems, one pattern remained the same: Masses of indigenous and African workers toiled for the benefit of small numbers of European elites and their descendants—called creoles (*criollos*) in Spanish America and *mazombos* in Brazil—who always constituted a small yet dominant minority of the population.

Initially, land was abundant, and Spaniards were more concerned about acquiring the labor to work it for them. However, competition for land intensified as more Spaniards immigrated to the colonies, the indigenous population recovered from the initial demographic collapse after the conquest, and lucrative markets for agricultural products developed both abroad and in the colonies. Ownership of land became a basis for wealth and prestige and conveyed power. From the beginning, the adelantados distributed land among their followers as a reward for services rendered.

In 1532, when Martim Afonso founded São Vicente, he distributed land with a lavish hand, quite contrary to prevailing custom in Portugal where, beginning in 375, kings sparingly parceled out the *sesmaria*, traditional land grants, so that no one person received more than could be effectively cultivated. Aware of the immensity of the territory in front of him, Afonso ignored such a precaution. Good coastal land was quickly divided into immense sugar plantations, and in a few decades, huge sesmarias for cattle ranches dominated the interior as well.

Many of the original land grants grew to gigantic proportions as astute landowners bought out their neighbors or simply encroached upon other lands. The declining indigenous population freed more land, which the Iberians rapidly monopolized. A series of legal devices favored the Spaniard in acquiring land: the *congregación, denuncia*, and *composición*. The congregación concentrated the indigenous in villages and thereby opened land for seizure. The denuncia required the indigenous to show legal claim or title to their property in order to retain it, and the composición used surveys to claim land, legalities for which their ancient laws had not prepared the indigenous. Over time, Spanish landowners steadily pushed the indigenous up the mountainsides and onto the arid soils of marginal lands.

The Portuguese monarchs were critical of the inefficiency of the large *fazendas* and later tried to limit the estate size. An eighteenth-century viceroy, the Marquis of Lavradio, complained bitterly that the huge estates were poorly managed, only partially cultivated, and retarded Brazil's development. He disparaged the unused fields held as symbols of prestige while farmers petitioned for land to till. Some regions imported food they were capable of producing themselves. Nonetheless, the *latifundia* remained a dominant characteristic in Brazil and Spanish America. There were *haciendas* in New Spain exceeding 1 million acres, and the Diaz d'Avila ranch in Brazil surpassed most European states in size.

The Luso-Brazilians quickly developed the prototype of the plantation economy, thanks to the profitable market in Europe for sugar, which grew well along the coast. By 1550, Pernambuco produced enough sugar in its fifty mills to annually load forty or fifty ships for Europe. Brazilian sugar plantations flourished from 1550 to 1650, establishing the pattern of monoculture, the dominance of a single crop. In contrast, plantation agriculture came later to Spanish America, where haciendas produced for local markets and to feed mining towns. Only as the eighteenth century neared did haciendas enter international trade on a scale comparable to Brazilian plantations.

The type of life exemplified by the hacienda or fazenda often has been termed *feudal*, a term with a strong emotional overtone connoting exploitation. Certainly, the classical feudalism of medieval society did not appear in the New World. Weak though his power might have been in remote areas, the king never relinquished the prerogatives of sovereignty to the landlords. Royal law prevailed. Neither does the self-sufficient manorial system properly describe the large estates because, for all their self-sufficiency, they were closely tied by their major cash crop to the world capitalist economy. Perhaps the *patrimonialism* defined by Max Weber is most accurate. Under patrimonialism, the landowner exerts authority as one aspect of his property ownership. Those who live on his land are under his control, and he uses force arbitrarily to enforce his personal authority on his estate. He controls all trade between his estate and the outside world, and through that trade, he participates in the capitalist marketplace.

The plantations, ranches, and mines provided a rich and varied source of income for Iberian monarchs, capitalists, and merchants. Sugar, tobacco, cacao, indigo, woods, cotton, gold, silver, diamonds, and hides were among the natural products the colonies offered the Old World. The Iberian Peninsula depended on New World products for its prosperity. For example, for many years Brazilian products constituted approximately two-thirds of Portugal's export trade. Iberian settlers searched for any means of making money, but they were restricted by geographical constraints, Iberian trade controls, and the desires of European markets. Often just one or a small number of products dictated whether a colony prospered or endured stagnation and misery.

For American elites, colonial wealth would be worth nothing without the European goods that it could buy. Spaniards disdained such indigenous products as maize and insisted on their familiar wheat, olive oil, and wine. From the earliest colonial days, merchants constituted a small but important class. They united into trade associations, *consulados*, and obtained formidable privileges. The crown authorized the first consulado for New Spain in 1592; the merchants of Lima received permission for one in 1613. Thereafter, the consulado spread through the empire, a significant indicator of increasing trade with the metropolis.

The Spanish fleet system shipped all goods between Spain and the colonies, starting in the 1520s. There were two fleets a year, one to New Spain and one to Cartagena and Panama. The system was successful in protecting its shipments, with only two losses: one to Sir Francis Drake in 1572 and the other to the Dutch in 1628. Weather was a more serious threat. The fleet system effectively protected Spanish trade until it was abandoned as inefficient in the late eighteenth century.

Simultaneously, there was tremendous contraband trade among the colonies and with other countries. The vast territory and limited number of officials made it impossible for Spain to fully control colonial trade. Ships would put in at the many uncontrolled coves, or would claim to be in distress, which gave them the legal right to put in to port at other locations. It is impossible to know the extent of contraband, but some scholars estimate that as much as two-thirds of the region's trade was illegal.

Although foreign trade was the motor of activity, the growth of mining centers led to local demand for food and goods, providing a market of lower-income Spaniards and indigenous. Local entrepreneurs served the new markets by carting, raising livestock, road building, processing imported goods, and creating *obrajes*, or workshops, for local manufacture. *Tratantes*, dealers in locally produced goods, had lower status than merchants linked to international commerce, but they managed a lucrative business. Nonetheless, most of the colonial world was a subsistence economy, operating outside all but local markets for surplus goods and artisanry.

The Iberians did not achieve notable efficiency in the exploitation of those natural products with which a generous nature endowed their lands. With little competition and cheap labor, there was no incentive to overcome old-fashioned, inefficient methods. For example, the Portuguese held almost a monopoly on sugar production for over a century. Between 1650 and 1715, the Dutch, English, and French increased sugar production in the Caribbean; they were efficient and organized, had new equipment and extensive financial resources, and enjoyed a geographic position closer to European markets. As a result, their sugar economies prospered, while Brazil's languished.

With quick and large profits as its goal, the economy of Latin America was largely speculative and hence subject to wide variations. The patrimonial system of land and labor contributed to economic inefficiency. In sum, the economy of the American colonies was not geared to their own best interests but to the making of immediate profits for the Iberian metropolises and a small, New World planter-trader elite, almost exclusively of European origin.

2.4: Colonial Administration

Spain and Portugal ruled their American empires for more than three centuries, a remarkable longevity that places them among the great imperial powers of all time. They owed that success to quite different concepts of imperial organization. The Spanish colonial administration was relatively well organized, the hierarchical ranks well defined, and the chain of command easy to recognize. The Portuguese empire was loosely organized with more transitory institutions. The different colonial systems were rooted in their own histories and the material conditions they encountered.

In both kingdoms, the monarch was the State. He ruled as the supreme earthly patriarch in the hierarchy of God, King, and Father, as envisioned in St. Augustine's concept of Christian monarchy. The mystique and tradition of the monarchy gave the institution such force that no one questioned the king's right to rule or refused his loyalty to the crown. It was the crown that legitimated all colonial arrangements.

However, the great distance between the metropolis and colonies and the slowness of communication and travel worked to confer considerable local autonomy on New World officials. In practice, kings could only hope to dictate the broad outlines of policy, leaving interpretation and implementation to colonial and local officials. *Obedezco pero no cumplo* ("I obey but I do not fulfill") became the accepted way for New World officials to manifest loyalty to the crown while bending the laws to suit local situations. This approach permitted flexibility that acknowledged imperial and colonial interests. As the third governor-general of Brazil, Mem de Sá, confided to the king, "This land ought not and cannot be ruled by the laws and customs of Portugal; if Your Highness was not quick to pardon, it would be difficult to colonize Brazil."

To keep their royal officials and subjects in check, the Iberian monarchs sent to the New World officials of unquestioned loyalty. At best, they suspected that the colonies increased in everyone "the spirit of ambition and the relaxation of virtues," and they hesitated to appoint many Americans to the highest colonial posts. But as the empires matured, Americans increasingly held influential ecclesiastical, military, and political positions. Despite royal frowns, Iberian officials married into distinguished American families, linking the local and Iberian elites. The crown's check on these officials was the *visita*, an on-the-spot investigation to which all subordinates could be subjected. And at the end of all terms of office, each administrator could expect a judicial inquiry into his public behavior.

Considering the size of the American colonies and handful of royal officials, almost all residing in the most populous cities, the extent of Iberian influence over the colonies was remarkable. The crowns maintained authority and control principally through the power of legitimacy or hegemony. Americans accepted the system, rarely questioned it, and seldom challenged it. Popular uprisings, mostly motivated by economic discontent, broke out periodically, but the populace reacted to specific grievances rather than adhering to any philosophy advocating systemic change.

The Spanish crown was assisted in administration of the colonies by the Council of the Indies (*Consejo de las Indias*), which was not formed until 1524. By contrast, the *Casa de Contratación* (House of Trade), in charge of colonial commerce, was founded in 1503. The Council comprised a group of ministers, named by the king, who controlled legislative, executive, and judicial decisions regarding the colonies. Like the king, however, the Council was an ocean away from the territories it sought to govern.

The king's principal representative was the viceroy. Columbus bore the title of viceroy, and the various conquistadores served as governors of the territories they conquered. Their administrative skills varied considerably, and their self-interest made them questionable crown representatives. In 1535, the crown appointed Antonio de Mendoza, a trusted diplomat of Charles V from one of Spain's foremost families, as first viceroy of New Spain. By 1551, he had imposed law and order, entrenching royal power. Control of Peru, the second viceroyalty, was more difficult. The first viceroy arrived in 1543 to find chaos, rivalries, and civil war. Not until the able administration of the fifth viceroy, Francisco de Toledo (1569–1581), was the king's authority firmly imposed.

The *audiencia* was the highest royal court and consultative council, with wide-ranging powers to make political and administrative decisions and serve as final arbiter for civil and criminal complaints. The first audiencias were established in Santo Domingo, 1511; Mexico City, 1527; Panama City, 1535; Lima, 1542; and Guatemala, 1543. The tenure of audiencia judges, *oidores*, exceeded that of the viceroy and overlapped each other, thereby providing continuity to royal administration.

Captaincies-general were the major subdivisions of the viceroyalties. Theoretically subordinate to the viceroys, in practice they communicated directly with Madrid and paid only formal homage to the viceroys. Ranking beneath them were the governors, *corregidores*, and *alcaldes mayores*, chief administrators of municipalities. Within their localities, these minor officials possessed executive, judicial, and limited legislative authority. Municipal government, known as the *cabildo* or *ayuntamiento*, initially provided the major opportunity for the creole, the American-born white. Parallel to the creole cabildos were indigenous cabildos; Spaniards relied on indigenous leadership to collect tribute and to organize labor for public works. In short, royal administration of the colonies rested on a handful of royal officials who oversaw Spanish settlers, their creole descendants, and the indigenous, who actually produced the wealth.

Although Brazil emerged as Portugal's most valuable overseas possession, until the nineteenth century it was governed by no special laws or institutions distinguishing it as a separate entity within the empire. From 1532 to 1534, to colonize Brazil without reaching into royal coffers, the Portuguese monarch distributed Brazil in the form of large captaincies to twelve

Map 2.1 Viceroyalties in the Colonial Americas, 1776

SOURCE: Craig, Albert M.; Graham, William A.; Kagan, Donald; Ozment, Steven M.; Turner, Frank M. *Heritage of World Civilizations Combined*, 7th ed. © 2006. Electronically reproduced by permission of Pearson Education, Inc., Upper Saddle River, NJ.

donataries who enjoyed broad powers in return for colonizing the land. By 1548, John III reversed that decision and reasserted his authority. In 1549, a central government under a governor-general brought some order. In 1646, the king elevated Brazil to a principality, and thereafter the heir to the throne was known as the Prince of Brazil. After 1720, all chiefs of government of Brazil bore the title of viceroy.

Portugal was well into the sixteenth century before the rulers distinguished between home and overseas affairs. No special body exclusively handled Brazilian matters. A variety of administrative organs exercising consultative, executive, judicial, and fiscal functions assisted the king. One of the most important was the Overseas Council (*Conselho Ultramarino*) created by John IV in 1642, to advise the king on various military, administrative, judicial, and ecclesiastical matters.

In Brazil, Salvador da Bahia, a splendid port, served as the first seat of the central government. The colony was divided into two states in 1621: Maranhão in the north, which was much poorer than the larger state, known as Brazil. The Dutch occupied a portion of the northeastern sugar coast from 1630 to 1654. By the end of the seventeenth century, Brazil faced a growing threat from the Spanish in the Plata region. In defense, in 1680, the Portuguese founded the settlement of Colônia do Sacramento on the bank of the Rio de la Plata. The Spanish challenge to Portuguese claims to the region caused a century and a half of intense rivalry and frequent warfare.

A growing colonial bureaucracy administered Brazil: the High Court (*Relação*), established in Bahia in 1609 and in Rio de Janeiro in 1751, had judicial responsibilities, while the Board of Revenue (*Junta do Fazenda*) handled taxes and treasury supervision. Captaincies were the principal territorial subdivisions, and governors or captains-general carried out the same responsibilities on a regional level as the governor-general or viceroy did on a broader scale. However, Brazilian leaders never exercised the same degree of authority as their counterparts in Spanish America.

The municipal government was the one with which most Brazilians came into contact and initially the only one in which they participated to any degree. In sparsely settled Brazil, the municipalities contained hundreds and often thousands of square miles, making the municipal council, or *senado da câmara*, the most important institution of local government. There was no Brazilian equivalent to Spanish America's indigenous town councils.

2.5: The Catholic Church

The Church was perhaps more important than royal administration. Ferdinand and Isabel were the Catholic monarchs, and they took their position seriously. The monarchs defended the faith within their realms, and in return the Pope conferred royal patronage, permitting the Iberian monarchs to exercise power over the Church in all but purely spiritual matters. They collected the tithe and decided how it should be spent; appointed and recalled ecclesiastical officials; and authorized construction of new churches. Royal patronage meant that the state dominated the Church, but it also allowed the Church to pervade the state—clerics often occupied top governmental posts, serving as ministers, captains-general, viceroys, and even regents. Cardinal Henry, after all, ruled the Portuguese empire in the sixteenth century.

The Catholic Church, one of the most important institutions of colonial Latin America, was central to the goal of turning the indigenous into Spanish subjects. The magnificent cathedral in the Zocalo of Mexico City was built on the ruins of an Aztec temple. (Library of Congress)

Columbus's instructions stated, "the King and Queen, having more regard for the augmentation of the faith than for any other utility, desire nothing other than to augment the Christian religion and to bring divine worship to many simple nations." Accordingly, twelve friars accompanied Columbus on his second voyage to the New World. Conversion was essential not only to give the indigenous the true faith and eternal salvation but also to make them loyal subjects to their Most Catholic Majesties. To be Portuguese or to be Spanish was to be Roman Catholic. Religion served as a proto-nationalism as the Catholic monarchs used faith to measure citizenship, starting with the expulsion of Jews and Muslims. Iberians were born, reared, married, and buried Catholics, and the Church touched every aspect of their lives.

Like the government, Church structure was hierarchical and territorial, with archbishops located in viceregal capitals. Below them were bishops in the other major cities, followed by clergy in cathedral chapters and urban parishes. At the bottom of the ecclesiastical structure were the priests in the countryside, instructing the indigenous in encomienda villages. These were the most recent arrivals, those with the least training and no connections. Like the encomienda, the rural parish was erected directly on the indigenous provincial unit, the Aztec empire's *altepetl* and the Incan *ayllu*. The priests used the authority of the caciques to get churches built and to get the indigenous to attend services. They also relied on the *fiscal*, an indigenous leader who served as a Church officer, and recruited sacristans and singers to participate in Church functions.

Priests behaved in the same way as other immigrants to the New World. They used ties of kinship and region to build their power bases, and they married their female relatives off to encomenderos, miners, and wealthy merchants. Their economic base was built on the pious donations given to them, which they used to buy income-producing property and to build elaborate churches. Elites tended to join the regular orders—Franciscans, Dominicans, and Augustinians—whereas those of lower rank became the secular priests. Not tied to an order, they participated in society much like any other Spaniard—they looked for ways to make money.

The colonial church earned income through tithes, the sale of papal indulgences, and parochial fees, and grew wealthy through the legacies left by the elites. The Church accumulated vast estates, quickly becoming the largest landowner in the New World. The Church also became an important lender, and ecclesiastical coffers grew with fees and interest rates on loans. Church wealth was by no means evenly distributed. In cities such as Lima, Salvador da Bahia, Antigua, and Mexico City, ostentatiously imposing churches crowded one another, whereas "shocking poverty" characterized hundreds of humble parish churches dotting the countryside. While some higher clergy lived in great wealth, impoverished clerics served the faithful in remote villages. Wealth reinforced the conservative inclinations of the Iberian Church. It was not from the masses that the Church drew its leadership. Generally the sons of the wealthy and/or noble became bishops and archbishops, positions in the New World that were dominated by the European-born. Thus, the highest ranks of the clergy, like those in the civil service, were filled by the elite.

The relationship between the Church and the indigenous was a complicated one. Initially, the clergy were responsible for evangelizing, converting the indigenous to Christianity. Their initial level of success is questionable; for example, the Nahuatl expression for baptism was to pour water over one's head—descriptive but not indicative of much understanding. A few clergymen, such as Bernardo de Sahagún, labored to understand the indigenous by teaching their leaders Spanish and Latin while simultaneously learning the native languages. Few went as far as Sahagún, who with indigenous assistance compiled a multivolume compendium of native history and tradition, the *General History of the Things of New Spain* (*Historia general de las cosas de Nueva España*, 1540–1585).

A religious people, the indigenous did not need to be convinced of the existence of deities. The indigenous were polytheistic and accustomed to adding the gods of conquerors to their pantheon of deities. After all, the god of the conqueror must be stronger; the symbol for conquest in Nahuatl, the language of the Aztec empire, was the burning temple. A more difficult problem was Catholic insistence on monotheism, and the proliferation of Latin American saints days can be attributed to the linking of Catholic saints with local deities. Each village had its own saint and *cofradia*, or lay brotherhood, devoted to festivals and commemoration of the saint.

The Catholicism that emerged among the indigenous was a mix of pre-conquest and indigenous beliefs. The indigenous appropriated symbols of Christianity, such as the cross and the rosary, and combined them with traditional spiritual symbols. For example, the indigenous continued to pile stones along roads to call on spiritual powers to protect travelers, but added a cross on top. They perhaps gave up idols from the past, but replaced them with statues of Mary.

The process of religious change was neither successful conversion by an imposing Church nor victorious resistance by indigenous who prayed to the idols that they hid behind altars. It was instead an ongoing conversation, one in which the indigenous were more likely to have the final say. Priests could not control the meanings that indigenous assigned to Christian messages, if for no other reason than the sheer impossibility of a small number of priests trying to control the private beliefs of a much larger indigenous population. As late as 1792 in the Viceroyalty of Peru, there were only 3,700 priests for about 1 million people.

Like the indigenous, Africans had their own belief systems, which continued in the New World alongside Catholicism. In most African belief systems, there was not one universal creator, but ancestor gods who created and looked after specific peoples. There was no concept of heaven or hell, but of temporal and spiritual worlds. Religion was practical, a way to understand and control events in the temporal world, and there was a constant dialog between the two worlds, facilitated by diviners, healers, and spirit possession. Africans added some Catholic elements to these spiritist beliefs, creating new mixtures such as Santería in Cuba, candomblé in Brazil, and vudún in Haiti. Catholic priests found their ability to reach Africans limited by a shortage of clergy, as well as differences in cosmologies, language obstacles, and lack of concern by slave owners.

Some clergy took seriously their role as protectors of their new congregants, a position exemplified by Antonio de Montesinos and Bartolomé de las Casas. In 1511, Montesinos chided parishioners in Santo Domingo. Calling himself "the voice of Christ crying in the wilderness of this island," he denounced colonists' abuse of the indigenous and demanded, "Are they not men? Do they not have rational souls? Are you not bound to love them as you love yourselves?" He was joined in his cause by Las Casas, once an encomendero himself, who became the most famous supporter of the indigenous cause. His *Short Account of the Destruction of the Indies* (*Brevísima relación de la destrucción de las Indias*, 1552) detailed, with some exaggeration and hearsay, the abuses of the Spaniards. Widely read in Europe, the work contributed to the "Black Legend," the claim that the Spanish conquest was more brutal than conquests in other parts of the world. This conquest, though brutal, was certainly not more so than the many conquests

Europeans who had never set foot in the New World imagined the conquest as particularly bloodthirsty, a view that fueled the "Black Legend" of Spanish cruelty. This engraving by Theodore de Bry shows conquerors watching as their dogs attacked the indigenous. (Library of Congress)

that characterized European history; the critique was motivated mostly by European jealousy of the Spanish.

Although many priests helped the indigenous, most clergy believed in the hierarchical social system that put the indigenous and African slaves at the bottom of society. The indigenous served as a workforce for the Church, as well as for secular society, tilling Church lands and building local churches and opulent urban cathedrals. Some clergymen were also guilty of abuse of native populations, as evidenced by many indigenous petitions to the crown.

Perhaps most importantly from the Iberian point of view, the clergy upheld the social order from the pulpit, where they preached resignation. If God had made them poor or slaves, it would be a sin to question why. Poverty was to have its reward in the next life. It was a message not just for the indigenous, but for the lower ranks of Iberian and mestizo society as well. The messages of passivity and acceptance legitimated a hierarchical society.

The Church maintained a careful vigil over that society. Alleged backsliders, especially Jewish converts, the New Christians, could expect to account for themselves before the Inquisition. Philip II authorized the establishment of the Holy Office in Spanish America in 1569, and it began to operate in Lima in 1570, in Mexico City in 1571, and in Cartagena in 1610. He exempted the indigenous from the jurisdiction of the Inquisition "because of their ignorance and their weak minds." The Inquisition often served a more political than religious end by targeting particular individuals in its vigilant efforts to purify society. The considerable power of the Inquisition lasted until the last half of the eighteenth century. As an institution it was never established in Brazil, but it operated there through the bishops and visitations from Inquisitors.

The Church was also responsible for much of the culture in the Spanish colonial world. While the Church censored books, it also educated Americans and fostered most of the serious scholarship in the New World. The Church exercised a virtual monopoly over education. Monasteries housed the first schools and taught reading, writing, arithmetic, and Catholic doctrine. The Spanish monarch encouraged the founding of universities in the New World, granting the first charters in 1551 to the University of Mexico and the University of San Marcos in Lima, mostly staffed by clergy. Before Harvard opened in 1636, a dozen Spanish American universities, drawing on medieval Spanish models, offered courses in law, medicine, theology, and the arts, mostly in Latin, but also taught theology students indigenous languages.

The Church also offered women an alternative to family life, a choice not always made voluntarily. Patriarchs sometimes placed their daughters in convents to guard their virginity or to prevent marriage. A marriage could mean a huge dowry of land, property, or capital, which the father preferred to pass on intact to a son. On the other hand, widows often retired to a convent from which they administered their wealth, estates, and property. A religious life was not dominated exclusively by prayer and meditation, nor were nunneries necessarily dreary houses of silence, service, and abnegation. Through the religious orders, women operated schools, hospitals, and orphanages. In some convents, the nuns, attended by their servants, entertained, read secular literature, played musical instruments, sang, prepared epicurean delights, and enjoyed a lively and comfortable life.

One of the most remarkable intellectuals of the colonial period was a nun, Sor Juana Inés de la Cruz (1651–1695), whose talent earned her fame in New Spain during her lifetime and whose complex and brilliant poetry ensured her literary legacy. A favorite of the viceregal court, she chose to enter the convent to pursue the kind of intellectual life that was not open to women in secular society. At the convent of San Jerónimo she had her own library and study, where she conducted *tertulias*, literary salons. Though famous for her poetry, she was equally interested in philosophy and natural science. With a change in the viceregal and bishopric hierarchy, she fell from favor and was criticized for her worldly studies and the antipatriarchal tone of some of her writing. Under pressure from the hierarchy, she sold her extensive library of some 4,000 volumes, as well as her musical and scientific instruments. She died in 1695 after contracting the plague while nursing her sick sisters.

The Virgin de Guadalupe

The melding of Catholic and indigenous tradition is most effectively embodied in the Mexican veneration of the Virgin of Guadalupe. According to legend, an indigenous man named Juan Diego was crossing the hill of Tepeyac on December 9, 1531, when a beautiful, dark-skinned woman appeared to him. She instructed him to tell the bishop, Juan de Zumarraga, that she "ardently wish[es] and greatly desire[s] that they build my temple for me here, where I will reveal . . . all my love, my compassion, my aid, and my protection."

Juan Diego followed her instructions, but the skeptical bishop demanded proof. The disconsolate man returned to see the Virgin, who instructed him to go to the top of the hill and gather flowers. The hill had been barren, but the amazed man found the hilltop in bloom. He gathered the flowers in his cloak and brought them to the bishop on December 12. When he opened his cloak to show the out-of-season flowers, they spilled out and left an image of the Virgin on the lining of the cloak.

Convinced and repentant, the bishop ordered the construction of the church. Conveniently, the very spot where the dark Virgin appeared was the site of a temple to the indigenous goddess Tonantzin, whose temple Zumarraga had ordered destroyed. Tonantzin had been an earth goddess, mother of gods, and protector of humanity.

Similarly, the Virgin de Guadalupe Tepeyac is seen as a protector, especially of the poor and oppressed. She is distinguished from the Spanish Virgin de Guadalupe de Estremadura by her location at what had been an indigenous holy site and by her appearance. Because she is dark skinned, *morena*, she provided a link with Catholicism for the indigenous.

The original church was replaced by a larger one in 1709, and a new basilica was dedicated at the site in 1976. Juan Diego's mantle is preserved and on display at the shrine, which draws thousands of devout Catholics each year, especially on December 12.

A statue depicts the dark-skinned Virgin of Guadalupe, whose miraculous appearance at Tepeyac is celebrated December 12. (Ingram Publishing/Newscom)

2.6: The Conquered Peoples

The arrival of the Spanish led to the greatest demographic disaster in history. While the exact number of pre-Columbian inhabitants is not known, it is clear they died in precipitous numbers—only 10–20 percent of their original numbers remained. Even discounting the actual conquest years, tribute records in New Spain show a decline of 80 percent from 1548 to 1595. In coastal areas and the Caribbean Islands, estimates have ranged closer to 100 percent. "The Indians die so easily," commented a German missionary in 1699, "that the bare look and smell of a Spaniard causes them to give up the ghost." The indigenous population began to grow in the mid-seventeenth century, but never recovered its previous numbers.

The deaths devastated the indigenous world. Often people weakened by one disease died from a secondary infection. So many fell ill that they were unable to nurse and feed each other, and therefore famine followed disease. There was also an enormous psychological toll as survivors saw people around them die by the thousands. In attempts to control the population, Spanish priests gathered the remaining people into newly formed villages, *reducciónes*, where disease spread even more rapidly.

The conquest was not felt uniformly across the region. The greatest impact was experienced in fully sedentary indigenous societies, which offered organized populations that the Spanish sought as laborers. The Spanish gave less attention to the nonsedentary and semisedentary indigenous, particularly in areas where there appeared to be no mineral wealth or other profitable product. The less-sedentary indigenous could more easily move to remote territories and frequently remained beyond Spanish control. For example, the Mapuche of Chile launched repeated attacks on the Spanish throughout the entire colonial period.

In the imperial societies, conquest meant that old leadership structures were gone, frequently along with the physical structures in which they were located. Spaniards destroyed the impressive city of Tenochtitlan and built Mexico City in its place. The cities built on the sites of old indigenous cities became Spanish centers, and indigenous life, with the exception of the indigenous intermediaries, moved entirely to the countryside and rural villages.

The ethnic unities that supported state organization declined, and native priests at all levels came systematically under attack: Lands that had been worked for their benefit were confiscated and redistributed; indigenous priests were killed, brought to trial, or harried into secrecy. The result was turmoil; but part of the turmoil enabled the lower classes and oppressed indigenous groups to free themselves from their previous bonds. Some indigenous commoners eagerly denounced arbitrary acts and pagan habits of their nobles. Some towns of landless farmers attached to indigenous noble estates declared their independence.

The political consequences of the conquest fell heaviest on indigenous elites above the city-state level, and provincial chieftains quickly lost their authority. The majority of the natives did not suffer a fall from social eminence because they were commoners to begin with. Indeed, some may argue that the new rulers were less demanding and less divine. There was an increasing blurring of class lines, with the breakdown of sumptuary laws, which had stipulated that only indigenous nobility could possess cacao or consume alcoholic beverages. In New Spain, there was a growing business in production of *pulque*, an alcoholic drink made by fermenting cactus juice, and the establishment of taverns and inns. What was once a custom reserved for ritual celebration became a common public practice.

Initially there was little contact between most indigenous and Spaniards, with indigenous leaders serving as intermediaries. Even late in the colonial period, indigenous leaders continued to serve as brokers between the communities. Eventually, as both populations grew and an increasingly complex economy and society drew the two worlds together, there was greater individual contact. The process was quite slow, even in the zones of greatest contact, as is shown by one useful indicator—the adoption of the Spanish language. For example, in what is now Central Mexico, it was not until 1650–1800 that many indigenous communities exhibited full bilingualism. As late as the Mexican Revolution in 1910, revolutionary leader Emiliano Zapata found it necessary to give speeches in Nahuatl, and non-Spanish-speaking populations are still significant today.

Indigenous communities displayed remarkable cultural resilience and were selective about what to incorporate and what to ignore. Spanish tools were frequently adopted, though indigenous styles of dress were maintained. Indigenous first names were followed by Spanish surnames. But basic preconquest community organization remained, as did household structures.

For some indigenous, the colonial world presented opportunities. Indigenous leaders bargained with colonists in exchange for organizing their peoples' labor, tribute, and passivity. The indigenous produced products for new markets, including subsistence products for laborers, such as corn and chile in New Spain and potatoes and coca leaves in Peru. They also saw opportunities in new markets, such as the production of cochineal.

New Spanish crops and especially the introduction of livestock changed the countryside, but there was enough land so that indigenous still held their own until late in the colonial period. They had to contend with stray cattle from the Spanish ranches wandering into subsistence plots and causing damage. In addition, the Spanish tried to monopolize trade, breaking up the old, indigenous commercial networks. The indigenous were able to trade, but not to dominate the old trade patterns as they had in the past.

The indigenous were not without resources—they used their internal unity, as well as intergroup conflict, to set the terms of interaction with the Spaniards. For example, communities underreported their numbers to lower tribute and service obligations. They recruited migrant indigenous to work community lands while community members were meeting their labor obligations. But they also called on Spanish authorities to enforce laws that would protect them from their own abusive leaders as well as from the excesses of the Spanish world.

The indigenous used the tools of the Spanish system, including petitions to higher authorities and lawsuits. And when those measures failed, communities turned to rebellion, endemic to colonial Latin America. Sometimes it was widespread, threatening entire regions, although more frequently it was restricted to the village level. After the initial resistance to conquest, later rebellions were usually reactions to changes in established relationships, such as increases in taxes or other obligations.

The indigenous came to view the crown as their protector, and indeed, the crown's attitude was that the indigenous needed protection: They were seen as *niños con barbas*, children with beards. Royal officials helped establish the corporate rights of indigenous villages, hoping to protect the indigenous from the worst features of Spanish ways and allow the best features to be carefully introduced. To that end, the Spanish colonies were divided legally and socially into two worlds: the *República de Indios*, Republic of Indians, and the *República de Españoles*, Republic of Spaniards. However, the separation of the two worlds was never complete, and race mixture blurred the boundaries as time went by.

2.7: Colonial Society

The basic organizational principle of the colonial world was hierarchy, characterized by concerns about status and honor. It was a patriarchal society, giving ultimate status to men over their dependents: women, children, and laborers. It was a system that was color coded: The hierarchy was topped by whites. Beneath them were the mixed groups—mestizos and mulattos—who emerged when those at the top mixed with those at the bottom—the indigenous and blacks, whether Africans or those born in the colonies.

While these categorizations are generally true, there were also divisions within each group. Iberian-born whites (Spanish *peninsulares* and Portuguese *reinóis*) considered themselves better than whites born in the New World (Spanish American *criollos* and Brazilian *mazombos*). The American-born alternated between admiration for peninsulares and scorn for latecomers who did not understand the New World. Social standing was not only linked to nobility but also to religion, wealth, occupation, and service to the crown. Peninsulares in the sixteenth century laughed at creole use of the honorifics *don* and *doña* (lord and lady) by anyone with an *encomienda*. In the eighteenth century, they scoffed that the creoles' only claim to status was that they were white.

Within the indigenous world, there was also differentiation based on nobility, economic status, wealth, and gender. Especially in the early colonial period, the Spanish recognized that some indigenous had claims to nobility—claims that were more valid than those of low-ranking Spaniards who came to the New World seeking their fortunes. In Peru, many indigenous *kurakas* were able to hold onto their status more successfully than heirs of the early conquistadores.

Africans might rank phenotypically at the bottom in the Iberian view, but they frequently had skills that the indigenous did not possess. Blacks born in the Americas were more acculturated to the Iberian world, and therefore ranked higher than new arrivals. Many blacks were able to buy their freedom, and a growing population of free blacks earned higher status.

Although Spaniards may have wished to see a colonial world neatly divided into Spanish and indigenous republics, the African presence challenged such duality. The first Africans to accompany the Spanish to the New World were viewed by the indigenous as just another type of Spaniard: After all, free and acculturated Africans from Spain, called *ladinos*, and African slaves participated in the conquests of the Caribbean and the Aztec and Inca empires. While the Spanish did not see blacks as their equals in the República de Españoles, they did try to keep them out of indigenous villages. But these attempts were no more successful than attempts to keep Spaniards out. Much of the historical record documents hostility between the indigenous and the African, but new research shows many exceptions to the pattern. Indigenous and Afro-Latin American people often came together voluntarily, particularly on the fringes of the colonies.

Africans and their descendants could be found in all parts of Latin America and formed a large part of the population. Their presence dominated the plantations on which they worked, and their influence spread quickly to the "big house," where African women served as cooks, wet nurses, and companions to the woman of the house, while black children romped with white children. But African influence also permeated the cities. In the sixteenth century, blacks outnumbered whites in Lima, Mexico City, and Salvador da Bahia, the three principal cities of the Western Hemisphere.

Slaves continued to focus on their original nations, and, particularly in urban areas, they elected kings and queens with great ceremony, celebrated traditional festivals, and formed mutual aid societies. Whatever cultural differences Africans may have had, they were also able to organize multiethnic runaway slave communities, known variously as *palenques* or *cumbes* in Spanish America and *quilombos* in Brazil. The most famous quilombo, Palmares, located in Alagoas, lasted roughly from 1630 to 1697; it included as many as 5,000 African slaves from different African regions, who spoke different languages but were united in the community. There were six attempts to conquer Palmares between 1680 and 1686, and the final assault was won only after a bitterly fought forty-four-day siege. Quilombos, in fact, were so established as communities that they competed with Brazilian farms and stores to sell supplies in the Rio de Janeiro area.

The threat of these separate African communities in some ways was not as great as the threat of the African presence within Spanish and Portuguese colonial society. The issue of purity of blood was a sensitive one to Spaniards already biased against the impurity of Jewish or Muslim heritage. Racial purity was added to the equation as a small population of Iberians was surrounded by a much larger indigenous and black population.

Racial purity would be complicated by the new mixtures. The position of mestizos and mulattos depended both on appearance and identification. Since the offspring of race mixture were most commonly the children of white fathers, they tended to be raised by their mothers in indigenous or black communities, with which they identified. Nonetheless, categories of mixture, as well as of indigenous and black were not fixed, but instead varied by place and time.

Although Spanish and Portuguese men mixed freely with indigenous and African women, they tried to maintain strict control over Spanish and Portuguese women to maintain a pure lineage. These concerns reinforced the Iberian tradition of the patriarchal family, sanctified by religion and serving as a model for the organization of society and polity.

During the early decades of conquest and colonization, more European males than females arrived in the New World. Thirty women were allowed on Columbus's third voyage in 1498, and in 1527, the crown licensed brothels in Puerto Rico and Santo Domingo "because there is need for it in order to avoid worse harm." In later conquests, Spanish women sometimes fought alongside the men. Doña Isabel de Guevara participated in the conquest of the region of La Plata in the 1530s. In seeking an encomienda years later, Doña Isabel recounted, "The men became so weak that all the tasks fell on the poor women, washing the clothes as well as nursing the men, preparing them the little food there was, keeping them clean, standing guard, patrolling the fires, loading the crossbows when the Indians came sometimes to do battle, even firing the cannon, and arousing the soldiers who were capable of fighting, shouting the alarm through the camp, acting as sergeants and putting the soldiers in order. . . . [O]ur contributions were such that if it had not been for us, all would have perished; and were it not for the men's reputation, I could truthfully write you much more and give them as the witnesses."

The paucity of Spanish women led many men to turn to indigenous women. From the beginning, the conquerors regarded indigenous women as part of the conquest, and rape was common. Indigenous men would frequently hide women from the Spaniards in the Caribbean islands. Other indigenous groups gave women to the Spaniards to cement alliances. The conquest of women is in many ways a metaphor for the conquest of the New World itself—the conquest of virgin territory. The most famous example of Spanish conquest of indigenous women is the story of Malintzin, known by the Spanish as *Doña Marina* or *Malinche*, a term that in Mexico has come to mean *traitor*. Malintzin was sold into slavery from the Nahua-speaking world to the Mayan territories, where she was later given as a gift to Cortés. Because she could speak Nahuatl, she became Cortés's translator. She bore him a son, but within a few years he married her off to one of his men, Juan de Jaramillo.

To increase stability in the colonies, the crown insisted that married men bring their wives to the New World and encouraged the export of eligible young women to marry off in the colonies. By the 1540s, there was one Spanish woman for each seven or eight men. As in Spain, marriages were arranged to benefit family wealth and position. Marriage linked encomenderos, miners, government officials, and merchants.

In this patriarchal organization, the male head of the family dominated the household and its businesses: In rural Brazil, that included plantation, slaves, and tenants. Although males of the household liberally expanded the basic family unit through their sexual escapades to include hosts of mestizo and mulatto children, the Iberian woman was expected to live a pure life of seclusion. Ideal models were set for the behavior of the women of the patriarch's family, who were destined either for matrimony or religious orders. They were to remain virgins until their marriage and to live separated from all men except their fathers, husbands, and sons. A high value was placed on their duties as wives and mothers, and control over their sexuality was essential to guarantee that the Iberian elites remained of pure blood. Although women of other economic strata could not follow such an elitist model, they were certainly influenced by it.

Despite the patriarchal bias, Iberian women enjoyed more freedoms than their European sisters. Women retained their family names when married, and both maternal and paternal surnames were passed to the children. Women owned half the property of their marriages and inherited equally with their brothers. Widows were frequently executors of men's wills and guardians of their children, and their freedom often resulted in great societal pressure for them to remarry. Sometimes the women would choose instead to retreat to convents as *beatas*, holy women, though not technically nuns. In the New World, women took on new roles: They were left in charge of encomiendas when their husbands traveled and proved to be sharp businesswomen. Some wealthy women were active in the business world, investing in real estate or manufacturing. They sometimes bought and sold slaves and took a direct role in the administration of their properties.

Whereas elite Iberian and creole women were expected to stay mostly in the home, lower-class women, of necessity, participated in many occupations. Women of the middle sectors tried to work in their homes, by renting rooms or taking in sewing. But lower-class women were very much in the public view, working as spinners, producers and sellers of brandy (*aguardiente*), bakers, healers (*curanderas*), midwives, seamstresses, potters, candlemakers, innkeepers, and landladies. They dominated the local market, selling food, produce, and handicrafts, and frequently worked as domestics in the homes of the upper classes. Women's roles were clearly linked to the issue of economic class and occupation, which in turn was coupled with race and ethnicity.

All of these groups were made up of individuals, who struggled with strategies for survival and advancement. Race, ethnicity, class, and gender shaped and limited their responses, but the boundaries between groups were porous. Slaves were freed; women became landowners. Not even race and ethnicity were stable. People shifted from one category to another by moving to new locations, marrying outside their community, renaming, and even buying documents that indicated a new racial status. There was constant negotiation for status within these hierarchical societies, but for 300 years, one concept was not questioned—that the Americas were part of the Spanish and Portuguese realms.

Questions for Discussion

1. What were the motives and methods of the Iberians who sailed to the New World and the indigenous people who greeted and fought them?
2. How was colonization a different challenge than conquest?
3. What institutions did the Iberians bring, and how were they affected by the colonial context?
4. What was the role of the Catholic Church in the colonies?
5. How did the categories of race, ethnicity, class, and gender interact in colonial society?
6. How were the Iberians able to maintain these colonies for more than 300 years?

3 Independence

Spain and Portugal managed to hold on to their sprawling overseas possessions for more than 300 years, an empire of impressive duration. But it is, perhaps, inevitable that distant subjects one day recognize first that they are capable of standing alone and subsequently that they need their independence to flourish economically and politically. As Spain fought European wars, Latin Americans found themselves at times experiencing the wealth of free trade and at others successfully handling their own military defense. They came to see themselves as the true rulers of their lands, and a feeling of inferiority before the Iberian born gave way to a feeling of equality and then to superiority. A greater appreciation of the regions where Latin Americans were born and raised gave rise to feelings of nativism, a group consciousness attributing supreme value to the land of one's birth and pledging unswerving dedication to it. Spain tried to gain better control of the colonies, but those attempts did more to provoke than to contain their colonies. The result was the struggle for independence. With the exception of Haiti and the initial efforts in Mexico, it was the creole and mazombo elites who directed the movements toward independence. But these elite leaders relied on the darker and poorer masses to win the wars. The majority populations had their own reasons for fighting, which were often at odds with the leadership, and the independence wars were simultaneously civil wars.

3.1: A New Sense of Self

Before the end of the first century of Iberian colonization, the inhabitants of the New World began to reflect on themselves, their surroundings, and their relations to the rest of the world. They spoke and wrote for the first time in introspective terms. Juan de Cárdenas, although born in Spain, testified in the *The Marvelous Problems and Secrets of the Indies* (1591) that in Mexico the creole surpassed the *peninsular* in wit and intelligence. Evincing a strong devotion to New Spain, Bernardo de Balbuena penned *The Grandeur of Mexico* (1604) in praise of all things Mexican. He implied that for beauty, interest, and charm, life in Mexico City equaled—or surpassed—most Spanish cities. Similarly, the Peruvian Franciscan Fray Buenaventura de Salinas y Córdova in his *Memoir of the Histories of the New World, Peru* (1630) sang the praises of Lima as "a holy Rome in its temples, ornaments and religious cults, a proud Genoa in the style and brio of those born in it . . . a wealthy Venice for the riches it produces for Spain . . . a thriving Salamanca for its university and colleges."

In 1618, Ambrósio Fernandes Brandão made the first attempt to define or interpret Brazil in his *Dialogues of the Greatness of Brazil*. In doing so, he exhibited his devotion to the colony, chiding those Portuguese who came to Brazil solely to exploit it and return wealthy to the peninsula. Poets, historians, and essayists reflected on the natural beauty of a generous nature. They took up with renewed vigor the theme extolled in the early sixteenth century that the New World was an earthly paradise.

The creoles were in a dialog with disdainful Spaniards. Royal chronicler Juan López de Velasco contended in his *Universal Geography and Description of the Indies* (1570) that the "creoles turn out like the natives even though they are not mixed with them [by] declining to the disposition of the land." His view was echoed by Bernardino de Sahagún (1590): "I do not marvel at the great defects and imbecility of those who are born in these lands because the Spaniards who inhabit them, and even more those who are born here, assume these bad inclinations."

The peninsular Spaniards called their colonial brethren *criollos*, an adaptation of the Portuguese *crioulo*, a term used to refer to black slaves born in Brazil. Its use to refer to Spanish Americans was derogatory and implied questionable racial purity. However, creoles chose to adopt the word and invest it with pride. In 1683, Juan Meléndez, a creole Dominican priest from Lima, asserted, "In order to distinguish ourselves from the Spaniards born in Spain, we call ourselves creoles, a term that Spaniards undoubtedly laugh at very much; but they laugh using the same logic as those who laugh at everything they do not understand. This is typical of stupid people who do not deserve to figure as human beings."

These nativist attitudes were given further impetus by Enlightenment philosophers who viewed the Americas as inferior to Europe. The denigration of all things American was most pointedly expressed in the *Philosophic and Political History of the Establishment of European Commerce in the Indies*, which first appeared in 1770 and again in fifty editions during the next thirty years. Americans were not only dismissed as less intelligent than Europeans, but even less virile: "The men there are less strong, less courageous, without beard or hair: degenerate in all signs of manhood. . . . The indifference of the males toward that other sex to which Nature has entrusted the place of reproduction suggests an organic imperfection, a sort of infancy of the people of America similar to that of the individuals in our continent who have not reached the age of puberty."

Latin Americans leaped to their own defense. Juan Vicente de Güemes Pacheco y Padilla, the creole son of a former viceroy of New Spain, insisted that American men were "extremely inclined to engage in sex." On a more lofty note, Francisco Javier Clavijero likened the ancient Mexican world to the classical societies of Greece and Rome in his four-volume history of New Spain, *Ancient History of Mexico* (1780–1781). The popular work was disseminated widely and in several languages, and the glorification of ancient indigenous societies would become a common trope. Juan José de Eguiara y Eguren, rector of the University of Mexico, showed that the subsequent inhabitants of New Spain were equally erudite by publishing the *Mexican Bibliography* (1755), a compendium of scholarly works. The peninsulares, however, remained unconvinced.

3.2: The Bourbon Reforms

Despite Iberian disdain for Americans, colonial rule under the Habsburg kings involved consensus between ruler and subject. Far from an absolute monarch, the Habsburg king needed cooperation from his subjects, who were able to negotiate for more satisfactory arrangements. The Habsburgs did not even use the word colony to refer to their American possessions. They were the kingdoms of the Indies, just as Spain was also composed of its kingdoms in Iberia and elsewhere in Europe. In the eighteenth century, however, the Bourbons, a French dynasty, came to power in Spain. Their hope was to create an absolutist monarchy and to put the Americas in their place as colonies.

The Bourbons attained the Spanish throne in 1700, after Charles II, the last Habsburg king, died without leaving an heir. Philip V's accession to the throne was contested by Archduke Charles of Austria, which led to the War of the Spanish Succession (1702–1713). It was the first of many European wars over control of both European and colonial territory that would impact the Americas during the next century. The new Bourbon leaders found a bankrupted kingdom, overextended in European affairs, with an inefficient system of colonial administration. It would take three kings and nearly a century for the Bourbons to address the problems.

The Bourbons intended to reform government, increase governmental control of the economy, and limit the power of the Church. These measures were necessary on the peninsula as well as in the colonies, and much of Philip V's administration (1700–1746) was spent securing control of Spain's provinces while the country remained embroiled in European wars. Philip met with limited success—in 1739, Spain shocked its creditors by suspending payments on its responsibilities, tantamount to declaring bankruptcy. By the time of Philip's death, his administration had achieved limited change.

Philip's heir, Ferdinand VI, used his father's strengthened state authority to make economic reforms. His ministers increased controls on colonial trade and levied heavier penalties for unregistered treasure, resulting in a substantial increase in colonial revenues reaching the crown. The administration also increased crown control by halting the sale of audiencia and corregidor offices. But Ferdinand's reign was brief, ending with his death in 1759.

It was primarily the work of the third Bourbon king, Charles III (1759–1788), whose measures would come to be characterized by historians as the Bourbon reforms. Charles was a more able monarch than his mentally unbalanced predecessors. Historian John Lynch classified him as "a prodigy among Bourbon misfits."

By the mid-eighteenth century, the Bourbons faced increased interest in their colonies on the part of other European monarchs, along with a hefty bill for the costs of participation in the Seven Years' War (1756–1763). Charles was determined to reform the imperial structure, with the goal of increasing administrative efficiency, political control, and profits. The power of the Council of the Indies waned as the ministers of the king took over many of its former duties. In the eighteenth century, the chief responsibility for the government of Spanish America rested in the hands of the Minister of the Indies. The Casa de Contratación also felt the weight of Bourbon reforms: The king's ministers absorbed so many of its powers that it was abolished in 1790.

Attempts to bring the Americas under tighter imperial control were led by José de Gálvez, who Charles sent to New Spain as visitor general from 1765 to 1771. Gálvez, who has been described as aggressive, ill tempered, and intolerant, was notorious for despising creoles. It fell to Gálvez to enforce the king's 1767 edict expelling the Jesuits from the colonies. The edict was greeted by uprisings in Guanajuato, San Luis Potosí, Pátzquaro, and Uruapan. Riots were suppressed by Spanish troops and local militia, and Gálvez meted out fierce punishment: 85 people were hanged, 73 flogged, 117 banished, and 674 imprisoned. Creole judges, priests, or officeholders who so much as questioned these measures were expelled. Far from frowning on these excesses, the king appointed Gálvez as Minister of the Indies in 1776, a post he held until his death in 1787.

The expulsion of the Jesuits asserted crown control over the Church, without challenging more powerful denominations, such as the Franciscans. The Jesuits controlled extensive wealthy estates in the Americas and were the object of jealousy by rivals in the church and in the community. About 2,600 Jesuits were expelled from Spanish America and an equal number from the peninsula. One priest in New Spain described how, without warning, soldiers arrived at his college and told the priests to pack. Two days later, after allowing one last visit to the shrine of Our Lady of Guadalupe, the priests were taken to Veracruz, shipped off to Cádiz, and then sent to the Papal States. Some elderly and infirm priests did not survive the arduous journey. The vast majority of Spanish American Jesuits were creoles—for example, 500 of the 678 from New Spain. When they left, the many Jesuit colleges were closed.

The Spanish troops and local militias who aided Gálvez in New Spain were another of the significant changes in the late colonial period. As European rivalries were frequently played out in colonial territory, Spain chose to maintain a standing army on the peninsula and in the Americas. A small number of soldiers were sent from Spain, but the majority was recruited in the colonies. In New Spain, regular army troops eventually numbered around 10,000, with another 20,000 in the militia. The idea was to transfer the cost of colonial defense to the Americans.

Initially, the militias were almost fictional and served little purpose other than to give titles to miners and merchants. The granting of military commissions to the creoles afforded them a new prestige. Usually high ranks were reserved for—or bought by—wealthy members of the local aristocracy. Hence, a close identification developed between high military rank and the upper class. Further, creole officers enjoyed a practical advantage, the *fuero militar*, a special military privilege exempting them from civil law. It established the military as a special class above the law, and the results would be increasingly disruptive for Latin American society. For the middle and lower classes, militia service was frequently less appealing, as they were called away from their shops and jobs to train, parade, and occasionally fight. Whereas the upper classes could pay for substitutes, the middle and lower classes lacked resources for such an option.

The most radical Bourbon innovation was the establishment of the intendancy system, an administrative unit used by the Bourbons in France and copied by their relatives in Spain. Philip V had tried in vain to establish the intendancy to control Spain's provinces (1711–1720). The institution was finally installed on the peninsula in 1749 and brought to the colonies starting in 1764 in Cuba. The intendants, royal officials of Spanish birth, with extensive judicial, administrative, and financial powers, supplanted the numerous governors, corregidores, and alcaldes mayores in hopes that a more efficient, uniform administration would increase royal revenues and end corruption. In financial affairs, the intendants reported directly to the crown. In religious, judicial, and administrative matters, they were subject to the viceroy. By 1790, the system extended to all the Spanish American colonies.

The creoles had always been denied the highest positions in the Americas. Of 170 viceroys, only four were creoles—and they were the sons of Spanish officials. Of the 602 captains-general, governors, and presidents in Spanish America, only fourteen had been creoles; of the 606 bishops and archbishops, 105 were born in the New World. But, by the late colonial period, creoles were accustomed to holding most of the remaining positions. Not only did creole families own most of the rural estates, they also provided nearly all parish priests and dominated the lower reaches of imperial government, particularly as secretaries and petty administrators. As the colonial viceroys brought fewer people with them, creoles filled out their entourages. The creoles not only dominated the cabildos, but they also held the majority in the audiencias.

Now, however, the crown began to restrict creole appointments to even those offices, reducing their ranks by about two-thirds. This blow came at a moment when the bureaucracy was expanding and creole population was increasing. The intendants, all peninsulars, formed a new layer of political power between creoles and the crown and fanned the flames of antipeninsular reaction. Lamented Antonio Joaquin de Rivadeneira y Barrientos, a Mexico City judge, "They come to govern a people they do not know, to administer laws they have not studied, to encounter customs with which they are not familiar, and to deal with people they have never seen before."

The Iberian suspicion of the New World elite in effect questioned their ability and loyalty. One Portuguese high official remarked of Brazil,

"That country increases in everyone the spirit of ambition and the relaxation of virtues." Intendant Francisco de Viedman in Upper Peru complained: "For appointments here natives of the country are not suitable, because they are extremely difficult to dissuade from the customary ways ingrained in them even in contravention of the laws. . . . even Spaniards who live for some time in these parts come to acquire the same or worse customs."

The new policies were aimed at making the colonies more profitable and bringing more income to Spain. The Bourbon kings infused a more liberal economic spirit into the empire, which at first blush would appear to favor colonial economic welfare, although the intent was to strengthen Spain by liberalizing trade, expanding agriculture, and reviving mining. However, the crown also set up a more efficient tax administration, which made sure that colonists actually paid their taxes. Cádiz had already lost its old commercial monopoly when Philip V permitted other Spanish ports to trade with the Americas and ended the fleet system in 1740. In the 1770s, Charles lifted the restrictions on intercolonial commerce. The opening up of trade, once restricted to trade fairs, fleets, and the ports of Cádiz and Seville, led to an explosion in trade volume. Mining production doubled in New Spain and Peru, and there were booms in the export of sugar, cochineal, dyewood, tobacco, cacao, and hides from throughout the colonies. But along with freer trade came more crown monopolies: on gunpowder, playing cards, salt mines, official stamped paper, coins, and mercury.

Obviously, the points of view of the Iberian and the American varied. The first came to the New World with a metropolitan outlook. He saw the empire as a unit that catered to the well-being of the European center. The latter had a regional bias. His prestige, power, and wealth rested on his mines and lands. His political base was the municipal government, whose limited authority and responsibility reinforced his parochial outlook. In short, he thought mainly in terms of his region and ignored the wider imperial views.

In Portugal, the Braganzas were as taken with ideas of reform as the Bourbons. Portuguese mercantilism was never as effective as that of its neighbor, particularly before 1750. Attempts were made sporadically to organize annual fleets protected by men-of-war to and from Brazil, but the highly decentralized Portuguese trade patterns and a shortage of merchant ships and warships caused difficulties. Between the mid-seventeenth and mid-eighteenth centuries, the crown partially succeeded in instituting a fleet system for the protection of Brazilian shipping. Still, it never functioned as well as the Spanish convoys. The four economic companies licensed by the crown fared little better. They were unpopular with residents of Brazil and with merchants in Portugal and Brazil. The Brazilians criticized them for raising prices with impunity. Merchants complained that monopolies restricted trade, and accused the companies of charging outlandish freight rates. Nonetheless, crown monopolies flourished for brazilwood, salt, tobacco, slaves, and diamonds.

Royal control over Brazil tightened during the eighteenth century. The plantations produced their major crops for export, demand dictated incomes, and incomes regulated production. Though Lisbon encouraged the capitalist trends toward higher production, it continued to prevent direct trade with European markets, much to the increasing frustration of Brazilian elites. Their desire to enter the capitalist marketplace of the North Atlantic, along with the imperial, mercantilist, and monopolistic policies of the Portuguese crown, charted a course of conflict prompting greater royal control.

The absolutist tendencies noticeable during the reign of John V (1706–1750) were realized under Sebastião José de Carvalho e Melo, the Marquis of Pombal, who ruled through the weak Joseph I (1750–1777). Pombal, the Brazilian equivalent of Gálvez, hoped to strengthen his economically moribund country through fuller utilization of its colonies. To better exploit Brazil, he centralized its government. He incorporated the state of Maranhão into Brazil in 1772. He dissolved the remaining hereditary captaincies, with one minor exception, and brought them under direct royal control and tried to restrict the independence of municipal governments. In 1759, he expelled the 600 Jesuits, accusing the order of challenging the secular government and interposing itself between the king and his indigenous subjects. Since many of the Jesuits were American-born, their expulsion was deeply resented in the colonies.

3.3: The Temptations of Trade

In the late colonial period, there was a new orientation in the colonies toward the Atlantic coast. In response, Spain created two new viceroyalties: New Granada in 1738, including modern Colombia, Venezuela, and Ecuador, and La Plata in 1776, encompassing parts of modern Chile, Argentina, Bolivia, and Uruguay. La Plata became important as a route between the Peruvian silver mines and Atlantic ports, and increased demand for hides brought prosperity to the region. Venezuela rose in importance as a wheat producer for Cartagena, where the fleet was supplied, and as a cacao producer. Cuba followed Brazil's lead, and between 1750 and 1770, the island converted to a sugar economy, with exports leaping from a mere 300 tons a year to 10,000 tons.

The privileged classes in the New World were most interested in the reform of commerce and trade. American merchants and planters chafed under crown monopolies and restrictions, especially as the more liberal Spanish commercial code of 1778 and burgeoning European contraband brought Latin America into ever-closer contact with Europe's more dynamic economies.

Americans were also consuming European ideas, and they were particularly interested in rationalism and scientific study of nature and economy. The works of René Descartes and Isaac Newton quickly entered the university curriculum. John Locke's conceptions of individual property rights dovetailed nicely with Adam Smith's liberal ideas of free trade and an unfettered market. American elites discussed the new ideas at *tertulias*, whereas the masses frequented cafes, where newspapers were read aloud and events discussed.

The revolution in scientific and economic ideas led to the formation of Economic Societies of the Friends of the Country (*Sociedades Económicas de Amigos del País*), which had developed in Spain and spread to the New World by the 1780s. Portuguese America had not a single university, but its counterpart of the Spanish economic societies was the six literary and scientific academies established between 1724 and 1794 in Salvador and Rio de Janeiro. Each had a short but apparently active life.

Physiocrat doctrine—the idea that wealth derived from nature (agriculture and mining) and multiplied under minimal governmental direction—gained support among Brazilian intellectuals. They spoke out in favor of reducing or abolishing taxes and soon were advocating greater freedom of trade. From Bahia, João Rodrigues de Brito boldly called for full liberty for the Brazilian farmers to grow whatever crops they wanted; to construct whatever works or factories were necessary for the good of their crops; to sell in any place and through whatever agent they chose without heavy taxes or burdensome bureaucracy; to sell to the highest bidder; and to sell their products when it best suited them.

Similar demands reverberated throughout the Spanish American empire. Chileans wanted to break down their economic isolation. José de Cos Iriberri, a contemporary of the Bahian Rodrigues de Brito, concluded, "Crops cannot yield wealth unless they are produced in quantity and obtain a good price; and for this they need sound methods of cultivation, large consumption, and access to foreign markets." Fellow Chilean Manuel de Salas agreed and insisted that free trade was the natural means to wealth. And countryman Anselmo de La Cruz asked a question heard with greater frequency throughout the colonies: "What better method could be adopted to develop the agriculture, industry, and trade of our kingdom than to allow it to export its natural products to all the nations of the world without exception?"

As the American colonies grew in population and activity, and as Spain became increasingly involved in European wars in the eighteenth century, breaches appeared in the mercantilist walls that Spain had carefully constructed around its American empire. British merchants audaciously assailed those walls and when possible widened the breaches. During the Seven Years' War, the Caribbean became a zone of military occupation. Havana was then the third largest city in the New World, smaller than Mexico City and Lima but larger than New York and Boston. When the island was taken by the British in 1762, Havana was opened to free trade. In ten months, more than 1,000 ships entered the port, compared with 1,500 ships in the preceding ten years. When the British left, the desire for more free trade remained.

The Spanish Bourbons tried to reinforce Spain's economic control by authorizing monopolistic companies. The Guipúzcoa Company in Venezuela best illustrates the effects of these monopolies and the protests they elicited from a jealous native merchant class. By the end of the seventeenth century, Venezuela exported a variety of natural products, most importantly tobacco, cacao, and salt, to Spain, Spanish America, and some foreign Caribbean islands. This trade expanded to

Simón Bolivar (1783–1830) is often considered the father of the South American independence movements. (Library of Congress)

England, France, and the British colonies during the early eighteenth century. A small, prosperous, and increasingly influential merchant class emerged. The liberator of northern South America, Simón Bolívar, descended from one of the most successful of these merchant families. The creation of the Guipúzcoa Company in 1728, designed to ensure that Venezuela stayed within imperial markets, evoked sharp protest from the merchants. They complained that the company infringed upon their interests, shut off their profitable trade with other Europeans, and failed to supply all their needs. Spain, they quickly pointed out, could not absorb all of Venezuela's agricultural exports, whereas an eager market in the West Indies and northern Europe offered to buy them. Finally, exasperated with the monopoly and unanswered complaints, the merchants fostered an armed revolt against the company in 1749. The revolt took four years to quell, but the resultant hostility lasted the rest of the century.

Two new institutions organized the activities of the disaffected Venezuelans. The Consulado de Caracas, established in 1793, brought together merchants and plantation owners and became a focal point for local protest. In 1797 the merchants formed a militia company to protect the coast from foreign attack. That responsibility intensified their nativism, which was transforming into patriotism. Nurtured by such local institutions, the complaints against the monopoly, burdensome taxes, and restrictions mounted in direct proportion to the increasing popularity of the idea of free trade. As one merchant expressed it, "Commerce ought to be as free as air." One result of intensifying complaints was a series of armed uprisings in 1795, 1797, and 1799. Great Britain continued to encourage such protests against Spain's commercial monopoly.

Economic dissatisfaction extended beyond the narrow confines of the colonial elite. Monopolies, such as tobacco, hurt consumers and marginal producers, as cultivation was limited to only areas that produced the highest quality product. Increases in sales taxes fell not just on elites, but on *campesinos* and workers. The discontent of the lower classes was manifested in a variety of protests. Even popular songs expressed those economic concerns:

> All our rights
> We see usurped
> And with taxes and tributes
> We are bent down.
> If anyone wants to know
> Why I go shirtless

It's because the taxes
Of the king denude me.
With much enthusiasm
The Intendents aid the Tyrant
To drink the blood
Of the American people.

3.4: The Impact of Ideas

The eighteenth century was a time of intellectual ferment in Europe, and those ideas were disseminated to Spain and the colonies. The printing presses in Spanish America—Lisbon rigidly prohibited the setting up of a press in its American possessions—contributed significantly to spreading ideas. Several outstanding newspapers were published, all fonts of enlightened ideas and nativism. Their pages were replete with references to, quotations from, and translations of major authors of the European Enlightenment.

Initially for Latin Americans, the Enlightenment ideas that were most important focused on reason and the belief in progress. Political ideas were of less importance, with many believing that increased liberty could be accomplished within the Spanish system rather than calling for independence. Indeed, the Spanish crown had become more liberal under the Bourbons; and Spaniards, too, looked forward to progress within the royal structure.

As creoles began to reconsider their relationship with the crown, they could look to classical Spanish political theory and the ideas of Francisco Suárez (1548–1617). Suárez argued that sovereignty was a contract based in popular origin and that the aim of society was to achieve the common good. Spanish intellectuals were also inspired by Benito Jerónimo Feijóo (1677–1764), who championed new knowledge and the pursuit of truth through reason and experience.

Latin American elites were no strangers to the ideas of European Enlightenment philosophers that were sweeping the continent. They encountered these ideas during their grand tours of Europe and in colonial publications, despite Spanish censorship. The men who would become independence leaders were influenced by Charles-Louis de Secondat, Baron de La Brède et de Montesquieu, who wrote in *The Spirit of the Laws* (1748): "The Indies and Spain are two powers under the same master; but the Indies are the principal, while Spain is only an accessory, it is in vain for politics to attempt to bring back the principal to the accessory; the Indies will always draw Spain to themselves." They were also inspired by Jean-Jacques Rousseau's 1762 *Social Contract*, and his view that "the general will alone may direct the forces of the State to achieve the goal for which it was founded, the common good. . . . The government's power is only the public power vested in it."

But there were limits to the usefulness of European Enlightenment thought. Most of these philosophers were cosmopolitan in their outlooks and did not think about nationalism or the rights of colonial subjects. An exception was the Englishman Jeremy Bentham, who chided the French revolutionaries at the National Convention of France in 1792: "Emancipate your colonies. . . . Justice, consistency, policy, economy, honor, generosity, all demand it of you."

Bentham advocated that France free its colonies, but Thomas Paine in *Common Sense* (1776) advocated that the colonies free themselves. "It is repugnant to reason, to the universal order of things, to all examples from the former ages, to suppose, that this continent can longer remain subject to any external power." In 1811, a translation of *Common Sense* circulated in Venezuela, passed by hand among its readers.

Paine's work was cited by Juan Pablo Viscardo y Guzmán, a Peruvian Jesuit exiled to Italy, who in 1791 wrote the first document to call for Spanish colonial independence: *Open Letter to the American Spaniards*. Viscardo contended that Spain had abrogated its treaties with the original conquerors and, therefore, forfeited its rights. He reached back to the *Siete Partidas*, the seven-part code adopted in 1265 under the guidance of Alfonso X, also known as Alfonso the Wise, of Castile. This major law code included a provision that "if the king violated rights and privileges of the people, the people had a right to disown him for their sovereign, and to elect another in place, even of the pagan religion."

Viscardo called for independence in no uncertain terms: ". . . Spain [is] constantly considering you as a people distinct from the European Spaniards, and this distinction imposes on you the most ignominious slavery. . . . [L]et us agree on our part to be a different people. . . . It would be a blasphemy to imagine that the Supreme Benefactor of man has permitted the discovery of the New World, merely that a small number of imbecile knaves might always be at liberty to desolate it."

The letter was sent to the colonies by Francisco de Miranda, a Venezuelan exile who spent most of his life in Europe, where he lobbied the governments of France and England to help liberate Spanish America. He also amassed an enormous library of philosophical and military tracts, and he made them available to the young men of Spanish America who came to Europe, among them future liberators Bernardo O'Higgins of Chile, José de San Martin of Argentina, and Simón Bolívar of Venezuela.

It would fall to Bolívar not just to become one of the main leaders of the independence movements, but also the one who would articulate the reasons for Latin American independence. He would need to move beyond the Enlightenment thinkers who stopped short of revolution, and to design an approach appropriate to Spanish American, rather than Anglo American, reality.

3.5: Early Warning Signs

The majority of the colonial population probably appreciated the economic motivations for independence more than the political ones. Popular antitax demonstrations rocked many cities in both Spanish and Portuguese America in the eighteenth century. Indigenous communities protested the *reparto* system, which forced them to buy goods from creoles and peninsulars. There were more than sixty revolts in the Andes alone during the 1770s. Most of these, however, were focused on local problems, without a larger vision of change in society. However, oppressive economic conditions helped to spark potentially serious uprisings in Peru, New Granada (Colombia), and Brazil.

The Tupac Amaru Revolt, which began near Cuzco in November 1780, was sparked primarily by the raising of the *alcabala*, or sales tax, from 2 to 4 percent in 1772 and to 6 percent in 1776. The tax hike came in the context of the continued burden of the *mita* and twenty-five years of increased demands: more efficient tax collection; legalization of the despised *reparto*; formation of the Viceroyalty of La Plata, which disrupted patterns of trade; and new as well as increased taxes. In addition, population was increasing and access to land decreasing, making it harder still to meet the new demands.

The revolt was led by José Gabriel Condorcanqui Noguera, a mestizo mule driver who also held the title of indigenous kuraka and took the name Tupac Amaru from the last Inca emperor. Tupac was comfortable in the creole, mestizo, and indigenous worlds, and he tried to appeal to all groups on their terms. His military command included sixteen Spaniards and creoles, seventeen mestizos, and nine indigenous. He claimed to wage his war in the name of Charles III and did not call for independence, but he also cast himself as a divine Inca. As he spoke against forced labor, invoking images of the ancient Inca empire, indigenous people responded en masse. A small minority wanted to eliminate Europeans and Africans, and their attacks frightened whites, who constituted barely 8 percent of the Peruvian population. Alarmed creoles joined the Spanish in brutally quelling the revolt in 1781. Tupac was captured, and after most of his family was killed in front of him, Tupac was drawn and quartered, his remains posted at several locations as a warning.

Nonetheless, the uprisings that he inspired continued until 1783. The movement spread across thousands of square miles and ended with the deaths of tens of thousands of people. The crown recognized it needed to address the cause of the uprising, abolished the reparto, and established a special audiencia to hear indigenous complaints. It is important to note, however, that not all indigenous followed Tupac Amaru's lead. At least twenty indigenous leaders stayed loyal to the crown. Their decisions were locally based, sometimes hinging on rivalries among individuals, villages, and ethnic groups.

The Comunero Revolt in New Granada (Colombia) in 1781 was also a protest against a tax increase as well as burdensome monopolies. Inspired by the Tupac Amaro revolt, they focused their wrath on the visitor general Juan Francisco Gutiérrez de Piñeres. The protest united elites and lower classes.

In addition to demands for tax cuts, the protestors called for returning salt mines taken from the indigenous and relief of tribute and ecclesiastical obligations. There were riots in Socorro in March, and by late May, the rebellion had spread throughout the Eastern Cordillera, with 20,000 armed protesters marching on Bogotá. Faced with the growing success of the rebels, the Spanish authorities capitulated to their demands for economic reforms. However, once the comuneros dispersed, the viceroy abrogated former agreements and arrested the leaders, who were executed in 1782 and 1783.

The Comunero uprising was not a movement for independence. Protestors commonly called out, "Long live the king and death to bad government." The protestors spoke of the *común*, the common good, not of a nation. They held fast to the tradition of compromise between government and governed and the tradition of popular approval of taxes. Eventually they called for creole self-government, but within the colonial structure.

In Brazil, there were two very different conspiracies that indicated unhappiness with Portuguese administration. The first occurred in Ouro Preto in the state of Minas Gerais in 1788–1789, when a group of elite conspirators planned to assassinate the governor and proclaim an independent republic. Most of the men were wealthy, and they had no plans for socioeconomic change. Word of the plot leaked out to the governor, and after a trial, six leaders were sentenced to hang. All were granted clemency but one: Joaquim José da Silva Xavier, an amateur dentist with the nickname Tiradentes (tooth puller). Tiradentes was hanged, decapitated, drawn, and quartered, and like Tupac Amaru in Peru, his remains were displayed in several places as a warning. For good measure, his home was destroyed and his lands salted to make them unusable.

Nevertheless, another, more disturbing plot would simmer in 1798 in Bahia. This time conspirators were from the popular classes: soldiers, artisans, mechanics, workers, and so large a number of tailors that the movement sometimes bears the title of the "Conspiracy of the Tailors." All were young (under thirty years old), and all were mulattos. The conspirators spoke in vague but eloquent terms of free trade, which they felt would bring prosperity to their port, and of equality for all men without distinctions of race or color. They denounced excessive taxation and oppressive restrictions. On the walls of the city, they posted manifestos denouncing the "detestable metropolitan yoke of Portugal." As in the Tiradentes conspiracy, the governor found out and the conspirators were tried. Three free mulattos were executed in the same manner as Tiradentes, and twenty-eight were exiled.

The Bahia plot was indicative of significant changes in Latin America in the late colonial period, particularly the ascension of castas to positions in society between the most humble classes and the elites. They began to make demands for their place in society, and while they were clearly potential allies for disgruntled elites, they also challenged the colonial social structure. Elites did not support the plot in Bahia, and fear of the popular classes would make elites throughout the region hesitant to challenge the crown that guaranteed social order.

3.6: International Examples

The turmoil of the late eighteenth century was, of course, not felt only in Spanish America. The New World colonies to the north chafed under similar economic and political controls exerted by Great Britain. The Spanish crown, however, felt so sure of its colonies that it did not see a threat in supporting the thirteen colonies in their struggle; Spanish support had more to do, however, with animosity to the British than with support of independence. Nonetheless, Spain was a signatory to the 1783 Treaty of Paris, which made the United States independent.

The ideas of the North American independence movement were no secret in Spain and its colonies. Newspapers published translated versions of the work of Thomas Paine and reported on the English colonial struggle. Although Latin Americans found many of the ideas appealing, they still tended to believe that their concerns could be worked out within the monarchy.

The French Revolution of 1789 at first seemed to broaden the ideas of U.S. independence. Many were impressed by the French idea that citizens should be politically active and that through action they could make change. But as the revolution moved further left, it became too radical for even the most liberal thinkers in Spanish America. As Francisco de Miranda commented in 1799, "We have before our eyes two great examples, the American and the French Revolutions. Let us prudently imitate the first and carefully shun the second."

The greatest repercussions of the French Revolution were felt in Saint Domingue, the French colony that occupied the western third of the island of Hispaniola. Sugar profits had soared in the eighteenth century as French planters exploited the good soil of the island, adopted the latest techniques for growing and grinding cane, and imported ever-larger numbers of African slaves to work on the land and to process the crops. A multiracial society had developed, a divided society, which by 1789 counted 40,000 whites and half a million blacks with approximately 25,000 mulattos. The Code Noir, promulgated in 1685, regulated slavery. Theoretically it provided some protection to the black slave, facilitated manumission, and admitted the freed slave to full rights in society. In reality, the European code but slightly ameliorated the slave's dreadful state. In general, slavery on the lucrative plantations was harsh. To meet the demand for sugar, the plantation owner callously overworked his slaves, and to reduce overhead, he frequently underfed them. An astonishingly high death rate testified to the brutality of the system.

The distant French revolutionary cry of "Liberty, equality, fraternity" echoed in the Caribbean in 1789. Each segment of the tense colony interpreted it differently. The white planters demanded and received from the Paris National Assembly a large measure of local autonomy. But in May 1791, the Constituent Assembly voted to give full French citizenship to free men of color in the colonies—provided their parents were free and they owned enough property.

The decision sparked fighting between white colonists and free blacks and mulattos.

Then, on August 22, 1791, the slaves demanded their own liberty and rebelled in northern Saint Domingue. More than 100,000 arose under the leadership of the educated slave Toussaint L'Ouverture, son of African slave parents. In pursuit of his goal of liberating his fellow slaves, he fought for the following decade against the French, British, Spaniards, and various mulatto groups. By 1801, L'Ouverture commanded the entire island of Hispaniola.

Meanwhile, the chaos of the French Revolution had yielded to the control of Napoleon, who took power in 1799. In 1801, Napoleon resolved to intervene to return the island to its former role as a profitable sugar producer. A huge army invaded Saint Domingue, and the French induced L'Ouverture to a meeting only to seize him treacherously. Imprisoned in Europe, he died in 1803. His two lieutenants, Jean-Jacques Dessalines and Henri Christophe, took up the leadership. A combination of the slaves' strength and yellow fever defeated the massive French effort. On January 1, 1804, Dessalines proclaimed the independence of the western part of Hispaniola, giving it the name of Haiti.

Latin American elites were deeply shaken by the slave uprising that won Haiti its independence from France. Haitian leader Toussaint L'Ouverture is pictured here in a military uniform with epaulettes, protecting a black woman and her two children while confronting a white man. (Library of Congress)

The surviving French planters fled to Cuba, Mexico, and Venezuela. They brought with them their horror stories of rampaging slaves, burning sugar fields, and bloody struggle. Haiti's independence was a potent symbol. For the exploited slaves of the New World, it represented hope. As news of the Haitian Revolution spread, whites throughout Latin America complained about their slaves' attitudes. One Cuban slave owner commented in 1795, "The insolence [of Havana's blacks and mulattos] no longer knows any bounds." A British traveler to Brazil in 1806 reported that "the secret spell that caused the Negro to tremble at the presence of the white man has been in a great degree dissolved [by Haiti's display of] black power."

For creole elites, it was a chilling example of what they had always feared—that the dark masses would rise against them. If that was the cost of independence, it was too high a price to pay.

3.7: Impetus from the Outside

Latin America watched in horror as the French Revolution devolved into the Jacobin terror. In 1793, Spain joined Britain in war against the French Republic, which ended in defeat for Spain. In humiliating treaties of 1795 and 1796, Spain was forced to ally with France, becoming an enemy of the British. Spain had been economically sound at the beginning of the war; by the end, the bankrupt crown confiscated Church property, raised taxes, and even taxed the nobility, all highly unpopular measures.

In 1799, Napoleon's coup ended the revolution. But more war was in store for Europe. In 1802, Napoleon became first consul for life, and he was crowned emperor in 1804. Although his troops failed to retake Haiti, he set out to bring Europe under his control. Because of the treaties, Spain was forced to fight on the side of the French in the ensuing Napoleonic Wars. In 1805, Britain defeated the Spanish navy at Trafalgar, and in 1806 a French blockade destroyed Spain's economy. In 1807, Spanish authorities gave Napoleon permission to cross Spain and conquer Portugal. The Portuguese royal family, the Braganzas, however, did not fall prisoner to the conquering French armies. Prince-Regent John packed the government aboard a fleet and, under the protecting guns of English men-of-war, sailed from Lisbon for Rio de Janeiro just as the French reached the outskirts of the capital. The transfer of a European crown to one of its colonies was unique. The Braganzas were the only European royalty to visit their possessions in the New World during the colonial period. They set up their court among the surprised but delighted Brazilians and ruled the empire from Rio de Janeiro for thirteen years.

After seizing control of the Portuguese metropolis, Napoleon immediately turned his attention to Spain. Spanish opponents of Charles IV forced him to resign in favor of his son, Ferdinand VII. Bonaparte invited the regents to France to discuss the dispute over the crown. There he either took the royal family hostage and forced Charles and Ferdinand to abdicate, or, depending on one's political point of view, they voluntarily chose exile in France, abandoning Spain to its fate.

The Spanish people rose up against the French forces on May 2, 1808, a battle later immortalized by the painter Francisco Goya. The Spaniards formed local juntas to rule in the name of the king under the principle that in his absence, sovereignty resided in the people, who bore the responsibility to defend the nation. It soon became clear that the Spaniards needed unity to defeat the French, and the regional juntas gave way to the Supreme Central Governing Committee of Spain and the Indies, the *Junta Central.*

Spanish Americans reacted with equal repugnance to the French usurper. Various juntas appeared in the New World to govern in Ferdinand's name. In effect, this step toward self-government constituted an irreparable break with Spain. By abducting the king, Napoleon had broken the major link between Spain and the Americas. The break once made was widened by the many grievances of the Latin Americans against Spain.

Most of Latin America achieved its independence during a period of two decades, between the proud declaration of Haiti's independence in 1804 and the Spanish defeat at Ayacucho, Peru, in 1824. Nearly 20 million inhabitants of Latin America severed their allegiance to France or to the Iberian monarchs. Every class and condition of people in Latin America participated at one time or another, at one place or another, in the protracted movement. But the slave uprising in Haiti proved to be unique. Most of the independence movements were led by creole elites.

3.8: Elitist Revolts

Between 1808 and 1810, Spaniards and Americans together formed juntas to rule in the name of the deposed king. Most agreed that their differences with the Crown could be resolved, and they hoped to remain within the Spanish monarchy. But then the Spanish resistance lost more ground to the French forces—Seville fell, and the Junta Central fled first to Cádiz and then to the Isle of León, the only Spanish territory still free from French control. In January 1810, the junta appointed a new Council of Regency to rule and then dissolved itself. The Regency made clear that it saw the colonies as part of and still subject to the rule of Spain.

In the colonies, more radical creoles took the reigns and ousted the Europeans. In 1810, the movements for independence began simultaneously in opposite ends of South America, Venezuela and Argentina. The struggle for Spanish America's independence fell into three rather well-defined periods: the initial thrust and expansion of the movement between 1810 and 1814, the faltering of the patriotic armies and the resurgence of royalist domination from 1814 to 1816, and the consummation of independence between 1817 and 1826.

When news of Spain's fall to Napoleon reached Venezuela in 1808, creole leaders sought to stay within the Spanish realm by asking authorities to allow the formation of a junta. Instead of granting their request, the authorities repressed the movement and jailed or exiled its leaders. When the central junta in Spain dissolved in favor of a regency, the Venezuelan elites—both Americans and Spaniards—overthrew the captaincy general and audiencia and used the cabildo as the basis for forming a new junta. The junta, however, asserted, "Venezuela has declared its independence, neither from the *Madre Patria*, nor from the sovereign, but from the [Council of] Regency, whose legitimacy remains in question in Spain itself." Such claims have been dismissed by some scholars as "false monarchism," a claim to rule in the name of the king only as a pretense for the goal of independence. Others argue that some creole elites sincerely intended to remain within the monarchy until circumstances made it appear impossible.

The new junta distrusted the Spanish regency, fearing that it was controlled by Iberian merchants whose interest collided with their Venezuelan counterparts. The junta also represented a sharply divided creole ruling class: conservatives versus radicals, and autonomists versus independists. In addition, the Caracas junta tried in vain to control the rest of Venezuelan territory, which had never been united. Alliances were not always what one might expect. A mostly Spanish junta in Cumaná province supported not the Regency in Spain but the junta of Caracas. But the town of Barcelona, seeking autonomy from Cumaná, created its own junta and recognized both the Junta Suprema de Caracas and the Council of Regency in Cádiz.

In a move to unite the area, the Caracas Junta called for elections to a congress to determine the region's future. The elections were indirect—locally chosen electors would then elect the deputies to a national congress. Voting was restricted to free males ages twenty-five years and older with at least 2,000 pesos in movable property. But before elections could be held, the council in Cádiz attempted to control the region by sending Captain General Fernando Miyares, former governor of Maracaibo, to take control. Cádiz also decreed an end to Venezuela's free trade, arguing that because Great Britain and Spain were not at war, such emergency measures were not necessary. The junta of Caracas refused to obey, and Cádiz sent troops based in Cuba and Puerto Rico, which blockaded Venezuela.

Younger, more radical Venezuelans rose up to challenge the Caracas leadership, which consisted of politically moderate merchants and *hacendados*. The radicals, with the exception of wealthy aristocrat Simón Bolívar, were generally middle class, including lawyers, notaries, journalists, small merchants, and lower-ranking officials. When news of a Spanish massacre in Quito reached Caracas, the radicals started a riot, which was mostly peopled by blacks and castas. The terrified moderates on the Caracas junta were more frightened by the thought of social disorder than by continued Spanish dominance. They responded by exiling radical leaders and barring the return of revolutionary Francisco Miranda from Britain. They also sent armed forces to subdue the city of Coro, but they were defeated by the radicals, who convinced the authorities to let Miranda return. With his leadership, they immediately began to press for independence.

The congress elected in 1811, however, had been chosen by property owners and was dominated by moderates. They appointed a new regime, a weak triumvirate to rule in the interests of the elites. Property requirements were imposed to vote or hold office, a national guard was established to regulate slaves, and new vagrancy laws took aim at the *llaneros*, the cowboys of the grasslands whose free lifestyle was seen by the urban elites as chaos. The radicals did not object to the new rules because they shared the same social and economic interests as the moderates. Despite their use of the dark masses, the radicals' sole issue was independence from Spain, not social and political change within the Americas.

The next year was spent in bloody battle as the royalists continued to fight against the so-called "First Republic." Because the white population of the region was only 22 percent, both royalists and rebels counted on the support of the blacks and castas. Venezuela descended into chaos, and both sides encouraged racial warfare. The radical independence forces won in August 1813, and Bolívar formed the "Second Republic," which was little more than a military dictatorship. Bolívar's victory was fragile and temporary. He faced the challenge of José Tomás Boves, a lower-class trader from Asturias who had lived in the llanos. He organized a cavalry of llaneros, many of them pardos and blacks, and promised them the lands of white republicans. His forces drove Bolívar from Venezuela, and whites fled in fear of the "dark hordes." Both peninsulars and creoles were relieved when a Spanish army arrived in April 1815. By then, one-fifth of the population, some 80,000 people, had died.

Events unfolded differently in Río de la Plata, which had been drawn into the Napoleonic Wars in 1806, when Buenos Aires was attacked by the British. The viceroy had fled, but the local militia gathered under the leadership of Santiago Liniers, a French officer who was serving in the Spanish forces. The *porteños* (residents of the port city of Buenos Aires), defeated the invaders, and Liniers was named temporary viceroy. The American forces were also successful against a second invasion attempted by the British in 1807.

When news of Joseph Bonaparte's ascension to the throne reached Buenos Aires in July 1808, Liniers and the provincial intendants proclaimed their loyalty to Ferdinand. In September, they recognized the junta central in Seville. In November, Liniers opened Buenos Aires to free trade, and trade with the now-allied British provided enough customs revenue to pay an 8,000 member militia.

In January 1809, the peninsulars plotted to depose Liniers, who they saw as too friendly to the creoles. Cornelio Saavedra, the creole militia leader who defended Liniers, characterized the Spaniards as men who, seeing the loss of Spain in Europe, "proposed the idea of forming another Spain in America in which they and the many whom they hoped would emigrate from Europe would continue to rule and dominate." The creoles successfully came to Liniers' defense, but in August 1809 the junta in Spain sent a new viceroy, Balthasar de Cisneros. Cisneros put an end to free trade and tried to reduce the size of the militia. The peninsulars expected Cisneros to restore their old influence, whereas the creoles struggled for supremacy by controlling the militia, which far outnumbered the Spanish garrison.

With the news in 1810 that the French had conquered Seville and set the junta central on the run to Cádiz, the creoles saw their opening. The creole militia rose up in May and demanded a *cabildo abierto*, an open town meeting—that is, open to the urban elite. The elites tried to impose a conservative junta, but they capitulated when threatened by the creole-controlled militia. The new junta was divided evenly between military leaders, who mostly came from the elites, and intellectuals, who were generally from the middle class. The creoles in charge drove the peninsulars from office, arrested, exiled, or executed them. A Committee of Public Safety was formed to catch counterrevolutionaries, who were summarily executed.

The leaders of the May revolution, however, were in for a surprise. The provinces outside of Buenos Aires saw the uprising as a local matter. Leaders in the Banda Oriental (Uruguay), Paraguay, and Upper Peru (Bolivia) felt no unity with the porteños. The interior of the region was largely conservative, based on the large haciendas that were worked by mestizos and indigenous, who were involved in patron–peon relationships. The provincial area of Buenos Aires was the wide open pampas, where there was a great deal of unclaimed land and the freedom offered by the mobile society of the gaucho. Neither region had any desire to be dominated by the port city, which was a commercial and bureaucratic center. Furthermore, the port wanted free trade, whereas the Andean provinces of Salta, Tucumán, Jujuy, and Catamarca, which produced food, livestock, and pack animals for the silver mines of Upper Peru, had begun to industrialize and required protection for the fledgling sugar, textile, and transport equipment factories. Mendoza, San Juan, and La Rioja in the far west, though mostly an area of subsistence agriculture, could sell wine and brandy, if protected.

The Buenos Aires junta tried to take the other regions by force, but a year later only a small territory was under their control. Political power was in the hands of a triumvirate headed by Bernardino Rivadavia, whose program called for a constitutional monarchy, the formation of a maritime insurance company and new meat-salting plants, and the colonization of the interior by Europeans. The colonization plans further antagonized the provinces, whereas urban Buenos Aires was unhappy with the idea of constitutional monarchy. The triumvirate was brought down in 1812 by forces led by José de San Martín. San Martín was born in Río de la Plata and was sent to Spain to train as a soldier, reaching the rank of lieutenant colonel. In 1814, San Martín was named commander of the rebel Army of the North, and Rio de la Plata's independence seemed momentarily safe.

While the colonies fought over which path to take, changes were also occurring in Spain. The Regency had called for a *cortes*, a parliament, and elections were held in the colonies to send representatives. The cortes decided that it was the repository of national sovereignty and created a government that consisted of the Regency as executive, a judiciary, and a legislative branch, which would dominate. When the Regency objected, its members were arrested and the cortes appointed a new Regency.

The cortes was divided on several issues: Spanish liberals wanted a modern constitutional monarchy, whereas conservatives wanted absolutism. Neither group wanted the Americans to have autonomy. After much debate, the cortes produced a constitution in 1812. It did not give the Americans full equality, but it did provide them with equal representation in the legislative branch. It recognized mestizos and the indigenous as citizens, but continued black slavery. Most important, it restricted the power of the king and made the legislative branch the most powerful.

After the defeat of Napoleon, Ferdinand VII was returned to the throne in 1814. He was met by General Francisco Elío, who pledged to support the king against the liberals. The king was also supported by sixty-nine deputies who presented him with a manifesto calling for absolutist rule. Ferdinand lost no time in abolishing the cortes and the constitution. He persecuted and jailed the liberals who had usurped his authority and he determined to take control of the colonies by dispatching a large army to South America in 1815. By 1816, independence movements had been crushed everywhere but Buenos Aires and Paraguay. Movements had not even begun yet in Peru, Central America, Cuba, or Puerto Rico. Not only were prospects of independence defeated, but there also seemed little hope for improved relations with Spain. Ferdinand's retaliatory measures, however, fueled new resentments and led to renewed fighting in the colonies.

In Buenos Aires, independence leaders called a congress in Tucumán in 1816 and wisely made overtures to the provinces. The same year, Bolívar, who had been in exile, returned to Venezuela. He took Angostura in 1817 and organized a new army with the help of José Antonio Páez. San Martín organized forces in Río de la Plata, along with the Chilean exile Bernardo O'Higgins, to liberate Chile in 1817 and 1818. In a truly amazing feat, San Martín's army of 5,000 crossed the Andes over six different passes, including several that were more than 10,000 feet high. Bolívar borrowed San Martín's strategy and crossed the Andes in the north to free New Granada in 1819. The Spanish forces were on the run, and they were desperately in need of reinforcements. But none would come.

In Spain, rebellions against Ferdinand's absolutism began in 1814. The monarch struggled to control the peninsula while holding on to his far-flung empire. In 1819, he began massing forces at Cádiz to send to the colonies. But there were problems: Many of Spain's ships were not seaworthy, and there was no money for repairs. Even worse, the troops became restless, amid wretched conditions, shortages of food and clothing, and the spread of disease. Junior officers complained, and on January 1, 1820, they were led in mutiny by Lieutenant Colonel Rafael del Riego. Riego had been a loyal Spanish officer who fought for Ferdinand and was even imprisoned in France. Not only did the troops refuse to go to the Americas, but Riego also demanded the restoration of the 1812 constitution. The rebellion was taken up by civilian protests that cut across classes. Ferdinand had no choice but to capitulate.

Royalist forces continued to fight in the colonies, and bloody battles ensued, but without reinforcements the royalists were doomed. The Americans triumphed in Venezuela in 1822 and in Ecuador in 1823. That same year, the forces of San Martín and Bolívar converged on Peru, and the decisive battle of Ayacucho in 1824, under the leadership of Antonio José de Sucre, effectively ended the war.

3.9: Popular Revolution in Mexico

New Spain initially started on the same path as the rest of Spanish America. Taking advantage of the political vacuum in Spain in 1808, the creoles maneuvered to form a local junta to govern the viceroyalty, a move calculated to shift political power from the Spaniards to the Mexican elite. Alarmed by the maneuvering, the peninsulars feared the loss of their traditional, preferred positions. They acted swiftly to form their own junta and thus pushed the creoles aside. The creoles plotted to seize power and enlisted the help of Father Miguel Hidalgo, the parish priest of the small town of Dolores. In September 1810, the peninsulars discovered the plan and jailed the leaders. The wife of one leader, Josefa Ortiz de Dominguez, got word to Hidalgo, who on September 16, 1810, rang the church bells to summon the mostly mestizo and indigenous parishioners.

But Hidalgo's ideas were different from the other creole leaders. Well educated, and profoundly influenced by the Enlightenment, Hidalgo professed advanced social ideas. He believed that the Church had a social mission to perform and a duty to improve the lot of the downtrodden. Personally, he bore numerous grievances against the peninsulars and the Spanish government—he had been educated by the Jesuits until their expulsion in 1767, and he had been investigated by the Inquisition on charges of mismanaging funds while he was the rector of the College of San Nicolás Obispo in Valladolid.

When his parishioners gathered, Hidalgo exhorted them to rise up and reclaim the land stolen from them 300 years before: "Long live Ferdinand VII! Long live America! Down with bad government! Death to the Spaniards!" Hidalgo's words fell on fertile ground. The town of Dolores was a mere ten miles from Guanajuato, making it dependent on the mining economy. In 1809, an unusually dry summer had decreased maize production, ruining many small farmers. Food prices had quadrupled, and miners could not feed the mules needed in mining, leading to layoffs of mine workers.

Hidalgo had unleashed new forces. Unlike the creoles who simply wanted to substitute themselves for the peninsulars in power, the mestizo and indigenous masses desired far-reaching social and economic changes. There were 600 people gathered in Dolores; as they marched toward Guanajuato, the numbers swelled to 25,000. At Guanajuato, the intendant, the local militia, the peninsulars, and some creoles barricaded themselves in the granary, leaving the city defenseless. The lower classes of Guanajuato joined the insurgents and burned and pillaged the city. By the time the mob reached Mexico City, it numbered 60,000–80,000, and in their rampage they made no distinction between creole and peninsular. With energies released after three centuries of repression, the indigenous and mestizos struck out at all they hated. The creoles became as frightened as the peninsulars, and the two rival factions united before the threat from the masses.

Hidalgo did little to discipline the people under him. Indeed, his control of them proved minimal. Poised before Mexico City, he hesitated and then ordered a withdrawal, an action that cost him allegiance of many of the masses.

The Spanish army regained its confidence and struck out in pursuit of the ragtag rebels. Once captured, the royalists tried Hidalgo before a nine-member panel that included six creoles. In mid-1811, ten months after the Grito de Dolores, a firing squad executed him. His severed head was mounted on the wall of the granary at Guanajuato, a clear warning to others.

Nonetheless, Hidalgo's banner was taken up by another parish priest, José Maria Morelos. Morelos came from a poor mestizo family and worked as a mule driver; through great personal effort he became a priest, always assigned to the poorest, backwoods parishes, where poor indios and mestizos labored. He had joined Hidalgo's movement, but after Hidalgo's death he determined that undisciplined hordes were not the answer. He trimmed the forces, organized them into a more disciplined force, and tried to appeal to the creoles while still carrying Hidalgo's banner of social reform. He defined his program: Establish the independence of Mexico; create a republican government in which the Mexican people would participate with the exclusion of the wealthy nobility and entrenched officeholders; abolish slavery; affirm the equality of all people; terminate the special privileges of the Church as well as the compulsory tithe; and partition the large estates so that all farmers could own land. At Chilpancingo, he declared that Mexico's sovereignty resided in the people, who could alter the government according to their will. He called forth pride in the Mexican—not in the Spanish—past. His program contained the seeds of a real social, economic, and political revolution and thereby repulsed peninsular and creole alike. He ably led his small, disciplined army in central Mexico for more than four years. In 1815, the Spaniards captured and executed him. The royalists immediately gained the ascendancy in Mexico and dashed the hopes of the mestizos and indigenous for social and economic changes. New Spain returned momentarily to its colonial slumbers.

When independence was won in Mexico, it was under conservative leadership, a reaction to Ferdinand's acceptance of the liberal 1812 constitution. The peninsulars and creoles in New Spain rejected Spanish liberalism just as they had earlier turned away from Mexican liberalism. In their reaction to the events in Spain, they decided to free themselves and chart their own destiny. They were joined by the Church hierarchy, who feared the loss of property and secular restrictions if the liberals in control of Spain had their way. The peninsulars and creoles selected a pompous creole army officer, Augustín de Iturbide, who had fought against Hidalgo and Morelos, first as their instrument to effect independence and then as their emperor. The most conservative forces of New Spain ushered in Mexican independence in 1821. They advocated neither social nor economic changes. They sought to preserve—or enhance if possible—their privileges. The only innovation was political: A creole emperor replaced the Spanish king, which was symbolic of the wider replacement of the peninsulars by the creoles in government. The events harmonized little with the concepts of Hidalgo and Morelos but suited creole desires. The Mexican struggle for independence began as a major social, economic, and political revolution but ended as a conservative coup d'etat. The only immediate victors were the creole elite.

3.10: Unsung Heroes and Heroines

When heroic tales of the wars were told, especially in the early national years as identities were constructed by a new independence narrative, the stories focused on a handful of leaders, especially Simón Bolívar, the father of Latin American independence, and José de San Martín, who dared to cross the Andes to help Bernardo O'Higgins liberate Chile before heading north to join Bolívar's forces in the liberation of Peru.

But the majority of those who fought in the wars were the lower classes, indigenous, black, and *castas*, freed slaves, cowboys of the plains (*llaneros*) and pampas (*gauchos*). The intellectual authorship of Latin American independence and most of the leadership came from the elites, who were more concerned with their own wealth and political power than the needs of the masses. But those small numbers could never win the war without mass support from those with their own grievances and who were promised greater freedom in new nations.

The independence movements also depended on the support of women, who served in various capacities throughout the struggle. Women were among the rural poor who, armed only with tools, flocked to the side of Miguel Hidalgo in New Spain. They helped defend Cuautla, Mexico, with José Maria Morelos, even singing and dancing to taunt their attackers. Juana Ascencio trained a mostly indigenous battalion to fight in Upper Peru.

Some were camp followers, cooking for the troops and tending their wounds. Others smuggled messages and weapons, hidden under their voluminous skirts. Women of the upper classes held *tertulias* in their homes at which intellectuals debated the issues—and planned rebellions—including Javiera Carrera in Santiago, Manuela Cañizares in Quito, Mariquita Sánchez in Buenos Aires, and Mariana Rodríguez de Toro in Mexico City.

Many women, of all classes, died fighting alongside men on both sides of the struggle. Some faced public executions, like Policarpa Salavarrieta of Bogotá, who used her position as a seamstress to spy on prominent royalists and to make uniforms for rebel soldiers, as well as recruiting, transporting, and hiding them, and Gertrudis Bocanegra, who recruited her husband to fight in New Spain, where she transported messages and provided supplies.

Most famous was Manuela Sáenz, stereotyped for many years as Bolívar's mistress. But Sáenz was much more—she earned the rank of lieutenant after fighting in the Battle of Pichincha, near Quito. She served in the cavalry at the Battle of Junín in the Peruvian highlands and was promoted to captain. She was promoted to colonel after aiding the rebel army in the final Battle of Ayacucho—a promotion suggested by Bolívar's second in command, Gen. José de Sucre. Bolívar would later nickname her the "liberator of the liberator" after she threw herself in front of him during as assassination attempt, giving him time to escape.

Map 3.1 Latin America in 1830

SOURCE: Craig, Albert M.; Graham, William A.; Kagan, Donald; Ozment, Steven M.; Turner, Frank M. *Heritage of World Civilizations Combined*, 7th ed. © 2006. Electronically reproduced by permission of Pearson Education, Inc., Upper Saddle River, NJ.

3.11: The Brazilian Exception

Portuguese America achieved its independence during the same tumultuous years. Brazil entered into nationhood almost bloodlessly, and following the trend evident in Spanish America, the mazombos clamored for the positions of the reinóis, although their ascendancy was more gradual than that of the creoles. The difference lay in the way Brazil achieved its independence.

Under the guidance of John VI, Brazil's position within the Portuguese empire improved rapidly. He opened the ports to world trade, authorized and encouraged industry, and raised Brazil's status to that of a kingdom, the equal of Portugal itself. Rio de Janeiro changed from a quiet viceregal capital to the thriving center of a far-flung world empire. The psychological impact on Brazilians was momentous. Foreigners who knew Brazil during the first decade of the monarch's residence there commented on the beneficial effect the presence of the crown exercised on the spirit of Brazilians. Ignacio José de Macedo typified the optimism of his fellow Brazilians when he predicted, "The unexpected transference of the Monarchy brought a brilliant dawn to these dark horizons, as spectacular as that on the day of its discovery. The new day of regeneration, an omen of brighter destinies, will bring long centuries of prosperity and glory."

When the royal court returned to Lisbon in 1821, after thirteen years of residence in Brazil, John left behind the Braganza heir, Prince Pedro, as regent. The young prince took up his duties with enthusiasm, only to find himself caught between two powerful and opposing forces. On the one side, the newly convened parliament, the Côrtes, in Lisbon, annoyed with and jealous of the importance Brazil had assumed within the empire during the previous decade and a half, sought to reduce Brazil to its previous colonial subservience; on the other, Brazilian patriots thought in terms of national independence. As the Côrtes made obvious its intent to strip Brazil of previous privileges as well as to restrict the authority of its prince-regent, Pedro listened more attentively to the mazombo views. He appointed the learned and nationalistic mazombo José Bonifácio de Andrada e Silva to his cabinet, making him the first Brazilian to hold such a high post. Bonifácio was instrumental in persuading Pedro to defy the humiliating orders of the Côrtes and to heed mazombo opinion, which refused to allow Lisbon to dictate policies for Brazil.

Princess Leopoldina, although Austrian by birth, dedicated her energies and devotion to Brazil after she arrived in Rio de Janeiro in 1817 to marry Pedro. She too urged him to defy Portugal. She wrote him as he traveled to São Paulo in September 1822: "Brazil under your guidance will be a great country. Brazil wants you as its monarch. . . . Pedro, this is the most important moment of your life. . . . You have the support of all Brazil." On September 7, 1822, convinced of the strength of Brazilian nationalism, Pedro declared the independence of Latin America's largest nation, and several months later in a splendid ceremony he was crowned "Constitutional Emperor and Perpetual Defender of Brazil."

The evolutionary course upon which Brazil embarked provided a stability and unity that no other former viceroyalty of the New World could boast. Contrary to the contractual political experiences of Spanish America, an unbroken patriarchal continuity, hereditary governance, flowed in Brazil. Similar to that of Spanish America, Brazilian independence affected and benefited only the elite. Like the elites everywhere, the Brazilians rushed to embrace foreign ideas and to continue, even to deepen, the colonial dependency on exports.

3.12: Aftermath

In Central America, where there was little fighting during the independence era, many people did not know that they were independent from Spain until months after the region had become part of the new nation of Mexico. But in other areas, where there had been heavy fighting, the devastation was extensive. Villages were destroyed, populations displaced, and economies ruined. Ill-fed, ragged armies became vectors of infectious disease, which they brought with them as they marched. Yellow fever swept over northern South America. Urban populations burgeoned as people fled the war-torn countryside. Mexico City became the fifth largest city in the world as it grew from 137,000 in 1800 to 169,000 in 1811. But there was also tremendous loss of life. From 1810 to 1816, some 600,000 Mexicans died. Venezuela lost 21 percent of its population from 1810 to 1825. The causes were not just the violence of the war itself, but also disease, famine, and unsanitary conditions, especially the lack of clean water and the inability to adequately deal with waste removal, leading to open sewage and garbage in the streets.

Furthermore, while many Latin Americans gladly joined in the independence wars, many others were pressed into service against their will. Soldiers arrived with orders from new republican leaders demanding food, clothing, horses, even the use of houses, for the war effort and fledgling republics.

The certainty of the royal order was gone, but no one knew what would emerge politically in its place. Only one thing was clear: The wars for independence, though fought by people from the lowest to the highest status, was directed by the creole elites, who expanded their already dominant position in Latin American society. Although they had courted the masses, they had no intention of sharing wealth and power with them. For the majority, there was little difference between colonialism and early nationhood. Inevitably, Latin American independence was characterized by continuity more than by change.

Questions for Discussion

1. What were the long-term problems that challenged the colonial order?
2. How did the Haitian Revolution impact Latin American independence?
3. What were the immediate causes of the independence movements?
4. How did the Mexican independence movement differ from those in South America?
5. Who would benefit and who would lose as a result of the political change from colony to nation?
6. Why did independence affirm rather than change the basic patterns of society?

4
New Nations

Most of Latin America had gained independence by the end of the first quarter of the nineteenth century. The protracted struggle elevated to power a few privileged elites who, with few exceptions, benefitted from the Spanish and Portuguese colonial systems and reaped even greater rewards during the early decades of nationhood. The independence of the new nations proved almost at once to be nominal because the ruling elites became economically subservient to Great Britain. The impetus to build a nation-state came from "above." In the apt judgment of Cuban independence leader José Martí, the elites created "theoretical republics." They tended to confuse their own well-being and desires with those of the nation at large, an erroneous identification as they represented less than 5 percent of the total population. That minority set the course upon which Latin America continued through the twentieth century.

The very idea of the nation-state was problematic. The countries were new creations, and each included many ethnic and racial divisions. Simón Bolívar suggested political definitions of nationalism, rather than European models that drew on a supposedly shared ancestry, history, and culture. But national leaders also searched for cultural ways to forge national identities and unite the fragile new *patrias*.

4.1: Monarchy or Republic?

Who would govern and how they would do so were fundamental questions facing the newly independent Latin Americans. They were questions previously unasked. For centuries, ultimate authority was concentrated in the Iberian kings, who ruled in accordance with an ancient body of laws and customs. For nearly three centuries, the inhabitants of the colonies had accepted their rule, and the monarchies provided their own hereditary continuity. The declarations of independence created a novel political vacuum.

Brazil alone easily resolved the questions, mainly because of the presence of the royal family in Portuguese America. On hand to lend legitimacy to the rapid, peaceful political transition of Brazil from viceroyalty to kingdom to empire were first King John VI and then his son, Prince Pedro, heir to the throne, who severed the ties between Portugal and Brazil in 1822 and wore the new imperial crown. The mazombo elite supported the concept of royal rule and thereby avoided the acrimonious debates between republicans and monarchists that split much of Spanish America. Obviously facilitating their decision was the convenient presence of a sympathetic prince, a Braganza who had declared Brazil's independence. By his birth, as well as through the concurrence of the Brazilian elite, Pedro's power was at once legitimate, the perfect unifier of the immense new empire.

Although there was a genuine consensus as to who would rule, the question remained open as to how he should rule. The emperor and the elite agreed that there should be a constitution, but the contents sparked a debate that generated the first major crisis in the Brazilian empire. Elections were held for an assembly that would exercise both constituent and legislative functions. The group that convened in 1823, comprising lawyers, judges, priests, military officers, doctors, landowners, and public officials, clearly represented the privileged classes. They came from the old landed aristocracy and the new urban elite, two groups that remained interlinked. Almost at once, the legislature and the executive clashed, each suspicious that the other infringed on its prerogatives. Furthermore, the legislators manifested rabid anti-Portuguese sentiments and thus by implication a hostility to the young emperor born in Lisbon. Convinced that the assembly scattered the seeds of revolution, Pedro dissolved it.

Nonetheless, Pedro intended to keep his word. He appointed a committee of ten Brazilians to write the constitution and submitted it to the municipal councils throughout Brazil for ratification. It provided for a highly centralized government with a vigorous executive. Power was divided among four branches—executive, legislative, judiciary, and the novel moderative power, which made the emperor responsible for maintenance of the nation's independence and the harmony of national institutions and the twenty provinces.

The emperor was given broad powers to unify a far-flung empire whose geographic and human diversity challenged the existence of the State. In the last analysis, the crown was the one pervasive, national institution representing all Brazilians. The General Assembly was divided into a senate, appointed by the emperor for life, and a chamber of deputies periodically and indirectly elected by highly restricted suffrage. The constitution afforded broad individual freedom and equality before the law. Proof of the viability of the constitution lay in its longevity: It lasted sixty-five years, until the monarchy fell in 1889. It has proven to be Brazil's most durable constitution and one of Latin America's longest lived.

The Brazilians gradually took control of their own government. At first, Pedro disappointed them by surrounding himself with Portuguese advisers, ministers, and prelates. The Brazilians had their independence but were tacitly barred from exercising the highest offices in their own empire. The mazombos accused the young emperor of paying more attention to affairs in Portugal than to the new empire. As Brazilians demanded access to the highest offices, the currents of anti-Portuguese sentiments swelled. Pedro's failure to understand those nationalistic sentiments was a primary cause of discontent leading to his abdication in 1831. After his departure for Europe, mazombo elites replaced the Portuguese-born who had monopolized the First Empire. A three-man regency ruled until 1840, when the fifteen year old Pedro II ascended the throne, bringing a Brazilian-born ruler to power. Thus, the mazombo ascendancy was more gradual than the creole. It began in 1808 when the royal court arrived in Rio de Janeiro and reached its climax in 1840 when a Brazilian-born emperor took the scepter.

Unlike Brazil, Spanish America experienced a difficult transfer and legitimation of power. The question of what form the new governments should take aroused heated debates, particularly over the issue of republic versus monarchy. Monarchy harmonized with the hierarchical, aristocratic structure of Spanish American society. However, a desire to repudiate at least the outward symbols of the Spanish past, an infatuation with the political doctrines of the Enlightenment, and the successful example of the United States strengthened the arguments for a republic. Only in Mexico did monarchists carry the day, when the creoles crowned Augustín de Iturbide, whose brief reign lasted from May 1822 until February 1823. Still, because Mexico included the territory from Oregon to Panama, it meant that together with Brazil, a majority of Latin America in late 1822 and early 1823 fell under monarchical sway.

Iturbide's reign was not a happy one, largely because he was trying to satisfy conflicting groups with the scant resources of an economy hard hit by the independence wars. To reassure merchants and capitalists, he cut taxes, which led to a decline in the revenue needed to maintain the army. To pay the army, which was crucial to uphold a government that had not yet achieved hegemony, he issued paper money, which led to inflation. He then took foreign loans that the government could not repay and demanded forced loans from the elites and the Church, alienating his strongest supporters. With growing political unrest, Iturbide closed Congress, which alienated his political allies. The final straw came when there were no funds to fight the Spanish royalist troops still holding out at the fortress of Veracruz. Iturbide responded to complaints by the Mexican force's commander, Antonio Lopez de Santa Anna, by firing him. Santa Anna then led a coup against Iturbide. The army banished Iturbide in early 1823, abolished the empire, and helped to establish a liberal, federal republic. With that, the principle of republicanism triumphed, at least as an ideal, throughout Spanish America.

4.2: Liberals or Conservatives?

Although most of Spanish America opted for republics rather than monarchies, the question of who should rule was still a thorny one. The immediate answer in many countries was to turn the reins of authority over to independence heroes, but their popularity quickly faded when they proved to be poor statesmen. Latin Americans found it more difficult to select their successors; efforts to fill presidential chairs unleashed bitter power struggles, which were conducive to despotism among various factions of the elite.

Latin Americans also debated how the new polities should be organized, with the key debate between federalism and centralism. Those opposed to change preferred centralism, which echoed the strong central control of the Spanish monarchy. But a host of local rivalries, and the apparent successful example of North American federalism, combined to persuade many Latin American leaders that regional autonomy was preferable. Federalism also drew the support of local and regional leaders who wanted to maintain their power. Further, most of the population identified with their immediate region, the *patria chica*, or small country, rather than the abstract idea of the nation. The patria chica had deep roots in the pre-Columbian past, when indigenous identity was based on the city-state, and in the Spanish tradition of identifying with their provinces.

Mexican General Antonio Lopez de Santa Anna, in his medal-bedecked uniform, was the epitome of the nineteenth-century caudillo who emerged from the independence wars to become a dictator. (INTERFOTO / Alamy)

The federalist and centralist factions became aligned with two main political currents that dominated Latin American politics during the nineteenth century—liberalism and conservativism. In some ways, it is difficult to specify the differences between the two political views because the elites had much more in common than in opposition. As one Mexican textbook put it, "The Liberals were long-haired young lawyers with modest incomes; whereas most of the Conservatives were prosperous members of the church or army, middle-aged and older, and regularly groomed at the barbershop."

Generally, Liberals looked to the United States as a model and favored, at least in theory, a wider form of democracy. The new Liberal parties urged republican institutions, formal equality before the law, an expanding national government, and secularization of society. Liberals often flirted with federalist schemes, which provided broader participation, and maintained a theoretical interest in the rights of man. They demanded an end to the Church's temporal powers, advocating a transfer of power to the new national governments. They embraced laissez-faire economic doctrine and professed a willingness to experiment with new ideas.

Conservatives looked to Europe's constitutional monarchies, defending centralism and a hierarchical society. They advocated a continuation of Latin American traditions: a strong ruler—either a president or constitutional monarch, large haciendas controlled by patriarchs, a docile and servile majority, all backed by the authority of the Catholic Church. They approved of the privileges of the Church and felt more comfortable with a controlled economic system, including forced labor. Conservatives wanted to limit political participation, fearing that openness would lead to chaos.

Despite these debates, Liberals and Conservatives agreed on the most important issues. Both wanted to see the economic transformation of their countries, although they debated how to go about it. And neither group wished to cede control to the masses. In Brazil, the Visconde de Albuquerque wryly remarked, "There is nothing quite so much like a Conservative as a Liberal in office." More distinctive than party labels were the personalities. The politician as a man and leader exercised far more strength than did the more abstract institution, the party.

The most significant issue to divide Liberals and Conservatives was the role of the Catholic Church. The Church was the one colonial institution to survive independence, and it enjoyed great popular support. In addition, as a result of efficient organization, able administration, and the generosity of the pious, the Church continued to amass riches. At the end of the eighteenth century, the Church controlled as much as 80 percent of the land in some provinces of New Spain. That, however, constituted only part of the Church's wealth. Although rural estates and urban properties of the Church accounted for half the total value of the nation's real estate, the Church's real wealth accrued from mortgages and the impressive sums of interest collected.

Church power came not just from its wealth. The clergy, one of the best-educated segments of society, enjoyed tremendous prestige. The clerics regularly entered politics, held high offices in the new governments, or endorsed political candidates. In almost all of the new countries, the clergy monopolized education from primary school through the university. Further, the masses had more contact with the Church than with officials of the new State. In addition to Sunday services, people looked to priests for the most important events of their lives: baptism of their babies, marriage ceremonies, last rites, and burial in a Church graveyard.

Criticism of the Church centered not on religion itself but rather on its secular power and influence. Liberals believed that the very existence of the State was threatened by the Church's temporal powers and argued for lay teaching; secularization of the cemeteries; civil marriage; establishment of a civil register for births, marriages, and deaths; and State control over religious patronage. The debates between Liberals and Conservatives, particularly about the Church, would come to dominate the first fifty years of political life.

4.3: Masses and Elites

The wars of independence and early nationhood were led by elites who mobilized masses, and there was an inherent tension in this scenario: The elites did not want to relinquish their perquisites, but the masses expected their interests to be addressed. Simón Bolívar had told the Congress of Angostura in 1819: "We need equality to recast, so to speak, into a unified nation, the classes of men, political opinions, and public custom." But in the same speech, in which he advocated a hereditary Senate to provide enlightened and stable leadership, he said, "It must be confessed that most men are unaware of their best interests and that they constantly endeavor to assail them in the hands of their custodians—the individual clashes with the mass, and the mass with authority."

Nonetheless, in the early years of the new nations, Liberal ideas held sway, and they were reflected in the emphasis on elections. In the first flush of freedom, most of the new nations granted broad suffrage rights to men. Free, independent men were given the right to vote, which included the indigenous and free blacks. In Buenos Aires, an 1821 law established universal male suffrage. Peru's 1823 constitution gave more limited voting rights to all Peruvian men who were married or over twenty-five years old and who were literate property owners, had a profession or trade, or worked in a "useful industry." The literacy requirement was postponed until 1840—presumably the new State would provide education—and later the indigenous and mestizos were exempted from that requirement. Not all of Latin America followed suit. For example, Chile's 1833 constitution only gave the vote to literate men who satisfied property or income requirements. Most regions provided indirect elections, with voters choosing an electoral college or voting for a legislature that in turn selected senators and presidents.

Women, however, were not included in these new democratic freedoms. The new political model established the male head of household as the conduit between state and family. The men would be active participants in these new polities; the women would be passive citizens. While reasons were given to deprive men of voting rights—illiteracy, lack of property—there was no rationale given for the deprivation of women's rights. Like children, women were assumed to be unable to reason. The new nations' leaders looked to British law and the Napoleonic Code, which limited women's rights in comparison to Spanish legal tradition, which was patriarchal but more flexible.

Although women were excluded from formal political roles, they joined informal political actions. For example, in Colombia poor black and mestiza women joined men in destroying fences that had been erected in the common land (*ejidos*) from 1848–1851, and in 1853, the police commissioner who planned to read the new Liberal constitution was met by a "mob of women" who threw "insults, sarcasm, water, and stones." Their access to political acts was limited by class and race. Some lower-class women could participate in demonstrations and even riots, although indigenous women were often limited by traditional, patriarchal community roles. Conservative elite women might play traditional roles in support of the church, while Liberal women were more likely to be excluded by the new "rational" republican ideas.

Elections for all levels of government were routinely held, indicating that it was an important aspect of legitimation. Even those who resorted to violence to attain power also took the electoral route, sometimes using violence to guarantee a victory or to confirm their position. Political contenders worked to mobilize voters to support them, which usually meant appealing to specific, local interests. Latin Americans still thought in terms of the *pueblos* (towns), the *comunidades* (communities), and the *vecino*, literally neighbor, or a resident of a particular place.

At rare times, elections drew a great deal of popular interest. The Arequipa, Peru, newspaper in 1833 reported on congressional elections: "We have witnessed the council chambers full of citizens casting their votes: and their patriotic enthusiasm on this occasion is worthy of imitation and praise." In the 1820s, voter turnout in Mexico City was at times as high as 70 percent. However, such substantial turnout was not common. Throughout Latin America, most people simply did not go to the polls. Voter turnout was usually below 5 percent, and, sometimes, it was even as low as 0.02 percent. Elites throughout Latin America lamented "the lack of civic spirit," and electoral campaigns often were aimed more at voter turnout than at assuring a particular outcome. There are several explanations for the lack of interest in voting: Voting and representation were new, abstract ideas; there was frequently violence at the polls; and people trusted their leaders to take care of business for them. The idea of placing trust in a leader, whether local or national, would become the bedrock for the *caudillos*, the strongmen who brought stability to the young nations.

Voting rights tended to become more restricted throughout the nineteenth century, as the elites wrote and rewrote their constitutions. The most popular models were the North American and French constitutions as well as the Spanish Constitution of 1812. The Latin Americans promulgated and abandoned constitutions with numbing regularity. It has been estimated that in the century and a half after independence, they wrote between 180 and 190 of them, a large percentage of them adopted during the chaotic period before 1850. Venezuela holds the record with twenty-two constitutions since 1811. Four major Latin American nations have a somewhat more stable constitutional record. Brazil's constitution, promulgated in 1824, lasted until 1889. After several attempts, Chile adopted a constitution in 1833 that remained in force until 1925. Argentina's constitution of 1853 survived until 1949 and was put into force again in 1956. Mexico promulgated a constitution in 1857 that remained the basic document until 1917.

Generally the constitutions invested the chief executive with paramount powers so that he exercised far greater authority than the other branches of government, which were invariably subservient to his will. By the mid-nineteenth century, all the Latin American governments shared at least three general characteristics: strong executives, a high degree of centralization, and restricted suffrage.

Municipal governments became the locus of struggle between the central government, which viewed the municipality as a unit of the new nation, and local inhabitants for whom the municipal government was embedded in local traditions and concerns. National leaders wanted municipal governments to carry out state and national laws, police the population, collect taxes, organize local militias, and provide information to the national level. Municipal government was funded in the same manner as local colonial administration—land rentals, taxes paid by vendors on market days, fines, and a head tax. The income paid for administrative costs, public works, schools (where they existed), and religious celebrations, usually of the local patron saints. In this way, elites sought to create new national identities by drawing on local ones.

But the outlook for the majority remained local, and they saw the municipal government not as a conduit for national control but as the manager of the traditional community. They also transformed traditional local organizations to meet the new demands of independence. For example, the *cofradía*, or religious brotherhood, grew to be not just a religious organization but one that provided for community well-being, as a location of community organizing, and as a landholder.

It has sometimes been imagined that the majority of the Latin American population, illiterate and rural, was left out of this nation-building process. Undoubtedly, for many people that was true, especially in areas relatively untouched by the independence wars. In some parts of Central America, it was months after independence before people learned that they were no longer part of the Spanish empire. But in regions where there had been a great deal of fighting, the masses were already mobilized around the Enlightenment rhetoric of the independence project. Would-be national leaders had to contend with politicized populations. The process of state building occurred not just at the national level but in all regions. And the direction that the new leaders took was not uncontested in the communities.

In the countryside, the indigenous and mestizo lower classes struggled with elites over control of property, particularly over municipal boundaries and the continued existence of community land. Local communities also fought against national programs that would adversely affect them. For example, the people of the Mexican state of Guerrero supported the overthrow of President Anastasio Bustamante because his government failed to protect local cotton growers and weavers from foreign imports. Throughout the 1840s, when Mexico was ruled by centralists, the people of Guerrero rebelled against increased taxes and the decrease in the number and autonomy of municipalities; they were aware of divisions at the national level, and they used that opportunity to press their demands.

Rural people learned to work the new systems to their advantage whenever possible. For example, indigenous people in Peru when in land disputes with estate owners emphasized that under an 1828 law, they had become individual proprietors. But when it came to taxation, they spoke of communal land because the law still provided for a lower level of taxation for the indigenous. In some areas, the indigenous joined with mestizos and formed a *campesino* identity. (Campesino is often translated as "peasant," but that term has questionable implications about relationship to land and markets. In general, Latin Americans use the term to refer to rural people who work primarily in subsistence agriculture and sometimes to refer to lower-income country people in general.) At times, popular groups sought the protection offered by Conservatives, at others the freedoms promised by Liberals.

This struggle between local autonomy and a unifying, national project was an ongoing challenge to national leaders. For example, many sovereign polities competed in the Argentine region, narrowing to two—the Argentine Confederation and the Free State of Buenos Aires—in 1852–1853, and finally to the nation of Argentina in 1861–1862. Originally the name *argentino* only referred to residents of Buenos Aires or the area around the river Plate; it was not accepted as a name for all residents of the country until the 1870s.

4.4: Threats to New Nations

The new nations were fragile constructions, and their leaders faced multiple threats to their continued existence. From the start, the new leaders feared that Spain or Portugal, alone or in union with other European governments, might try to recapture the former colonies in the New World. The conservative monarchies of Russia, Austria, and Prussia formed the Holy Alliance, which numbered among its goals the eradication of representative government in Europe and prevention of its spread to areas where it was previously unknown. The Alliance boldly intervened in a number of European nations to dampen the fires of liberalism. At one time, it seemed possible that the Holy Alliance might help Spain in reasserting its authority over its former American colonies. The possibility alarmed their rival, England, which was interested in Latin American markets, as well as the United States, which was concerned about Russian settlement advancing down the western coast of North America.

The English urged the United States to join in a statement discouraging foreign colonization of the Americas. Instead, an independent-minded President James Monroe in 1823 issued what became known as the Monroe Doctrine, which declared that the Americas were no longer open to European colonization and that the United States would regard any intervention of a European power in the Americas as an unfriendly act against the United States. Most Latin American elites welcomed the possibility of help from their northern neighbors as they took their first shaky steps as new nations. However, they found the doctrine to be empty rhetoric: The United States did not come to the aid of Latin American countries until the end of the nineteenth century.

Latin American fears of European intervention were not unfounded. Spain invaded Mexico and Central America in 1829 and 1832. The French occupied Veracruz, Mexico, in 1838 to force Mexico to pay alleged debts; and they blockaded Buenos Aires in 1838–1840 and again in 1845–1848, this time in conjunction with the British, to discipline Argentine dictator Juan Manuel de Rosas. During the 1860s, Spain made war on Peru, seizing one of its guano-producing islands, and bombarded the Chilean port of Valparaíso.

But it was not only the Europeans that the new countries had to fear. Although in 1823 the United States claimed to be Latin America's defender, by 1846 the United States had become an aggressor in its westward march. The central

Map 4.1 Mexico, 1824–1853

Adapted from Spodek Howard, *The World's History Combined*, 3rd ed. © 2006. Electronically reproduced by permission of Pearson Education, Inc., Upper Saddle River, NJ.

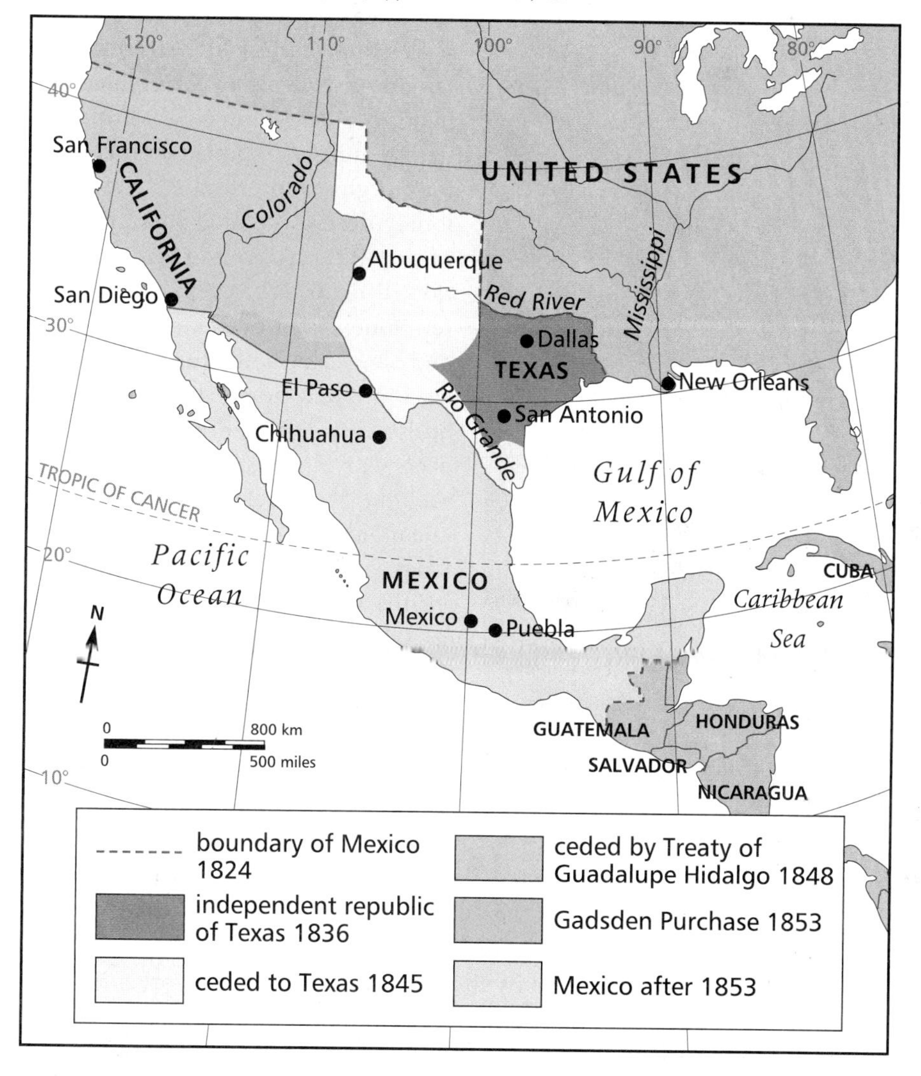

conflict began as a dispute about the western boundary of Texas, which had just been annexed by the United States. The boundary had always been the Nueces River, but Texans and the United States now claimed the boundary was the Río Bravo (or Río Grande as it was called in the United States). The result was a disastrous war in which the better armed and trained U.S. forces pushed all the way to Mexico City. In the Treaty of Guadalupe Hidalgo (1848), which ended the war of North American Invasion (known in the United States as the Mexican–American War), the United States won the huge California and New Mexico territories.

The new Latin American nations recruited the indigenous to man their armies. This 1868 photograph shows a Peruvian soldier and his wife. (Library of Congress)

The threats to the new Latin American nations were by no means all external. The unity of the new nations was extremely weak, and the forces that shattered it were most often internal. Geography provided one major obstacle to unity. Vast tracts of nearly empty expanse, impenetrable jungles, mountain barriers, and lonely deserts isolated small pockets of population, making communication difficult. Transportation was often nonexistent or at best hazardous and slow, especially during the rainy season. Such poor communications and transportation complicated the exchange of goods, services, and ideas on a national basis. It was easier and cheaper to ship a ton of goods from Guayaquil, Ecuador, to New York City via the Straits of Magellan than to send it 200 miles overland to the capital, Quito. Rio de Janeiro could import flour and wheat more economically from England than from Argentina. Similarly, the inhabitants of northern Brazil found it easier to import from Europe than to buy the same product from southern Brazil, even though most people lived near the coast, which was served by sailing vessels.

To penetrate the interior, the Brazilians relied on inland waterways, in some areas generously supplied by the Amazon and Plata networks, or cattle trails. A journey from Rio de Janiero to Cuiabá, capital of the interior province of Mato Grosso, took eight months in the 1820s. The situation was comparable in Spanish America. The trip from Veracruz, Mexico's principal port, to Mexico City, a distance of slightly less than 300 miles over the nation's best and most used highway, took about four days of arduous travel when Frances Calderón de la Barca made the journey under the most favorable conditions in 1839. She described the road as "infamous, a succession of holes and rocks." In the 1820s, a journey from Buenos Aires to Mendoza, approximately 950 miles inland at the foothills of the Andes, took a month by ox cart or two weeks by carriage, although a government courier in an emergency could make the trip in five days on horseback.

Distance, difficult geography, slow communication and transportation, and local rivalries, in part spurred by isolation, encouraged the growth of a regionalism hostile to national unity. Experiments with federalism intensified that regionalism and reinforced the loyalty to the patria chica. As a result, soon after independence the former territories of the Spanish viceroyalties disintegrated. None splintered more than the Viceroyalty of New Spain. In 1823, Central America, the former Kingdom of Guatemala, seceded. In turn, in 1838–1839, the United Provinces of Central America broke into five republics. Texas left the Mexican union in 1836, and after the war of 1846–1848, the United States won California, Arizona, and New Mexico. Gran Colombia failed to maintain the former unity of the Viceroyalty of New Granada: Venezuela left the union in 1829, and Ecuador followed the next year. Chile and Bolivia felt no loyalty to Lima, and consequently the Viceroyalty of Peru disbanded even before the independence period was over. In a similar fashion, Paraguay, Uruguay, and part of Bolivia denied the authority of Buenos Aires, thus spelling the end of the Viceroyalty of La Plata. By 1840, the four monolithic Spanish viceroyalties had split, giving rise to all of the Spanish-speaking republics of the New World except Cuba, which remained a Spanish colony until 1898, and Panama, which split from Colombia in 1903.

None of the eighteen new nations had clearly defined borders, a problem destined to cause war, bloodshed, and ill will. In some cases, commercial rivalries added to the difficulties. Further, the new nations raised trade barriers, which in turn intensified those rivalries. The resultant suspicion and distrust among the eighteen countries exacerbated the tensions felt within the new societies. On occasion, these tensions gave rise to war, as neighbor fought neighbor in the hope of gaining a trade advantage, greater security, or additional territory.

Argentina and Brazil struggled in the Cisplatine War (1825–1828) over possession of Uruguay and in an exhausted stalemate agreed to make the disputed territory independent. Chile attacked Peru and Bolivia in 1836 to prevent the federation of the two neighbors, and during the War of the Pacific (1879–1883), the three fought for possession of the nitrate deposits of the Atacama Desert. Chile won and expanded northward at the expense of both Peru and Bolivia. The Dominican Republic battled Haiti in 1844 to regain its independence. Throughout the nineteenth century, the five Central American republics challenged each other repeatedly on the battlefield. This catalog of conflicts is only representative, not comprehensive.

The wars constituted another arena in which local identities intersected with the national project envisioned by elites. For example, autonomous campesino guerrilla bands organized to protect their villages in Peru from Chilean invaders during the War of the Pacific. They were particularly opposed to leadership in northern Peru that took a conciliatory approach to the invading Chileans. Thus, local indigenous groups envisioned themselves as fighting for the nation in contrast to abandonment by the northern leadership.

Mexico's Early National Woes

Mexico exemplified Latin America's early national problems: foreign invasions, internal divisions, and caudillo rule.

In 1829, Spanish troops invaded in hopes of reclaiming the territory. In 1838, the French invaded to collect unpaid debts and settle claims by French citizens, among them a baker whose claim dubbed the invasion the Pastry War. U.S. invasion (1846–1848) was the most devastating, leading to the loss of half of Mexico's territory.

Geographic divisions also plagued Mexico. Central America seceded from Mexico early on, in 1824, with little fanfare. A more serious loss was the secession of Texas in 1835. Mexican authorities had authorized foreign settlement in the remote province, but English-speaking Protestants resented Mexican taxation, Catholicism and the tithe, and prohibition of slavery. Texas won independence in 1836; U.S. annexation (1845) set the stage for invasion and loss of Mexico's northwest.

Throughout this period, Liberals and Conservatives jockeyed for power, and Mexicans turned repeatedly to colorful caudillo Antonio Lopez de Santa Anna. Santa Anna fought for Mexico's independence and helped overthrow Agustín Iturbide. He defended Mexico's first elected president, Guadalupe Victoria, against Conservative rebellion and was rewarded with the governorship of Veracruz, where he built a lasting political base. His heroic role against Spanish invasion in 1829 catapulted him to national prominence, and in 1833 Santa Anna was elected president. It is often said that he held the presidency eleven times. More accurately, Santa Anna repeatedly took leaves from office—frequently to fight against rebellions and invasions—then resumed the presidency. He did so four times between 1833 and 1835.

Santa Anna was always more of a military than political leader, and Mexicans sought him in wartime, celebrating and rejecting him based on his battlefield results. He left office in disgrace after losing Texas in 1836 but was exalted again after the Pastry War—especially when he lost a leg but continued to lead troops while carried on a litter. He returned to the presidency for four months in 1839 and held a funeral for his amputated leg. Subsequent presidential terms were brief: October 1841–1842; March to October 1843; June to September 1844; and March to April and May to September 1847, leaving office to battle U.S. forces. Defeat at U.S. hands seemingly sealed his fate, but Conservatives restored him to the presidency in 1853. He was finally turned out for good in 1855.

Santa Anna was first elected as a Liberal, reflecting early national preferences for constitutionalism, representational government, and legal equality. But early Liberal governance was accompanied by unrest, which Conservative elites blamed on widened popular participation. Conservative reaction was embodied and eventually discredited by Santa Anna's rule.

The Santa Anna era ended with the Revolution of Ayutla (1855), which brought Liberals to power. Reform was spearheaded by Minister of Justice Benito Juarez and Minister of Development Miguel Lerdo. The Juarez law (1855) abolished fueros, limiting military and church courts to military discipline and theology. The Ley Lerdo (1856) broke up corporate landholdings, particularly church lands and indigenous ejidos, to expand market relations and private property. The Church Law (1857) removed church control over registries for births, marriages, adoptions, and deaths. The laws were incorporated in the 1857 constitution. The Catholic Church condemned those who swore allegiance to the new constitution, and the government refused jobs to those who would not.

Conservative reaction led to the War of the Reform (1858–1861), a bloody struggle finally resulting in Liberal victory and Juarez's 1861 election. But Juarez faced the usual problems: an empty treasury; army and police who needed to be paid; European creditors clamoring for debt payment; lack of funds to pay the state bureaucracy; and a huge monthly deficit. Confiscated church lands brought in less than expected, and customs receipts suffered because of the war. Already in arrears, the government could not arrange new loans.

Unpaid debts became the perfect excuse for Louis Napoleon Bonaparte, nephew of Napoleon I, to try to create a new French empire by invading Mexico.

The war between Mexico and France (1861–1863) featured one spectacular Mexican victory—the Battle of Puebla, May 5, 1862, known as Cinco de Mayo. While Mexicans won that battle, led by General Ignacio Zaragoza and the young General Porfirio Díaz, they lost the war. The result was the monarchy of Emperor Maximilian, an Austrian archduke kept on the throne by 20,000 French troops. His coronation led almost immediately to renewed warfare, and he was finally toppled in 1867, when the republic was restored.

The major conflict of the century pitted tiny, landlocked Paraguay against the Triple Alliance—Argentina, Brazil, and Uruguay—in a clash of imperialistic pretensions in the strategic La Plata basin. It took the allies five years, 1865–1870, to subdue Paraguay. That war solved two difficult problems that had troubled the region since independence: First, it definitively opened the Plata River network to international commerce and travel, a major concern of Brazil, which wanted to use the rivers to communicate with several of its interior provinces. Second, it freed the small nations of Uruguay and Paraguay from further direct intervention from Argentina and Brazil, which came to see the independence of the two small Platine countries as buffer zones.

Relations between the new nations and the Roman Catholic Church created tensions of another sort. Latin American chiefs of state claimed the right to exercise national patronage as heirs of the former royal patronage. The Pope in turn announced that the patronage had reverted back to the papacy with the declarations of independence. In their open sympathy with the Spanish monarchy, the popes had antagonized the Latin American governments. In 1824, Pope Leo XII issued an encyclical to archbishops and bishops in America to support Ferdinand VII. This confirmed to Latin American leaders that the Vatican was in league with the Holy Alliance. The new governments in turn expelled a number of ranking clerics who refused to swear allegiance to the new State.

Out of consideration for the feelings of Madrid, the Vatican for a long time refused to recognize the new American nations, much to the chagrin of their governments, fearful that the discontent of the American Catholics with their isolation might endanger independence. Rome began to change its attitude toward the Latin American nations in 1826, when the Pope announced his willingness to receive American representatives strictly as ecclesiastical delegates, in no way implying political recognition. The next year the Roman pontiff began to approve candidates presented by the American governments. With the death of Ferdinand VII in 1833, the Pope no longer felt any Spanish constraints on his policy in the New World. In 1835, the Vatican recognized New Granada (Colombia) and the following year accredited to Bogotá the first papal nuncio. The recognition of New Granada signified the end both of the problem over political recognition and of Spanish influence over the Vatican's relations with Spanish America.

4.5: Economic Instability

Economic freedom had been one of the main concerns of the creole elites who fought for independence. But the destruction caused by the wars was devastating, particularly in Mexico, Venezuela, and Gran Colombia. Mining and manufacturing suffered the most during the decades after independence. With some mines flooded and machinery destroyed during the fighting, the labor system in flux, and investments lacking, the production of the once-fabled mines plummeted. The decline continued steadily in Mexico and Peru until midcentury. The Bolivian mines did not revive until around 1875. It would take substantial investment to refurbish these industries, but capital vanished along with the Spaniards who fled at the end of the wars.

Great Britain was waiting in the wings. Foreign Secretary George Canning mused in 1824, "Spanish America is free, and if we do not mismanage our affairs sadly, she is English." As soon as Portugal and Spain fell to Napoleon, eager British merchants began to move in large numbers into Latin America to capture the markets they had so long craved. The British immediately sold more to Latin America than anyone else and almost monopolized the imports into certain countries.

The British government successfully wrested from the Latin Americans agreements favorable to its merchants, traders, and bankers. Brazil's experience was classic. The new empire provided the English merchants and manufacturers with their most lucrative Latin American market. Exports to Brazil in 1825 equaled those sold to the rest of South America and Mexico combined and totaled half those sent to the United States. Naturally the British wanted to keep their Brazilian market. In exchange for arranging Portugal's recognition of Brazilian independence in 1825, London exacted a highly advantageous commercial treaty from Pedro I. It limited the duty placed on English imports to 15 percent and bound Brazil not to concede a lower tariff to any other nation. The treaty thereby assured British manufacturers domination over the Brazilian market and postponed any Brazilian efforts to industrialize.

London supplied most of the loans and investments to the new nations. Already by 1822, four Latin American loans had been floated. In 1824 there were five more, and the following year an additional five. Foreign investors had particular interests in these economies. During the first half of the century, Europe had entered a period of rapid population growth and accelerated industrialization and urbanization. They needed raw products: food for the urban centers and materials for the factories. In turn, they sought markets in which to sell growing industrial surpluses. Latin America exported the raw materials required in Europe and imported the manufactured goods pouring in from distant factories.

Between 1800 and 1850, world trade tripled, and Latin America participated in that growth. While two or three ships a year handled trade between Chile and

England from 1815 to 1820, more than 300 carried Chilean exports to England in 1847. The value of exports leaving Buenos Aires nearly tripled from 1825 to 1850. The sector of the economy that recovered most quickly after the wars was agriculture, and Europe provided a ready market. The sale of agrarian products provided the basis for any prosperity that Latin America enjoyed in the decades before 1850.

Trade was facilitated by improved international transportation. Faster sailing vessels and the introduction of the steamship, used successfully in North Atlantic crossings in the 1830s, were responsible. The steamships appeared in Brazil in 1819 and Chile in 1822. In 1840, the British chartered the Royal Mail Steam Packet Company to provide regular twice monthly steamship service to the Caribbean. That same year, the Pacific Steam Navigation Company initiated steamship service along the western coast of South America. For the Atlantic coast, the Royal Mail Steam Packet from England to Brazil began service in 1851. The United States expanded its international steamship service, reaching Latin America in 1847 with the Pacific Mail Company. These improved communication and transportation systems further meshed the economies of Latin America with those of the United States and Great Britain.

To produce goods for export, Latin American elites needed to control land and labor. But the labor question was not as easily resolved as in the colonial period. Commitments had been made in order to mobilize the masses during the war, and to some extent, those promises had to be kept. During the independence struggle, creoles had used the metaphor of slavery to discuss their colonial plight. Bolívar told his troops in 1824, "You are going to complete the greatest task that heaven has been able to entrust to man—that of saving the entire world from slavery." The independence leader, of course, was speaking metaphorically about colonial subjects as slaves. However, this rhetoric was adopted by the real Latin American slaves. When Ecuadoran slave Angela Batallas appealed to the new nation for her freedom, she said, "I do not believe that meritorious members of a republic that . . . have given all necessary proofs of liberalism, employing their arms and heroically risking their lives to liberate us from the Spanish yoke, would want to pledge to keep me in servitude." She even appealed directly to Bolívar who, despite his elitism, had commented, "It seems to me madness that a revolution for freedom expects to maintain slavery."

The madness, however, did not immediately end. Although all of the new nations ended the African slave trade and freed children born to slaves, full abolition came more slowly. Only the Dominican Republic, Central America, Chile, and Mexico abolished slavery in the 1820s, and by midcentury, many Latin American blacks were still enslaved. Indigenous tribute also lingered, not ending legally until 1854 in Peru, 1857 in Ecuador, and 1874 in Bolivia.

Enforcement of the laws was uneven at best. As usual, resourceful landowners found ways of observing the letter of the law while changing only slightly the patterns of labor employment. They developed systems of apprenticeship and debt peonage. In some relatively isolated areas, such as Chiapas, debt

Table 4.1 Years when Latin American Countries abolished Slavery

Abolition of Slavery	Year
Dominican Republic	1822
Chile	1823
Central America	1824
Mexico	1829
Uruguay	1830
Bolivia	1851
Colombia, Ecuador	1852
Argentina	1853
Venezuela	1854
Peru	1855
Paraguay	1862
Cuba	1886
Brazil	1888

peonage was fairly successful in tying workers to haciendas via inheritable debt arrangements. But in much of Latin America, inadequate police forces and transportation systems simply made it impossible to find or return workers.

Most large estates survived intact from the turmoil of the independence period. In fact, many multiplied in size during those turbulent years. In northern Mexico, the Sanchez Navarro family, through astute business practices and shrewd political maneuvering, managed to preserve everything it had amassed during the colonial period. Their landholdings reached a maximum between 1840 and 1848, consisting of seventeen haciendas that encompassed more than 16 million acres. In Argentina, the Anchorena family, wealthy merchants in Buenos Aires, began to invest in ranches in 1818. Four decades later, they were the largest landowners in the country with 1.6 million prime, amply watered acres.

Indeed, the times were exceptionally propitious for landlords to extend their holdings. The governments put the lands of the Church, indigenous communities, and public domain on the market. From the old *ejidos* the new governments authorized small plots for subsistence, while hacendados and *fazendeiros* added to their already large landholdings. By 1830 in Argentina, approximately 21 million acres of public land had been acquired by 500 individuals.

Meanwhile, the rapid consolidation of large but often inefficient estates raised serious social and economic questions, which were rarely addressed. In Mexico, shortly after independence, Francisco Severo Maldonado warned that national prosperity required widespread land ownership. He advocated the establishment of a bank to buy land from those who owned large, unused parcels and to sell it "at the lowest possible price" to those without land. One observer of Brazil's economy, Sebastião Ferreira Soares, concluded in 1860 that for

Brazil's economy to develop, it would require that uncultivated land be put in the hands of people who work it. An editorial in the Buenos Aires newspaper *El Rio de La Plata*, September 1, 1869, lamented, "The huge fortunes have the unfortunate tendency to grow even larger, and their owners possess vast tracts of land which lie fallow and abandoned. Their greed for land does not equal their ability to use it intelligently or actively."

One inevitable result of agrarian mismanagement was the increase in the price of basic foodstuffs. In 1856, the leading newspaper of Brazil's vast Northeast, *Diário de Pernambuco*, sharply condemned large landholdings as a barrier to development. Their owners withheld land from use or cultivated the land inefficiently, resulting in scarce, expensive foodstuffs. Better use of the land, the newspaper editorialized, would provide more and cheaper food for local markets as well as more exports. At the same time in the southern province of Rio de Janeiro, a region undergoing a boom in coffee production for export, Ferreira Soares came to similar conclusions. He observed with alarm the rapid extension of the export sector accompanied by declining production of food for internal consumption. He noted that foods that had been exported from Rio de Janeiro as late as 1850 were being imported a decade later. Prices of basic foodstuffs—beans, corn, and flour—rose accordingly.

Despite the occasional criticism, there was unlikely to be any change in economic organization. It was the landed gentry who controlled the Latin American governments. The chiefs of state owned large agricultural estates or were intimately connected with the landowning class. Representatives of that privileged class filled the legislatures. The increasingly restrictive voting requirements of property ownership and/or literacy almost limited the franchise to that class. The courts represented them—from the ranks of the elite came the lawyers and judges—and usually decided cases in their favor.

The elites tended to romanticize the large estate, one useful means of enhancing their ideology. No one better idealized the mid-nineteenth-century hacienda than Jorge Isaacs in his highly acclaimed novel *Maria* (1867). The author created a patriarchal estate in the Cauca Valley of Colombia that served as a model: Orderly, hierarchical, harmonious, the novel's well-run estate centered on the comfortable "big house" and patriarchal authority, which extended from the doting family to the devoted slaves. The novel's popularity arose mainly from the ideal but tragic romance it depicted, although it must have also come from its appealing portrait of the idyllic country life in which people and nature intertwined, in which social roles were well defined and unquestionably accepted, and in which alienation apparently was unknown. This vision complemented a system that doubtless seemed less perfect to the *campesinos* than it did to the elites.

Prosperity, however, was elusive. Unsettled conditions during the early decades of the national period inhibited economic growth. Wars damaged infrastructure, interrupted trade, and frightened off wary investors, as did the chronic political instability. Politics rather than economics absorbed most of the attention of the new nations. At the same time, the quality of public administration deteriorated. Many trained public administrators departed with the defeated Spanish armies or returned to Lisbon with John's court. Recruitment was seldom based on talent; rather, positions in the civil service came as a political reward, and the frequent changes of government hindered the training of a new professional civil service. The national treasuries lay bare. Public financing was precarious, and the fiscal irresponsibility of the governments notorious. By 1850, the Latin American governments had defaulted on most of their loans, and no new investment would come into the region for some twenty years.

But the economic problems did not induce Latin American elites to rethink their economic policies. Just as they were captivated by foreign political ideologies that bore little relevance to local conditions, the Latin American elite also showed a penchant for economic doctrines more suitable to an industrializing Europe than underdeveloped Latin America. Adam Smith mesmerized many Latin American intellectuals, who embraced free trade as a solution to their nations' economic problems. Of course, Smith was writing about England in 1776, at a time when the country enjoyed the natural protection of being the world's only industrialized power. But, in the words of Mexico's *El Observador* in 1830, the country needed "absolute and general freedom of commerce" to promote prosperity.

In reaction to the former mercantilism they had deplored, the Latin Americans adopted policies of economic liberalism that they associated with the triumph of the Enlightenment but that bore no relation to the requirements of Latin America. Consistently modest tariffs deprived the new governments of sorely needed incomes and facilitated the flood of European manufactured articles inundating the New World, to the detriment of local industrialization. Mexico, for example, opened its ports in 1821 to all foreign goods at a uniform tariff of 25 percent ad valorem. Artisan manufacturing immediately declined. A petition to the national government in 1822 from Guadalajara for protection blamed the Liberal tariff for putting 2,000 artisans in that city alone out of work.

The elites were also taken with the ideas of David Ricardo, who argued that each country should emphasize its comparative advantage. For example, perhaps both England and Spain could produce wine and textiles. Nature, however, gave Spain much greater endowments for wine production, making it able to produce a higher quality product at a lower cost. Similarly, Ricardo argued, the already industrialized English could produce higher quality and lower-priced textiles. If each focused on its strengths, each would do well. There are, of course, several problems with this theory. The comparative advantage of Latin America was in primary products, which England wanted to buy at the lowest possible cost. Those products were then transformed into finished products and sold back to Latin America at a higher price. Furthermore, Latin America could become as strong a manufacturer as England with the proper protections for its industry, the path that the United States took. But the elites, who were getting rich via mining and farming, followed Ricardo uncritically.

The economic system that the elites obviously associated with progress was capitalism. It could not be otherwise when their primary models were England,

France, and the United States. The constitutions, laws, and political practices they put into effect complemented the penetration and growth of capitalism. They abandoned the protective but restrictive neocapitalism and mercantilism of the Iberian empires to try their fortunes with the dominant capitalist nations of the century. The landowners produced their crops on as large a scale as possible for sale to an external market from which they expected a satisfactory profit. The wealth from the countryside and mines—shared increasingly with middlemen in the ports—also brought prestige to landowners and politicians and ultimately conferred power. The growing merchant class, dependent first on the flow of primary products from the countryside and mines and second on the ability of the landed gentry and political leaders to afford costly European imports, exhibited scant inclination to challenge a system that also benefited them.

In catering to the caprices of an unpredictable market shaped by their trading partners, Latin Americans encouraged the growth of a reflex economy, little different than the previous colonial economy. The economic cycle of boom and bust repeatedly reoccurred in all regions of Latin America, condemning most of the area to the periphery of international capitalism. Peru was one of the most successful countries in Latin America as a participant in the growing export trade. But it, too, would display the classic boom-and-bust pattern.

The source of Peru's wealth was enormous deposits of guano, bird droppings, found off the coast on the Chinchas, Ballestras, Lobos, Macabi, and Guanape islands. Peru's guano was particularly attractive because of its high nitrate content, due to the diet of fish consumed by the birds and to the coastal weather conditions. The Humboldt Current brings cold water from Antarctica to the Equator, which mixes with the warm air on the coast and prevents rainfall. The hot air bakes the droppings, keeping the nitrates from evaporating.

Guano was in great demand as fertilizer in Europe and the United States, and Peru quickly capitalized on the product. Exports went from zero in 1840 to 350,000 tons a year by the 1850s, constituting 60 percent of the country's exports. It has been estimated that during the guano boom, 1840–1880, more than 21 tons of guano were exported, creating $2 billion in profits. By the 1880s, however, the boom was over. First, Peru lost important guano territories to Chile in the War of the Pacific. Then, guano lost its appeal as the market switched to nitrates as fertilizer. Chile, having won the Atacama Desert from Peru, cornered the market on nitrates and began its own boom—and eventual bust—cycle.

4.6: A Clash of Cultures

When the elites rejected Spanish domination, they also found themselves culturally adrift. They tried to define new American nations, not just in political and economic but in cultural terms as well. They struggled to create histories and symbols that would support a new identity. In the early days of independence, they drew on the noble, and largely mythologized, indigenous past as their own heritage. The Lima Congress in 1822 explained to the indigenous: "Do not be surprised that we call you brothers. We truly are; we descend from the same fathers, we form a single family." The Chilean historian Miguel Luis Amunátegui commented wryly in 1849: "Creoles in whose veins Spanish blood circulated declared themselves the heirs and avengers of the Aztecs, the Incas and the Araucanians who had been massacred centuries earlier by their fathers." Images of idealized indigenous nobility appeared in national anthems and on state seals, new coins, and even the curtain of the refurbished Lima theater. The new leaders reached to the past for names: New Spain became Mexico, evoking the Mexica, and the provinces of Nuevo Santander and Nueva Galicia became Tamaulipas and Xalisco.

But these elites did not see the glorious indigenous past when they looked at contemporary indigenous communities. Instead, they saw people they described as dirty, backward, and degraded. The elites blamed the inferior condition of the indigenous on Spanish colonial repression. Because the elites rejected both the Spanish heritage and indigenous culture, they turned instead to the ideas of the Enlightenment. They hoped that by embracing the ideas of French and British thinkers, they could reshape their nations, fostering modernity and prosperity. Because these ideas shaped official attitudes and institutions and pervaded the thinking of the elites, these ideas also exerted profound influence on the lives of all Latin Americans, no matter how humble.

The ideology of progress that emerged from the elites' flirtation with the Enlightenment was nowhere better expressed than by Argentina's Generation of 1837, an exceptionally articulate group of Liberal intellectuals. Their ideas reached far beyond the Argentine frontiers to shape much of the thinking of modern Latin America. Impressive urgency was achieved by those intellectuals because they were united in their opposition to popular caudillo Juan Manuel de Rosas, who dominated Argentina from 1829 until 1852. The Generation of 1837 regarded its conflict with Rosas as a struggle between "civilization" and "barbarism," a dialectic repeatedly invoked by intellectuals throughout the century. In defining civilization, the Generation of 1837 identified the Argentina they intended to create—and in fact did create—as a copy of Europe.

Associated with the port of Buenos Aires, the intellectuals looked with horror on the rest of the nation as a vast desert in need of the civilizing hand of Europe. Buenos Aires would serve, according to their blueprint, as a funnel through which European culture would pass on its civilizing mission to redeem the countryside—if it was redeemable. Many of the elite finally concluded it was not and advocated European immigration as the best means to "save" their country. They aspired to govern Argentina by means of a highly restrictive democracy. Esteban Echeverria summed up that aspiration in his influential *Dogma Socialista* (*The Socialist Doctrine*, 1838): Men of reason should govern rationally to avoid the despotism of the masses. Domingo Faustino Sarmiento forcefully set forth the dialectic in his *Facundo, or Civilization and Barbarism* (*Facundo o civilización y barbarie en las pampas argentinas*, 1845): The conflict was between the progress of the Europeanized city versus the ignorance, barbarism, and primitivism of the countryside, exemplified by the rule of caudillo Juan Facundo Quiroga and the lifestyle of the *gaucho*, cowboys who lived a nomadic lifestyle based on herding cattle and hunting. The independence of the gaucho, or of the self-supporting indigenous village, conflicted with elite desires to control labor in service to a new, competitive economy that would bring progress to Latin America.

The ideology of progress pursued by dominant elites extolled a type of liberty and democracy sanctioning individualism, competition, and unchecked pursuit of profit. The elites spoke constantly of "progress," which implied admiration for the latest ideas, inventions, and styles of Europe and the United States and a desire to adopt—rarely to adapt—them. The elites believed that "to progress" meant to recreate their nations as closely as possible in the shape of their European and North American models. Because the elites would benefit, they assumed that by extension their nations would benefit as well. They always equated their class well-being with national welfare.

The ideology of progress favored the wealthy, resourceful minority over the majority with limited resources. In fact, the values the elites placed on abstract liberties and democracy did not benefit most of the population, whose experiences were rooted in the New World. They drew from a past that had its conflicts but also generally fostered a culture of interdependence and reciprocity. The masses could not hope to gain much from elite values, and in fact, democracy as it took form in nineteenth-century Latin America quickly became a superficial rationale excusing or disguising the exploitation of the many by the few.

As they did in the colonial era, the popular classes drew cautiously from European sources. They chose to carefully mediate outside influences and continued to embrace their own cultures, based on common language, heritage, beliefs, and means of facing daily life. These folk cultures were based on traditional moral dictates, community obligations, and kinship relations. Economic decisions were embedded in social considerations prioritizing community and reciprocity.

The first priority for the indigenous was to protect their traditional communities. They even were willing to continue to pay the head tax, which had defined the subordination of the indigenous in the colonial world. Independence leaders initially saw the head tax as a despicable colonial holdover. José de San Martín abolished the tax in Peru by decree in 1821, calling it the "shameful tax." He also decreed that "henceforth the aborigines will not be called Indians or Natives; they are sons and citizens of Peru and will be known as 'Peruvians'."

But the desire of the independence leaders to integrate the indigenous into the new nations came up against unexpected challenges. New national leaders found they needed to rely on the head tax because they were desperate for revenue. With production languishing in the wake of independence destruction and postindependence conflicts, there was little in the way of customs duties. As elites, they were loath to tax their own class; and even if they wanted to do so, weak states lacked the authority to compel elites to pay taxes.

The limits of state authority also came up against the desire of the indigenous to guard their communities, clinging to colonial burdens that also guaranteed benefits. The head tax marked the indigenous as an "Indian," which also guaranteed community land as well as exemption from other taxes and from military service. When the national authorities of Ecuador attempted to collect a new sales tax in 1834, the indigenous immediately protested. Frightened regional authorities sent out the militia, but the national government quickly ruled that local officials should not attempt to collect the tax from the indigenous. The head tax was not eliminated until 1854 in Peru and 1857 in Ecuador.

The indigenous and folk communities also maintained their traditional way of life in terms of food, clothing, and housing. In Mexico and Central America, diets consisted primarily of beans, rice, and maize, which was used to make tortillas, as well as chiles and squash. In the Andean highlands, meals centered on several varieties of potatoes as well as maize and various root crops. Brazilians relied on black beans and manioc flour. Homes were huts built from dried mud bricks (*adobes*), or stones covered with mud. Roofs were thatch, sometimes over wooden boughs, and floors were dirt. Clothing was of simple cotton cloth, and most people went barefoot in warmer climates; in colder areas, woolen ponchos, sweaters, and blankets carried traditional patterns. Indigenous weaving identified various ethnic communities, and the styles were those handed down for generations.

Elites complained at length about the indigenous style of life, commenting particularly on their miserable huts and poor clothing. Their real concern was that the indigenous seemed to have so little need of the commodities that local manufacture could provide, if there were a market. Limited indigenous needs also meant that there was little in their lifestyles to compel the indigenous to work on neighboring haciendas. The elites, meanwhile, created a market for imported European goods, adorning themselves in the latest styles from London and Paris, however inappropriate they might be for local climates.

4.7: Control by Caudillos

The weakness of new political institutions gave rise to the rule of the *caudillo*, or strong leader. He was a product of the lingering violence of the independence wars and the fractured nature of the new nations. Most caudillos began as military leaders who built a following during independence or subsequent wars. These leaders were charismatic and appealed to a mass base of support. They built political coalitions and used government administration to collect taxes and to pass and enforce laws. But the authority of the *caudillo* came from his personal authority rather than institutional legitimacy.

Most caudillos advocated a selective Europeanization and enjoyed the support of the elites. Frequently they offered a continuation of the patterns of the past: large landed estates, servile-labor systems, export-oriented economies, and highly centralized political power. They were often dictatorial in their methods and controversial for the violent methods employed. But they were also beacons of stability amid the uncertainty that plagued the young nations.

The caudillo could not rule without support, and he usually sought it from the rural aristocracy, the Catholic Church, and the army. The early caudillos were more often than not members of or related to the rural landowning class, men notable for their desire to preserve their class's prestige, wealth, and power, and for their opposition to land reform, extension of the suffrage, and popular government. The caudillos usually represented these interests and tended to suppress the influence of more liberal urban elements. The Church as an institution was a conservative force suspicious of reforms and, with a few notable exceptions, rallied to endorse any caudillo who protected Church interests.

The army, the only truly national institution, immediately emerged as a political force. Its strength revealed the weakness of political institutions. In many countries, it was the dominant political force and so remained well into the twentieth century. Since few of the republics developed satisfactory means to select or alter governments, palace coups—in which the military always had a role—became the customary means to affect political shifts throughout the nineteenth century. The military then exercised the dual role of guaranteeing order on one hand and changing governments on the other. No caudillo or president would willfully alienate the military. Consequently, officers enjoyed generous salaries and rapid promotions.

The Latin American armies became and have remained top heavy with brass. Thus, armies not only retarded the growth of democracy through their political meddling but also slowed down economic growth by absorbing a lion's share of the national budgets. On average, prior to 1850, the military received more than 50 percent of national budgets. Mexico provided one of the most shocking examples: Between 1821 and 1845, the military budget exceeded the total income of the government on fourteen occasions. Caudillos often arose from the ranks of the army; thus they understood and commanded the major institution for maintaining order and power. Chile was the first, and for a long time the only, Spanish-speaking nation to restrict the army to its proper role of defending the nation from foreign attack. After 1831 and until the civil war of 1891, the army kept out of Chilean politics, a Latin American rarity.

A few caudillos, however, championed the needs of the dispossessed majority and can be considered "popular" or "folk" caudillos. A complex group, they shared some of the characteristics of the elite caudillos, but two major distinctions marked them as different. They refused to accept unconditionally the elites' ideology of progress, exhibiting a preference for the American experience with its Indo–Afro–Iberian ingredients and, consequently, greater reservations about the post-Enlightenment European model. Further, they claimed to serve the popular classes rather than the elite.

The popular classes expected their leader not just to represent them, but to personify their values and protect the community. Their caudillo understood the folk's distinctive way of life and acted in harmony with it. In the eyes of the people, he inculcated the local, regional, or national values—traditional values—with which most of the people felt comfortable. He was a natural, charismatic leader of the majority and ultimately a patriarch in whom they entrusted their interests. In his discussion of leadership and folk, the twentieth-century Peruvian political philosopher José Carlos Mariátegui ascribed to the leader the roles of "interpreter and trustee." Mariátegui concluded, "His policy is no longer determined by his personal judgment but by a group of collective interests and requirements."

Argentine intellectual Juan Bautista Alberdi, probably more than anyone else in the nineteenth century, studied the psychology of the relationship of popular caudillos with the masses and concluded that the people regarded a popular caudillo as "guardian of their traditions," the defender of their way of life. He insisted that such leaders constituted "the will of the popular masses . . . the immediate organ and arm of the people . . . the caudillos are democracy." The relationship between the folk and these caudillos resembled the traditional patron–client relationship of reciprocity. The folk were expected to obey the caudillo and provide him service, and the leaders bore the obligation to protect and to provide for the welfare of the people. This reciprocal arrangement was a personal relationship, challenged in the nineteenth century by the more impersonal capitalist concept that a growing gross national product would best provide for all.

The popularity of those caudillos is undeniable. Their governments rested on a base of popular culture, drew support and inspiration from the people, and expressed, however vaguely, their style. Under the leadership of such caudillos, the masses apparently felt far more identification with government than they ever did under the imported political solutions advocated by elite. On many occasions, they displayed support of their caudillos by fighting tenaciously to protect them from the Europeanized elites or foreign invaders. Few in number at the national level, the populist caudillos had disappeared by 1870. But during the first half of the nineteenth century, these few leaders showed the possibility of a different kind of leadership and development.

One of the most controversial popular caudillos was Juan Manuel de Rosas, who enjoyed the support of the Argentine gauchos from 1829 until his exile in 1852. He appeared in Argentine history at the exact moment that Argentina, submerged in anarchy, threatened to split apart. The old viceroyalty of La Plata had disintegrated in the early nineteenth century, and Argentina itself dissolved into squabbling regions. Sharpest was the rivalry between the prosperous port and coastal province of Buenos Aires, rich through trade with Europe, and the interior provinces, which were impoverished and wracked by civil wars.

The Argentine gaucho, the cowboy of the pampas, became the symbol of backwardness for such Europeanized elites as Domingo Faustino Sarmiento. In his 1845 *Facundo: Civilization and Barbarism,* Sarmiento launched a thinly veiled attack against caudillo Juan Manuel de Rosas and a direct critique of the culture of the Argentine countryside. (Library of Congress)

Buenos Aires elites advocated a centralized government—which it fully expected to dominate—whereas the interior provinces favored a federalized one that would prevent the hegemony of the port. The bitter struggle between Buenos Aires and the interior delayed Argentine unification until Rosas strode onto the political stage. He had lived and worked in *gaucho* country before being elected governor of Buenos Aires in 1829, and he judiciously pursued the federalist idea of a pastoral economy. Understandably, the cattle breeders and the hide and meat producers supported Rosas because his economic inclinations favored them. Indeed, he was one of them—he owned extensive estancias, and his cousins were the Anchorenas, the largest landowners in Argentina. In his full exercise of political control, however, the caudillo acted as a centralist. Suspicious of Europe, he defied on occasion both England and France, deflecting their economic penetration of Argentina.

Rosas did not hesitate to use terror to control the country. But he also created stability and consolidated national territory, while many Latin American countries were torn apart. His method was to combine authoritarianism with a primitive populism. "For me the ideal of good government would be paternal autocracy," Rosas opined. ". . . I have always admired the autocratic dictators who have been the first servants of their people." Rosas cultivated support from the masses, offering rewards that were both symbolic and material. Afro-Argentines constituted more than 20 percent of the Buenos Aires population, and Rosas wooed them by abolishing the slave trade, promoting Afro-Argentines to military offices, and eliminating restrictions on Afro-Argentine dances.

Rosas grew up on the pampa and prided himself on absorbing the gaucho culture of riding, roping, and fighting. He did not try to enforce limitation on the gauchos' freedom of movement, as called for by such Liberal elites as Sarmiento. To stimulate occupation of the land, in 1840 Rosas initiated a program to distribute land to soldiers. The policy, however, owed more to Rosas's need for loyal soldiers than to a commitment to land reform. There was no large-scale redistribution of land under Rosas, and vagrancy laws were initiated toward the end of his rule.

It is clear, however, that the masses—both the rural gauchos and urban blacks and mulattos—demonstrated their loyalty to Rosas by their willingness to fight for him for nearly a quarter of a century. While undoubtedly many worked for Rosas and his allies and owed their patron their support, such direct relationships cannot account for the large numbers who fought on his behalf. In contrast, such caudillos as Peru's Agustín Gamarra (president 1829–1833, 1838–1841) failed miserably in trying to recruit a mass fighting force, relying instead on coercion and ending up with soldiers who were wholly unreliable. But Rosas suffered defeat and exile only when the Argentine elites enlisted the Brazilian and Uruguayan armies to overthrow him. Defeated at the Battle of Monte Caseros, Rosas left for European exile. The fall of Rosas in 1852 opened the door to the promulgation of a Liberal constitution, the growth of capitalism, a flurry of land speculation, and the commercial expansion of the cattle industry on an unprecedented scale. Argentina merged with the capitalist world and intensified its dependency in the process.

An intriguing example of the caudillo who preferred local models to imported ones is Rafael Carrera, who governed Guatemala from 1839 until his death in 1865. At least half indigenous, Carrera negated many of the Enlightenment reforms applied by previous Liberal governments and ruled, at least in part, for the benefit of the indigenous, the vast majority of the Guatemalans. The elites regarded him as a barbarian; the indigenous exalted him as their savior.

Carrera led the indigenous revolt of 1838–1839. Among the many changes that popular rebellion represented was the refusal of the indigenous to countenance any further exploitation through Europeanization. They wished to be left alone by the elites of Guatemala City so that they could follow their own culture. They rejected Europeanizing education, culture, economy, and laws that would integrate them into a capitalist economy centered in Europe. They chose instead to withdraw, a common reaction of the indigenous to the Europeans. But, in regions where the elites depended on those indigenous for labor and taxes, withdrawal signified rebellion.

Carrera understood the indigenous position, sympathized with their desires, and rose to power on their strength. As Carrera wrote in his memoirs, "When attempts are made suddenly to attack and change the customs of the people, it provokes in them such emotion that, no matter how sound the intention of those who seek to change their traditional ways and institutions, they rise in protest." Carrera regarded it as his principal duty to allow "the people to return to their customs, their habits, and their particular manner of living."

The Carrera government was unique to Latin America for encouraging the political ascendancy of the indigenous. One could even argue that under Carrera the government was "Indianized." Indigenous and mestizos, all of relatively humble classes, participated in the government, holding, in addition to the presidency, such high-ranking offices as the vice presidency, heads of ministries, governorships, and high military ranks. The army became nearly an indigenous institution.

To lift some of the economic burden from the impoverished majority, Carrera reduced taxes on foodstuffs and abolished the head tax. Further, he excused the indigenous from contributing to the loans the government levied periodically to meet fiscal emergencies. By removing many taxes on the indigenous, which were paid in the official currency, the government lessened the need for the indigenous population to earn money, thus reducing the pressure on them to work on the estates. The indigenous could instead devote their time to their own agricultural plots.

Most significant was Carrera's protection of indigenous lands, return of land to indigenous communities, and settlement of land disputes in their favor. The government declared in 1845 that all who worked unclaimed lands should receive them. It was decided in 1848 and again in the following year that all pueblos without ejidos were to be granted them without cost, and, if population exceeded available lands, then lands elsewhere were to be made available to any persons who voluntarily decided to move to take advantage of them. In 1851, Carrera decreed that "the Indians are not to be dispossessed of their communal lands on any pretext of selling them."

The indigenous victory under Carrera proved as transitory as the gauchos' under Rosas. The death of Carrera in 1865 reinvigorated elite efforts to wield power, and they succeeded under the leadership of another and different type of caudillo, Justo Rufino Barrios (1873–1885), who emphasized order and material progress. Under the Liberal reforms of the Barrios period, capitalism made its definitive entry into Guatemala, which meant large-scale exportation of coffee with all the attendant consequences for the agrarian economy.

During the Carrera era, a popular caudillo also came to power in Bolivia, another overwhelmingly indigenous country. Manuel Belzú combined the forces of populism, nationalism, and revolution. He built an effective power base of campesino and urban-artisan support, which brought him to the presidency in 1848 and sustained him until he peacefully left office in 1855.

The impoverished artisans and campesinos rallied to Belzú probably because his novel rhetoric spoke directly to their needs and certainly because of a series of wildly popular actions he took. He encouraged the organization of the first modest labor unions, ended some free-trade practices, terminated some odious monopolies, abolished slavery, permitted the landless indigenous to take over lands they worked for the latifundista elite, and praised the indigenous past. Often vague and frequently unsuccessful, his varied programs nonetheless won popular support. To his credit, Belzú seemed to have understood the basic problems bedeviling Bolivia: foreign penetration and manipulation of the economy and the alienation of indigenous land.

Under the intriguing title "To Civilize Oneself in Order to Die of Hunger," a series of articles in a weekly La Paz newspaper in 1852 highlighted a vigorous campaign denouncing free trade and favoring protectionism. The paper argued that free-trade-deprived Bolivian workers of jobs while enriching foreigners and importers. It advocated "protectionism" as a means to promote local industry and benefit the working class, goals that had the obvious support of the president. Indeed, free trade bore some responsibility for the nation's poor agricultural performance. A chronic imbalance of trade between 1825 and 1846 had cost Bolivia nearly 15 million pesos, much of which was spent to import food the country was perfectly capable of producing. La Paz, for example, imported beef, mutton, and potatoes.

Although taking no legal steps to reform the land structures, Belzú never opposed the indigenous occupation of their former community lands. Landlords, fearful of the restive indigenous masses, found it prudent to move to the safer confines of the cities, abandoning their estates, which the indigenous promptly occupied. Two major consequences of the de facto land reforms were the greater supplies of food entering the marketplaces and the drop in food prices. Belzú further delighted the campesinos by relieving them of some taxes.

If rhetoric were the measure of government, Belzú's administration was revolutionary. These examples from his public speeches serve as a yardstick:

> Comrades, an insensitive throng of aristocrats has become arbiter of your wealth and your destiny; they exploit you ceaselessly and you do not observe it; they cheat you constantly and you don't sense it; they accumulate huge fortunes with your labor and blood and you are unaware of it. They divide the land, honors, jobs and privileges among themselves, leaving you only with misery, disgrace, and work, and you keep quiet. How long will you sleep? Wake up once and for all! The time has come to ask the aristocrats to show their titles and to investigate the basis for private property. Aren't you equal to other Bolivians? Aren't all people equal? Why do only a few enjoy the conditions of intellectual, moral and material development and not all of you?
>
> Companions, private property is the principal source of most offenses and crimes in Bolivia; it is the cause of the permanent struggle among Bolivians; it is the basis of our present selfishness eternally condemned by universal morals. No more property! No more property owners! No more inheritances! Down with the aristocrats! Land for everyone; enough of the exploitation of man. . . . Aren't you also Bolivians? Haven't you been born to equality in this privileged land?

For the great mass of dispossessed campesinos, Belzú's heady words did not fall on idle ears. Some seized the estates, and where the landlords resisted, Belzú's followers attacked and defeated them.

The president cultivated popular identification with his administration. From the balcony of the presidential palace, Belzú assured his listeners, "I am one of you, poor and humble, a disinherited son of the people. For that reason, the aristocrats and the rich hate me and are ashamed to be under my authority." The president frequently reminded his followers that all power originated in the people who had conferred it on him. "The popular masses have made themselves heard and played their role spontaneously; they have put down rebellions and fought for the constitutional government. The rise to power of this formidable force is a social reality of undeniable transcendence."

In the last analysis, Belzú was too Europeanized to feel comfortable for long as a folk caudillo, for he insisted on codifying his government within the confines of a Europeanized constitution. His political reforms and Constitution of 1851 reduced the presidential term to a specific period and prohibited reelection. Elections in 1855, classified by one Bolivian historian as "the cleanest ever held," brought a constitutional end to the Belzú presidency, awarding the office to the president's preferred candidate—his illegitimate son, Jorge Córdoba—a man unequal to the tumultuous task. In August of 1855, at the height of his power, Belzú stepped down, unwilling to follow the well-established precedent of *continuismo*, turned over the presidency to his elected successor, and temporarily left Bolivia. To the indigenous masses, he remained their "tata Belzú," friend and protector, whose short, unique government had benefited them.

The indigenous had every reason to be apprehensive of the electoral process in which they had played no role. With Belzú in Europe, the old elites quickly seized power. At the same time, they took possession of their former lands and returned the campesinos to subservience. In the years that followed, the indigenous often revolted with cries of "Viva Belzú!" on their lips, but as the elites became increasingly integrated into international trade and consequently strengthened, they were not about to repeat the previous political errors that had permitted a popular caudillo to govern. When, for example, the Huaichu of Lake Titicaca rebelled in 1869 to regain communal lands, President Mariano Melgarejo dispatched the army to massacre them.

The populist caudillos constitute an intriguing chapter in nineteenth-century Latin American history. They appeared and disappeared within a sixty-year period. Yet, in 1850, three folk caudillos ruled at the same time: Carrera, Belzú, and Rosas. Two of those nations had large indigenous populations with traditional cultures, while the Argentine gauchos boasted an equally well-defined folk culture. The popular caudillos identified with the majority and vice versa. In all three cases, foreign investments were low; the governments and majority expressed strong views against foreigners and shunned foreign influences. At the same time, land became available for the majority and pressures on the campesinos diminished. Subsistence agriculture dominated export agriculture. More food was available for popular consumption.

Although those popular governments obviously appealed to large elements of society that had been customarily neglected, such governments were few in number and had disappeared by 1870. The elites succeeded in imposing their will on the masses and on Latin America. Nonetheless, those few popular governments serve as useful reminders of possible alternatives to the Europeanized governments that the elites imposed. If, indeed, the people enjoyed a more satisfactory quality of life under these caudillos, those governments then suggest possible roads to development that were denigrated or ignored after 1870.

4.8: Change and Continuity

The continuity between the thirty years after independence and the colonial period is remarkable. Economic changes were few. Agriculture and the large estate retained their prominence, and the new nations became as subservient to British economic policies as they once had been to those of Spain and Portugal. The wars of independence had weakened some of the foundation stones of society, but the edifice stood largely intact. A small, privileged elite ruled over muted although sometimes restless masses. Fewer than one in ten Latin Americans could read, and fewer than one in twenty earned enough to live in even modest comfort. Land remained the principal source of wealth, prestige, and power, and the few owned much of the land.

There were also, however, some significant changes. The first and most obvious was the transmission of power from the Iberians to the creole and mazombo elites. Political power no longer emanated from Europe; it had a local source. The second was the emergence of the military in Spanish America as an important political institution destined to play a decisive role in Latin American history. The military was the elite's only guarantee of order, and initially it provided prestigious employment for some sons of the rich as well as a means of upward mobility for ambitious plebeians. Early in the national period, the Liberals challenged the status of the military, thus alienating the officers and driving them into the welcoming embrace of the Conservatives. The remarkable early stability of both Brazil and Chile can be explained in part by the close relationship between Conservatives and the military.

Some of the changes masked continuity. For example, the formal categories regarding race divisions and noble birth were eliminated. But the reality remained: A small, mostly white, group of elites maintained control over wealth, power, and the lives of darker masses. Elites were still concerned with *calidad*, but quality was no longer as limited to racial and hereditary considerations. Quality now referred to respectability, which was earned by family background, the organization and location of one's household, formal training and education, occupation, economic resources, and perceived color. At the other extreme were *pleybeyos*, or plebeians, those who were basically coarse and common. Other terms that were used for this increasingly binary division were *gente alta*, literally high people, and *gente baja*, low people, and *gente decente*, decent people, versus *gente de pueblo*, common people. The majority were also referred to as *las masas* (the masses) and *las clases populares* (the popular classes). The desire of the *gente alta* to maintain their position could be seen in the change in constitutions and laws that narrowed suffrage rights as well as continued attempts to control land and labor.

The change was summed up in a Mexican folk expression: "Same horse, different rider."

Questions for Discussion

1. What were the debates about the kind of government that the new Latin American nations should install?
2. How did economic weakness influence the exercise of sovereignty?
3. To what extent did colonial patterns continue in early nationhood?
4. Who were the caudillos, and why did they exercise such an important role?
5. What were the challenges that the new nations faced from foreign governments?
6. What were the internal divisions that led to instability?

5 The Emergence of the Modern State

By the late-nineteenth century, the chaos of the early independence years gave way to the emergence of a new political stability and economic prosperity. Foreign threats had diminished, the republican principle had triumphed everywhere but in Brazil, and centralized government had been gradually accepted. Nationalism became a better-defined force as more citizens expressed greater pride in their homeland, appreciated its uniqueness, and sought its progress. Feuding elites realized that they would profit most from stable governments that would encourage foreign investment and trade. Positivist ideology dominated government circles, complementing capitalist expansion. However, whereas elites overcame many of their earlier divisions to concentrate on the national project of "progress," large numbers of the population felt marginalized and threatened by the changes. They responded with protests and at times violence.

Although political change and instability marked the first fifty years of nationhood, economic and social innovations as well as political stability characterized Latin America in the late nineteenth and early twentieth centuries. An accelerating prosperity—at least for the favored classes—encouraged material growth and attracted a wave of immigrants, particularly to Argentina, Brazil, and Chile. The combination of stability and prosperity helped to accelerate three trends: industrialization, urbanization, and modernization, which in turn threatened to alter some of the established patterns inherited from the colonial past. As always, the elites benefited more than the masses.

5.1: Political Stability

The conflict among elites that characterized the early national period gave way to agreement on a new economic project: the large-scale export of primary products and import of foreign capital and manufactured goods, accompanied by limited industrialization. This new economic order required political order as well—without political stability, foreigners would not invest in Latin America.

With civil disorder on the wane, the chiefs of state consolidated and extended their authority. They governed with few if any checks from the congresses and the courts, which they customarily dominated. In some cases, they selected their own successors, who were assured impressive electoral victories. Nonetheless, greater respect for legal forms prevailed, and some caudillos even appeared to be more legally conscientious than their predecessors. They paid more lip service to constitutional formalities, and some even showed an occasional indication of heeding the constitution. In somber frock coats, representatives of the elite discussed and debated the political issues of the day.

The term *elite* can be hard to define. First, one must recognize that no single elite existed in any Latin American nation. Rather, a plurality of elites combined in various ways to dominate each nation. At best, elite is a shorthand expression signifying those in social, economic, and political control. The elites used their economic power, prestige, and education to exert authority over a society whose formal institutions were of European inspiration. They made the major decisions affecting the economic and political life of their nations.

Two countries were unusual in achieving political stability early in the century: Brazil and Chile. Brazil had enjoyed stability during the First Empire (1822–1831), but during the regency established after the abdication of Pedro I, regional forces threatened the empire. One province after another, from the far north to the far south, rebelled against the government in Rio de Janeiro. To save national unity, the Brazilians proclaimed the young Pedro II emperor in 1840, four years before he was legally of age to ascend the throne. As expected, the monarchy, the only truly effective national institution in Brazil, provided the ideal instrument to impose unity on the disintegrating nation. Brazil's nearly disastrous experiment with federalism ended when Pedro II reimposed a high degree of centralism on the nation. In 1847, the emperor created the post of President of the Council of Ministers, and a type of parliamentary system developed. The emperor saw to it that the two political parties, Liberals and Conservatives, alternated in power.

While domestic stability was restored, international tensions arose in 1864 when a Brazilian ship on the Paraná River was captured by Paraguayan forces for violating its waters. Paraguay resented Brazil's attempts at domination of the region, particularly the recent invasion of Uruguay. Brazil, which needed to use the river to reach the province of Mato Grosso, declared war, expecting to win quickly. Instead, the War of the Triple Alliance dragged on from 1865 to 1870, and the allies—Argentina and Uruguay—withdrew their troops before Brazil finally won.

Because the war was not quickly won, the emperor found himself challenged at home and abroad. In 1868 the emperor overruled the Liberal prime minister's choice of an appointee to a Senate vacancy. When Prime Minister Zacarias de Góes e Vasconcelos resigned, the entire Liberal ministry followed him. After the Conservatives won the following elections, the Liberals issued a reform manifesto calling for a federalist program that would have weakened the central government in Rio de Janeiro. The federalist call was taken up by the new Federal Republican Party, which in a December 1870 manifesto called for substitution of a republic for the monarchy. Additional calls came for expansion of the political base and even abolition of slavery.

The foundations of the monarchy slowly weakened. In the early 1870s, a Church–Crown conflict alienated Conservatives allied with the Church. The dispute began when the Bishop of Olinda carried out a papal order, unapproved by the emperor, to expel Masons from the lay brotherhoods of the Roman Catholic Church. Pedro II ordered the bishop to remove the penalty. His refusal to comply directly challenged the emperor. The government arrested and jailed the churchman along with another bishop, and the royal courts found both guilty of disobeying civil law. Further Conservative discontent came in 1888 when the manumission of the slaves without any compensation to the owners caused an important sector of the landowning class to abandon the emperor.

Emancipation had been brought about largely by urban groups who did not identify closely with the monarchy, which they felt did not represent their interests. They viewed the aging emperor as a symbol of the past, anathema to the modernization they preached. It was among these urban groups that republican doctrine spread. That doctrine, as well as the urban mentality, also pervaded the army officer corps. Pedro had ignored the military officers, an increasingly restless group in the 1880s. In them, the disgruntled clergy, landowners, and urban dwellers, as well as the coffee planters, found their instrument for political change. The republican cause won many military converts, particularly among the junior officers who equated a republic with progress. The fate of the monarchy was sealed when the principal military leader, Marshal Deodoro da Fonseca, switched his allegiance and proclaimed in favor of a republic. Under his leadership on November 15, 1889, the army overthrew the monarchy and declared Brazil to be a republic. The old emperor abdicated and, like his father before him, sailed into European exile. The transition was bloodless. A new constitution—presidential, federal, democratic, and republican—was promulgated on February 24, 1891.

Chile reached stability even earlier. By 1830, a small, powerful landowning class had consolidated quickly in the relatively small geographic area of central Chile, hemmed in by the arid Atacama Desert to the north and the unconquered Mapuches to the south. From the ranks of the relatively homogenous Conservative elites emerged one of Chile's most skillful leaders of the nineteenth century, Diego Portales. Through his force and efficiency, he imposed Conservative rule in 1830, and it lasted until 1861. Although he never served as president, he held various ministerial portfolios and ruled behind the scenes until his assassination in 1837. Portales was concerned with power, discipline, stability, and order, not social or economic reforms. In sharp contrast to the rest of Spanish America, he succeeded in subordinating the army to a civilian government, thereby removing the military from nineteenth-century politics. He also framed the Constitution of 1833, which lasted until 1925.

Presidential mandates in Chile lasted five years and could be renewed for an additional five years. Three Conservative leaders each held the office for ten years, enforcing the remarkable stability imposed by Portales. Manuel Montt, who served from 1851 to 1861, accepted a moderate, José Joaquín Pérez, to succeed him. The thirty years, 1861–1891, were a period of Liberal rule tempered by Conservative opposition, in direct contrast to the preceding three decades. The major reforms enacted by the Liberals indicated the direction and degree of change they advocated: private liberty of worship in houses and schools, the establishment of cemeteries for non-Catholics, the abolition of the privileged Church courts, civil marriage, freedom of the press, no reelection of the president, a modification of the electoral reform law to substitute a literacy test for property qualifications, greater autonomy for municipal governments, and the power of Congress to override a presidential veto. As is obvious, none of these altered the power structure in Chile; none attempted to shift the social and economic imbalance of the country. However, Chile's record of stability encouraged economic prosperity and material progress.

Argentina and Mexico experienced far greater difficulty than did Chile and Brazil in their search for political stability. After Rosas fell in 1852, the city and province of Buenos Aires refused to adhere to the new Argentine union, fearing that they would have to surrender too much power. The old struggle between the port and the provinces continued until 1862, when the powerful governor of Buenos Aires, Bartolomé Mitre, was able to impose his will on the entire nation. His leadership reunited the nation. A succession of strong, able presidents, each of whom served for six years, followed Mitre. These presidents wielded almost total power. In 1880, the thorny question of the city and province of Buenos Aires was finally solved. New legislation separated the city from the province, federalized the city, and declared it the national capital, a role it had previously played. The rich province then went on its own way with a new capital, La Plata. Meanwhile, the nation was investing the energy once given to political struggles into economic growth. In 1879–1880, General Julio Roca led a war that became known as the Conquest of the Desert, although it really constituted the conquest of the Araucanians. With the invention of the repeating rifle, using the same pattern of annihilation carried out on the U.S. plains, the Argentines were able to finally succeed in overpowering the nonsedentary Indians who had daunted their Spanish forefathers. For his efforts, Roca went on to become president.

Mexico's search for stability was one of the most difficult in Latin America, complicated by questions of federalism and the position of the Church in the new nation. The drama that began in 1810 lasted well over half a century and exhausted the nation. Finally a mestizo strongman, Porfirio Díaz, who had risen through the military ranks to become general and had fought against Santa Anna, the French, and Maximilian, brought peace to Mexico in 1876—an iron peace, as it turned out. For the next thirty-four years, he imposed a Conservative, centralized government on Mexico while ruling under the Liberal, federal Constitution of 1857. His government brought stability to a degree unknown since the colonial period.

The political longevity of Díaz rested on the powerful supporting alliances he created. The Church, the army, foreign capitalists, and the great landowners found it beneficial to back his regime, and they reaped substantial material rewards for their allegiance. On the other hand, the wily Díaz manipulated them at will for his own ends; for example, he promoted potential rivals to positions that required they relocate, undermining any chance of building a local power base. Under his guidance, Mexico made outstanding material progress and witnessed a prosperity surpassing the best years of colonial mining. The supporting alliance that Díaz formed proved to be the most effective combination of forces to promote political stability. To varying degrees the chiefs of states of other nations made use of similar combinations to buttress their power. The two key groups obviously were the army and the landowners.

In country after country in the late-nineteenth century, Latin American elites followed the Argentine and Mexican patterns. They realized that years of political infighting had done nothing for the country and little for their personal wealth. Latin America offered investment opportunities for the wealth of Great Britain's expanding industry. But foreigners were not eager to invest in countries where governments changed hands on an annual basis and the security of investments could not be guaranteed. Latin American elites cast aside their political differences in the interests of economic unity. The political unity that resulted was in the interest of a stability that would attract foreign investment.

In some ways, the masses lost more than they gained with the new stability. The chaos of the first fifty years frequently gave people room to maneuver. With stability, however, came greater State control, which was enhanced by the latest technology. News of an uprising could be sent to the capital by telegraph, and the government could respond by sending troops on the train.

5.2: Positivism and Progress

Latin American elites looked longingly at the material progress being made in Great Britain, France, Germany, and the United States. Many of them had mastered French, the second language of the elite, and a few had knowledge of English or German, which gave them direct access to literature from the nations whose progress impressed them. The newspapers carried full accounts of what was happening in the leading nations of the Western world, and the programs of the learned societies featured discussions of the technical advances of the industrializing nations. Many members of the elite traveled abroad and were exposed to the innovations firsthand. They returned to their quieter capitals in the New World with nostalgia for Paris and the irrepressible desire to mimic everything they had seen there. Of course, the imports bore testimony of the manufacturing skill and ingenuity of those technically advanced societies to larger segments of the population.

The elites closely followed the intellectual trends of Europe. In fact, they could more readily discuss the novels of Émile Zola or Gustave Flaubert than those of Jorge Isaacs or Joaquim Maria Machado de Assis, and they paused to admire European painters, while they ignored the canvases of their compatriots.

Not surprisingly, some members of the elite became knowledgeable about the new philosophy of positivism, formulated in Europe during the second quarter of the nineteenth century by French philosopher and sociologist Auguste Comte. They eventually imported it into Latin America, where governments warmly welcomed it.

Many of the ideas on progress that the Latin Americans extracted from the Enlightenment, Charles Darwin, and Herbert Spencer seemed to converge in Comte's positivism. Comte maintained that the rules of scientific observation and experimentation could also be applied to human societies, which could advance through scientific, rational processes. He defined three stages of social development: the theological, characterized by belief in the supernatural or superstition; the metaphysical, a belief in ideas that would be used to destroy the theological era; and the positive stage, in which phenomena are explained by the scientific method of observation, hypothesis, and experimentation, leading to material progress.

Comte applied his ideas to an analysis of the French Revolution: the *ancien régime* was a theocratic monarchy, which provided order but no progress. The Revolution was motivated by new ideas—a metaphysical reaction—but led to chaos. The positive period that followed brought both political order and material progress. Latin American elites adapted Comte's historical model to their own reality: the theocratic stage of Iberian colonialism, destroyed by the Enlightenment-inspired (metaphysical) independence wars that led to the chaos of the early national period, and the new modern, positive stage of order and progress. This model had the advantage of dismissing the Church as a brake on progress, while also dismissing liberal ideas of equal rights as leading to chaos. Positivism, then, could be used to support both authoritarian government and material progress, especially as manifested in railroads and industrialization.

With its emphasis on material growth and well-being, positivism ideally suited the trends of the second half of the century. It favored a capitalist mentality, regarding private wealth as sacred. Indeed, private accumulation of wealth was a sign of progress as well as an instrument for progress. Because of the weakness of domestic, private institutions, the State had to assume the role of directing progress. Deferring to the role of the State over the individual, positivism complemented the patriarchal experience of the Latin Americans. Of course, to promote capitalism and to direct progress, the State, had to foster stability. With its special emphasis on order and progress, positivism reached the height of its influence between 1880 and 1900. It became an official doctrine of the Díaz regime in Mexico; some of his principal ministers, who fittingly were known as *científicos*, had imbibed deeply of Comte's doctrines and tried to offer a scientific solution to the problems of organizing national life. As Mexico's 1990 Nobel Prize–winning poet Octavio Paz would reflect, "Positivism offered the social hierarchies a new justification. Inequalities were now explained, not by race or inheritance or religion, but by science."

In Venezuela, President Antonio Guzmán Blanco governed directly or indirectly from 1870 to 1888 under the influence of positivism. First, he imposed order and then set out in pursuit of elusive progress. Order meant the continuation of the class and racial inequities of the past, allowing minimal social change, whereas progress signified adoption of the outward manifestations of European civilization. Dreaming of Paris, Guzmán Blanco laid out wide boulevards for Caracas and constructed an opera house as well as a pantheon for national heroes. New railroads and expanded port facilities speeded Venezuela's exports to European markets to pay for the material progress.

Positivism also attracted many adherents in Brazil, particularly among the new graduates of technical and military schools, members of the fledgling middle class. They favored the abolition of slavery, establishment of a republic (albeit not a democratic one), and separation of Church and State, changes that did occur in Brazil between 1888 and 1891. Material progress absorbed much of their concern. The flag of the Brazilian republic, created in 1889, bears to the present the positivist motto "Order and Progress."

Where political stability and economic prosperity existed, notable material progress occurred. It is not surprising that Brazil and Chile were among the first Latin American nations to experiment with the innovations correlated with progress offered by Europe and the United States. The steam engine made an early appearance in these two nations. In 1815, Bahia boasted its first steam-driven sugar mill. By 1852, there were 144 in operation. Similar technological changes took place throughout Latin America, albeit more slowly in the smaller countries.

None of the innovations had more impact than the railroads. Its steel rails had the potential of linking formerly divided territories. Penetrating distant regions, the railroads tapped new sources of economic prosperity. Bulky and perishable products could be rushed over long distances to eager markets. Just as important, troops could be sent on the railroad, bringing State power to the most distant corners. Railroad expansion was supported by such government incentives as duty-free entry of materials and equipment, interest guarantees, and land grants. British capital readily responded and dominated rail construction. Britain supplied technicians and engineers and then sold English coal to the railroad companies to run the steam engines.

Cuba can boast of the first railroad in Latin America. The line from Havana to Güines, approximately thirty miles, began operation in 1838. Argentina's railroad era began in 1857 with the inauguration of a seven-mile line, and by 1914 Argentina boasted the most extensive railroad network in Latin America and the eighth largest in the world. In total, South America's railroad mileage grew from 2,000 miles in 1870 to 59,000 in 1900.

Along the newly laid rail lines, at rail junctions and at railheads, sprang up villages and towns. Older settlements took on a new life. The railroad brought the countryside and the city into closer contact, and as a result the isolated, paternalistic patterns of rural life were challenged as never before. The railroads

No technology had a greater impact on Latin America than the railroad. The train, like the one pictured here in Mexico circa 1884, created markets and strengthened the State, which could now send troops to formerly remote areas.
(Library of Congress)

opened new markets for the incipient industries of the large cities and brought into the hinterlands a greater variety of goods than those populations had hitherto seen. Railroads thus helped to promote the fledgling industries. They also provided some amusement during their early years. When the railroad finally reached Guatemala City in the 1880s, it became a custom for the capital's inhabitants of all social levels to congregate at the railroad station for the semiweekly arrivals and departures of the train. On those occasions, the national band played in the plaza in front of the station for the further entertainment of the crowd.

But the railroad was also disruptive, as can be clearly seen in the case of Mexico. When Porfirio Díaz took power in 1876, only 400 miles of track had been laid. By the end of his regime in 1910, there were approximately 15,000 miles of rail. Speculators bought land along the railroad routes, frequently driving the indigenous from their landholdings. The indigenous, however, did not passively acquiesce as the railroad came through their lands. There were fifty-five agrarian rebellions from 1877 to 1884, many of them so significant that they required the deployment of federal troops to suppress them. Of those, fifty took place within twenty-five of a railroad route, and thirty-two of them took place within twelve miles.

But the elites were more concerned with the prosperity that the railroads fueled. A national market was created as products could be moved even from the most distant regions. The railroads tapped new minerals for export. Whereas Mexico previously had sold only silver abroad, the railroads made it possible to export zinc, lead, and copper, intensifying the flow of Mexico's natural resources to the industrialized nations. It is noteworthy that Mexican exports increased eight and one-half times between 1877 and 1910, coinciding with the period of intensive railroad construction. U.S. interests completed the line linking the border with Mexico City so that in May of 1880 it was possible to travel by rail from Chicago to Mexico City for the first time.

Although railroads perfectly illustrated Latin America's desire for progress, they exerted an ironic negative influence on development. They deepened Latin America's dependency, strengthened neocolonial institutions, and impoverished the governments. The primary explanation for those adverse effects lay in the fact that the foreigners who built and owned the railroads did so where they would best complement the North Atlantic economies rather than Latin America's. In Argentina, as but one example, railroads that the English financed, built, equipped, and administered carried the resources of the rich pampas to the port of Buenos Aires for their inevitable export. As this example aptly suggests, the railroads served export markets by lowering transportation costs of bulky items, incorporating new regions into commercial agriculture, and opening up new lands and mines to exploitation. More often than not, the railroads expanded and strengthened the latifundia wherever the rails reached because they often conferred considerable value on lands once considered marginal. Whereas previously campesinos had been tolerated on that marginal land, its new value caused the landlords first to push the campesinos off the land and then to incorporate them into the estate's labor force. Commercial agriculture, much of it destined for export, replaced subsistence agriculture, and the numbers and size of small landholdings declined.

Bolivia provided a sobering example of the contribution of railroads both to the dependency on a single export and to the reduction of food production for local markets. By the end of the nineteenth century, rails connecting the highlands with Pacific ports accelerated tin ore exports. The rail cars descended the Andes loaded with the ore. In order to fill the otherwise empty cars on their return, the trains carried agricultural products imported from Peru, Chile, and the United States. The importation of food wrought havoc on the agrarian economy of the fertile Santa Cruz region, depriving it of its national market. Production declined sharply. Bolivia became locked into a double dependency status: dependent on foreign markets for its single export and on foreign producers for a part of its food supply.

Costa Rica illustrated in a slightly different but even more disastrous way the effects of railroads on the economy of a small nation. To market larger quantities of coffee and thus earn the money to modernize, the government encouraged the construction of a railroad (1870–1890) that stretched from the highlands, where the coffee grew, to Puerto Limón for shipment abroad. The onerous loans to finance the project overburdened the treasury as the government paid outrageous interest rates to unscrupulous foreign moneylenders.

Further, the government bestowed on chief engineer Minor Keith 800,000 acres, land that fronted on the railroad and later became the center of the plantations of the United Fruit Company. Meanwhile, the railroad remained in foreign hands. British investors also controlled the ports, mines, electric lighting, major public works, and foreign commerce as well as the principal domestic marketplaces. In short, Costa Rica surrendered its economic independence and mortgaged its future before 1890 to obtain some of the physical aspects of modernization.

Most of the other Latin American governments also contracted heavy debts to pay for their railroads. Those debts often led to foreign complications and always to some sacrifice of economic independence. One brazen example of foreign intervention triggered by railroad funding occurred in Venezuela. In 1903, German gunboats appeared off the coast to force the government to pay 1.4 million pounds sterling for loans associated with railroad building.

With the outsized importance of the railroad to Latin America in the nineteenth century, it should not be surprising that it became a key actor in fiction of the era. In Clorinda Matto de Turner's *Birds Without a Nest* (*Aves Sin Nido*, 1889), the railroad serves to transport people from the primitive countryside to modern Lima. En route, the train is derailed when it collides with a herd of cows that refuse to move from the track, an apt metaphor for the struggle between rural tradition and urban modernity. Another sign of the times is that the train is righted by Mr. Smith, the engineer from the United States. The railroad was the central actor in Argentine Ricardo Güiraldes's novela, "Rosaura," (1922) the tale of a provincial young woman who is wooed and abandoned by a Europeanized young man from the city. Güiraldes characterizes the railroad's impact on the town:

> The soul of Lobos was simple and primitive as a red bloom. Lobos thought, loved, lived, in its own way. Then came the parallel infinities of swift rails, and the train, marching armored to indifference from horizon to horizon, from stranger to stranger, brushed its passing plume over the settlement. Lobos fell ill of that poison.

The telegraph had a more subtle impact, but it, too, furthered the communications revolution in Latin America. The telegraph contributed to national unity—at least among the elites—and helped to bring neighboring nations closer together. Both Chile and Brazil began service over short lines in 1852. In 1866, a cable connected Buenos Aires and Montevideo; and the following year the United States opened cable communication with Cuba. Other countries followed suit. Transatlantic cables put Latin America in instantaneous communication with Europe. Pedro II dictated the first message to be cabled from Brazil to Europe in 1874. Significantly, Rio de Janeiro was linked to Europe by cable long before it could communicate by telegraph with other parts of its own empire.

International communication was also enhanced by the increasing number of international steamship lines serving Latin America, putting the region in direct and regular contact with the principal ports of Europe and the United States. Sailings became increasingly frequent, ships carried more cargo, and service improved. The major ports underwent renovation with the addition of new and larger warehouses, faster loading and unloading machinery, larger and sturdier wharfs, and the dredging of deeper channels.

Clearly not all of Latin America shared fully in the progress. One astonished visitor to the highlands of Ecuador and to Quito in 1885 gasped, "The country does not know the meaning of the words progress and prosperity." At that time, the only communication between Quito, the capital high in the Andes, and Guayaquil, the Pacific port, was still by mule path, a route impassable during six months of the year because of the rainy season. Under optimum conditions, it was possible to make the journey in eight or nine days over the same route in use for centuries. As a port, Guayaquil was in contact with the world and showed signs of modernization—public streetcar, gas lighting—long before the isolated capital did. Not until the end of the first decade of the twentieth century did a railroad link Guayaquil with Quito. The 227-mile railroad climb from the port to the capital took two days. The slower growth of Ecuador was not unique; it characterized many of the nations that lagged behind the leadership of Argentina, Brazil, Chile, and Mexico.

The concern with improved transportation and communication symbolized the dedication to progress. To a great extent, material advances measured modernization: How many miles of railroads, how much horsepower generated by steam engines, how many tons handled per hour in the ports, and how many miles of telegraph wires? The higher the number given in response to these questions, the greater the progress the nation could claim. In their satisfaction with these material advances, the elites seemed oblivious of another aspect of modernization: The very steamships, railroads, and ports tied them and their nations ever more tightly to a handful of industrialized nations in Western Europe and North America, which bought their raw products and provided manufactured goods in return.

The elites failed to see the significance of the fact that many of the railroads did not link the principal cities of their nations but rather ran from the plantations or mines directly to the ports, subordinating the goal of national unification to the interests of the industrial nations for agricultural products and minerals. As the amount of loans increased to pay for the new material progress, the governments had to budget increasing amounts to pay interest rates. As foreign investment rose, the outflow of profit remittances multiplied. Increasingly, the voices of foreign investors and bankers spoke with greater authority in making economic decisions for the host countries. Local economic options diminished. In short, progress as it took shape in nineteenth-century Latin America adversely prioritized the needs and desires of a tiny Latin American elite and its foreign economic partners.

5.3: Economic Prosperity

The elite pursuit of progress had economic prosperity as the goal, and they were not disappointed. Thanks to loans, investments, and rising exports, a growing economy characterized parts of Latin America during the last half of the nineteenth century. That growth came from the consolidation of the export economy.

The demands of the industrialized nations for raw products rose impressively throughout the second half of the nineteenth century. Not only was there a rapidly increasing number of consumers in Western Europe and the United States, but also their per capita purchasing power was ever greater. As the industrial centers bought more agricultural and mineral products from Latin America, the region's trade underwent a dramatic expansion. Plantation owners and miners produced larger amounts of grain, coffee, sugar, cotton, cacao, bananas, livestock, copper, silver, tin, lead, zinc, and nitrates for export. The natural products—such as palm oils, nuts, woods, rubber, and medicinal plants—also found a ready market abroad, and their exportation rose sharply. Similar to well-established patterns of the colonial past, the export sector of the economy remained the most active, the dynamic focus of investment, technological improvement, official concern, and demand for labor. More often than not, foreigners, with their own agendas, dominated that sector.

In order to focus on growing exports, the elites tried to monopolize the land. They had two goals: to transfer land from subsistence to export crops and to entice the campesinos to leave their small plots and work for the large estates. To that end, most Latin American governments adopted laws aimed at divesting indigenous communities of their land as well as requiring those without sufficient land to work on the large export-oriented estates.

The success of such measures was mixed, depending to a great extent on the strength of the State and its local units of control. Where force was strong, the elites triumphed. In many places, however, the folk were able to continue manipulating the system. One of the most successful governments was the Díaz regime in Mexico. Legislation to commercialize land had been passed during the Reform administrations in the 1850s, but the laws could not be enforced because of the War of the Reform and French intervention and occupation. Díaz's rule, and his *Guardia Rural* (rural guard), made enforcement possible. By 1894, 20 percent of Mexico's land had changed hands. By the early twentieth century, most rural villages had lost their ejidos. By 1910, half of all rural Mexicans lived and worked on haciendas.

Table 5.1 Export Commodity Concentration Ratios, 1913

Country	First Product	Percentage	Second Product	Percentage	Total
Argentina	Maize	22.5	Wheat	20.7	43.2
Bolivia	Tin	72.3	Silver	4.3	76.6
Brazil	Coffee	62.3	Rubber	15.9	78.2
Chile	Nitrates	71.3	Copper	7.0	78.3
Colombia	Coffee	37.2	Gold	20.4	57.6
Costa Rica	Bananas	50.9	Coffee	35.2	86.1
Cuba	Sugar	72.0	Tobacco	19.5	91.5
Dominican Republic	Cacao	39.2	Sugar	34.8	74.0
Ecuador	Cacao	64.1	Coffee	5.4	69.5
El Salvador	Coffee	79.6	Precious metals	15.9	95.5
Guatemala	Coffee	84.8	Bananas	5.7	90.5
Haiti	Coffee	64.0	Cacao	6.8	70.8
Honduras	Bananas	50.1	Precious metals	25.9	76.0
Mexico	Silver	30.3	Copper	10.3	40.6
Nicaragua	Coffee	64.9	Precious metals	13.8	78.7
Panama	Bananas	65.0	Coconuts	7.0	72.0
Paraguay	Yerba mate	32.1	Tobacco	15.8	47.9
Peru	Copper	22.0	Sugar	15.4	37.4
Puerto Rico	Sugar	47.0	Coffee	19.0	66.0
Uruguay	Wool	42.0	Meat	24.0	66.0
Venezuela	Coffee	52.0	Cacao	21.4	73.4

SOURCE: Victor Bulmer-Thomas, *The Economic History of Latin America Since Independence*, 2nd ed., New York: Cambridge University Press, 2003, 58. Reprinted with the permission of Cambridge University Press.

Workers were more frequently tied to the haciendas by debt peonage, which had been tried with much less success during the colonial and early national eras. Peons would be given cash advances but were required to work until they were repaid. Workers often were paid in scrip rather than money, usable only at the *tienda de raya*, or hacienda store. Workers were charged exorbitant prices and routinely cheated on their payments. If the worker died, his debts passed to his children. Debt peonage in Mexico was particularly successful on the henequen plantations of the Yucatan and the lumber camps of Chiapas. B. Traven, the German novelist who captured the reality of Mexico in the early twentieth century in such classics as *The Treasure of the Sierra Madre* (1927), wrote in *The Rebellion of the Hanged* (1936) about men who escaped from floggings in Yucatan only to descend into the hell of the Chiapas mahogany camps, where workers were disciplined by hanging them from the trees for days, exposed to the elements and the animals.

In many regions, however, elite measures failed dismally. Between 1834 and 1923, Nicaraguan legislators adopted twenty-five laws attempting to control labor, and the issue was addressed in the constitutions of 1905 and 1911. The plethora of laws is clear testament to their futility. Workers routinely accepted cash advances at coffee haciendas and then ran off to other haciendas. In the coffee-growing region of Carazo, hacendados complained that local authorities were incapable of tracking down rebellious workers. The Nicaraguans concluded that the best way to keep a steady workforce nearby was to provide them with small plots of land—just enough to provide subsistence, but not enough to prevent campesinos from needing the extra income of harvest work.

Foreign observers marveled at the rapid rate of increase of Latin American trade in the last half of the nineteenth century. By 1890, Latin America's foreign commerce exceeded $1 billion per year and had increased roughly 43 percent between 1870 and 1884. In comparison, British trade increased by 27.2 percent during the same period. The principal nations engaged in foreign commerce were Brazil, Argentina, Cuba, Chile, and Mexico, which together accounted for more than three-quarters of Latin America's trade. With the exception of Argentina, they exported more than they imported. Trade statistics for the individual nations were impressive. Argentine exports jumped sevenfold between 1853 and 1873 and doubled again by 1893. Mexican exports quadrupled between 1877 and 1900. The smaller nations benefited too. Exports of coffee from Costa Rica increased fourfold between 1855 and 1915.

A number of inventions, among them the railroad, barbed wire, the canning process, and the refrigerator ship, facilitated the exploitation of Argentina's hitherto untamed pampas, potentially one of the world's most fertile regions. The sailing of the first primitive refrigerator ship from Buenos Aires to Europe in 1876 changed the course of Argentine economic history. The successful voyage proved that chilled and frozen beef could be sold in lucrative European markets. Measures were taken at once to improve the quality of the beef. By 1900, 278 refrigerator ships sailed between Great Britain and Argentina. During the 1870s, Argentina also made its first wheat shipments to Europe, a modest twenty-one tons in 1876. In 1900, wheat exports reached 2,250,000 tons. This extraordinary economic boom made Argentina the most prosperous nation in Latin America.

Chile offers an excellent but by no means exceptional example of the development of a mineral-exporting economy. Already in the first decades of independence, it exported increasing amounts of silver and copper. By midcentury, Chile enjoyed a well-balanced program of exports divided between various minerals and agricultural products. That diversification ended after 1878 when the new nitrate exports grew increasingly important until, within a few years, they dominated the export sector. Dependent on the world's need for nitrates, the Chilean economy declined or prospered according to demand and price for one product. The vulnerability of the economy was obvious.

Most governmental leaders paid far more attention to the international economy than to the national, domestic economy. The export sector of the Latin American economy grew more rapidly than the domestic sector, and income from foreign trade contributed an unusually high percentage of the gross national product. Foreign trade emphatically did not mean commerce among the Latin American nations. They were strangers in each other's marketplaces, frequently competitors producing the same export products. Their economies complemented the demands of the distant major capitalist economies in Western Europe and the United States.

Mounting foreign investments also characterized the Latin American economy in the late-nineteenth century. Europe generously sent capital, technology, and technicians to Latin America to assure the increased agricultural and mineral production that its factories and urban populations required. The more pronounced stability of Latin America engendered greater confidence and generosity among foreign investors. The politicians of Latin America discovered the advantages of foreign investment. It created new wealth, which caudillos, politicians, and elite alike enjoyed. In a pattern established throughout the hemisphere, Porfirio Díaz meticulously paid Mexico's foreign debts and decreed laws favorable to foreign investors. Foreign investment poured into Mexico. As investments rose, profit remittances flowed out of Latin America. The foreign capitalists invested in part to obtain the raw products they needed, but they also expected a rich return in the form of profits, interest payments, patent fees, and commissions.

To handle the new flow of money, banks, both national and foreign, sprang up with amazing rapidity in the major Latin American cities. Only one bank existed in Brazil in 1845, but in the following twelve years, twelve new banks were founded. Similar patterns were followed in the rest of the region. The banks facilitated international trade and investment, but the majority of people continued to borrow small amounts from individuals in their own communities.

By the eve of World War I, foreign investments totaled $8.5 billion. The English invested the heaviest, $5 billion, or 20 percent of British overseas investment. The French were second with $1.7 billion, followed very closely by the United States with $1.6 billion. Germany was fourth with somewhat less than $1 billion. The largest share of the money went to Argentina, Brazil, and Mexico, and the most popular investments were railroads, public utilities, and mining. British investments predominated in South America; those of U.S. capitalists dominated in Mexico and the Caribbean area.

With sources of capital, markets, and headquarters abroad, the foreign investors and business people identified neither with their host countries nor with local needs. Their ability to direct capital investment, their hostility to tariffs, and their preference for free-trade policies exerted an unfavorable influence on Latin America's economic development.

5.4: Progress on the Periphery

Few parts of Latin America were left untouched by the drive for progress of the late-nineteenth century. Even areas that had been poor backwaters in the colonial period followed the same patterns as the rest of the region by orienting their economies and polities to the export market. The small Central American nation of Nicaragua is one example.

As early as 1835, the Nicaraguan government tried to promote exports by giving ten-year tax exemptions to the production of indigo, cochineal, and coffee. It soon became clear that the most likely success was coffee production on Nicaragua's fertile highlands. By 1846, the government exempted coffee planters, their families, and workers from military service and exempted coffee farms from taxes. Production was encouraged in 1858 by allowing the growers to import duty-free goods equivalent to the value of the coffee they exported. The government undertook transportation and communications improvements, with an eye toward the coffee market. The first telegraph lines were established in 1876, and in 1882, underwater telegraph lines connected Nicaragua to the world's major cities; by 1890, there were 1,539 miles of telegraph lines, and by 1905 there were 2,691 miles. Construction of the railroad, Ferrocarril del Pacífico de Nicaragua, began in Corinto in 1878, and by 1886, it had reached the port of Momotombo on Lake Nicaragua. By 1909, there were 171 miles of track. Furthermore, coffee planters were not charged for shipping their produce on the government-owned railroads. Clearly, Nicaragua's economic and political leaders were interested in developing the economy and, starting in the mid-nineteenth century, saw coffee as the way to do it. Both the Conservative governments (1862–1903) and the Liberal government of José Santos Zelaya (1893–1909) threw their support behind the new export commodity.

Nicaraguan farmers responded to the incentives provided by the government and the market: 26 million trees were in production by 1892. Exports rose from 4.5 million pounds in 1880 to 11.3 million by 1890. From 1904 to 1924, coffee exports averaged 23 million pounds a year. In 1871, coffee represented only 10 percent of the value of Nicaragua's exports, in fourth place behind indigo, rubber, and gold. By the 1880s, coffee had become the country's most important export, and in 1890, coffee constituted at least 60 percent of exports.

The larger producers of cacao, indigo, and sugar quickly turned to coffee production. They sought more land to bring into cultivation, and elites in government responded with laws intended to commercialize property. A small proportion of nationally owned land was sold to large growers, and a small percentage of municipal ejidos and indigenous communal land ended up in coffee production. However, despite decrees in 1877 and 1881 that authorized the sale of communal lands, municipal governments continued to receive large tracts of land from the federal government for use as ejidos. And they gave plots of land in the ejidos to small farmers, most of whom continued to produce subsistence crops. The donation of small plots of land continued apace, and rental plots were available at nominal fees well into the twentieth century.

By providing small plots of land, the government guaranteed a seasonal workforce for the large coffee estates that formed. Coffee requires minimal upkeep during the year, but during the harvest there are massive labor

Coffee became one of Latin America's most important exports in the late-nineteenth century. This stereographic image of Nicaraguan women sorting coffee in 1903 also shows how images of Latin America were exported as something exotic for foreign customers. (Library of Congress)

demands. The government had adopted a multitude of vagrancy laws and requirements that Nicaraguans work in the harvests; most of the laws were flouted with regularity as workers ran off to competitors. The easiest solution was to provide workers with land that enabled them to grow food for subsistence, with the lure of the nearby haciendas for seasonal labor to earn the needed additional income.

As large coffee farms spread across most of the countryside, they were joined by an even greater number of small and medium-sized coffee farms. These smaller farmers produced significant amounts of coffee and became an important part of the coffee economy, lending legitimacy to the economic and political system. Men who owned even relatively small farms had the right to vote, and they participated in heated municipal and regional elections. They also made use of the legal system to protect their rights.

The various indigenous communities of Nicaragua responded differently to the new commercial incentives. In Carazo, some indigenous who had produced sugar in the early national period easily switched to coffee and joined the Hispanized, or *ladino*, world. In parts of neighboring Masaya, they lost their lands and largely became a workforce for the powerful interests in the old colonial capital of Granada. In the northern province of Matagalpa, where coffee production came relatively late, many indigenous communities were more successful in keeping their land and indigenous identity.

Similar patterns appeared in Guatemala and El Salvador, where small coffee farms were created alongside the huge plantations and where some indigenous communities managed to hold onto their lands into the twentieth century. Some profited, others lost, but few were untouched by the spread of the export economy and the growth of government institutions that accompanied it.

5.5: Modest Industrialization

During the second half of the century, the larger, more stable nations began to industrialize in order to meet growing internal demands for manufactured goods, to develop more balanced economies, and to protect the national economies from extremes of fluctuation in international trade. Some farsighted statesmen believed that without such industrialization Latin Americans would be doomed to economic dependency and backwardness. The Chilean Manuel Camilo Vial, Minister of Finance during the administration of Manuel Bulnes (1841–1851), preached, "Any nation in which agriculture dominates everything, in which slavery or feudalism shows its odious face, follows the march of humanity among the stragglers.... That future threatens us also, if we do not promote industry with a firm hand and a constant will." The governments raised tariffs, particularly during the final decades of the century, to encourage and protect the new industries, along with tax incentives and permission to import machinery duty free. Industrialization grew gradually in the last decades of the nineteenth century, despite such disadvantages as limited capital, unskilled labor, low labor productivity, lack of coal, limited markets, and a mentality emphasizing continued reliance on exploitation of mineral and agricultural possibilities.

The burgeoning cities provided a ready labor supply for the new industrialization. Women—and all too many children—worked in the factories. Women had always played a major role in agriculture. While traveling in Central America in the mid-nineteenth century, William V. Wells observed, "I have always found the women of the lower classes in Central America simple, kind-hearted, and hospitable, generally performing the most laborious part of the work, and never tiring under their ceaseless tasks. They are truly the hewers of wood and drawers of water." In the second half of the century, increasing numbers of women tended the new industrial machinery, working the same long hours as men but for lower wages.

In the beginning, industrialization was primarily focused on processing natural products for local consumption or export. Flour mills, sugar refineries, meat-packaging plants, tanning factories, lumber mills, wineries, and breweries developed wherever the requisite resources were at hand. Then service industries appeared: gas and electric utilities, repair shops and foundries, and construction enterprises. Finally, protected industries began to manufacture other goods for home consumption, principally, textiles and processed food.

In Brazil, the textile industry was by far the most important. The nine cotton mills in 1865 multiplied into 100 before the fall of the empire. The new republican government that came to power in 1889 visualized an industrial expansion for the nation. Symbolic of their ambitions, the Republicans changed the name of the Ministry of Agriculture to the Ministry of Industry. As further encouragement, they promulgated a protective tariff in 1890, raising to 60 percent the duty on 300 items, principally textiles and food products, which competed with nationally produced goods. Conversely, they lowered the duty on primary goods used in national manufacturing. Fundamental to any industrialization, the government established four new engineering schools in the 1890s.

The Argentine government began to turn its attention seriously to the encouragement of industrialization during the 1870s. In 1876, it enacted a high protective tariff. Industrialization concentrated in and around Buenos Aires, which by 1889 counted some 400 industrial establishments employing approximately 11,000 workers. During the last decade of the century, larger factories appeared and industry began to penetrate other regions, although Buenos Aires would always remain the focal point. The census of 1895 listed the number of factories and workshops as 23,000 with 170,000 employees (small workshops employing fewer than ten workers were by far the most common); the next census, 1914, raised the number of factories and workshops to 49,000 and the number of employees to 410,000. By that time, Argentina manufactured 37 percent of the processed foods its inhabitants consumed, 17 percent of the clothing they wore, and 12 percent of the metals and machinery they used.

Under Porfirio Díaz, Mexicans also used their success in export production to invest in manufacturing. By 1902, there were 5,500 manufacturing interests in Mexico. The most important industries were iron and steel, developed with a combination of foreign and domestic finance. New industries included production of cement, textiles, cigarettes, soap, bricks, furniture, flour, and beer, including José Schneider's 1890 founding of the Cerveceria Cuahtemoc, which produced Carta Blanca. Industrial production tripled during the Porfiriato, and the value of manufacturing rose by 6 percent a year. Despite these increases, manufacturing was still dwarfed by production of primary products, and industrial labor did not exceed 15 percent of the nation's workforce.

Most of the progress in manufacturing took place in Argentina, Brazil, Chile, Mexico, and Peru. Among the five nations, domestic production of consumer goods reached from 50 to 80 percent. However, the goods were largely of low quality, protected by tariffs, and no progress was made toward the creation of manufactured goods for export.

5.6: The Growth of Cities

Although most of the population remained in the countryside, the export economies of the late-nineteenth century fostered the growth of cities. For purposes of definition, most of the Latin American nations classified as urban those localities that had some type of local government and a population of at least 1,000–2,000 inhabitants.

The cities played ever more important roles in each nation. The government and administrative apparatus, commerce, and industry were located in the cities. Increasingly, they served as hubs of complex transportation and communications networks. Further, they provided important recreational, cultural, and educational services. Rapid urban growth resulted from the arrival of greater numbers of foreign immigrants, a constantly increasing population (Latin America counted 60 million inhabitants in 1900), and an attraction the city exerted over many rural dwellers. Promises of better jobs and a more pleasant life lured thousands each year from the countryside to the city, a road that became increasingly more heavily trodden. On the other hand, even where that promise was lacking, the grinding poverty and modernization of the countryside pushed many desperate folk into urban areas.

But cities failed to play the role they might have in encouraging national development. The urban facade of modernization deceived. The high concentration of land in a few hands dictated the function of cities just as it molded other aspects of life. The export-oriented economy encouraged the prosperity of a few ports, a transportation system focusing on them, the expenditure of export wealth to beautify the capital city, and the concentration of absentee landlords in the capital to be near the center of power. The capital and the ports absorbed the wealth of the export economy. The large estates, overreliance on an export economy, and the resultant dependency help to explain why only one or two major "modern" cities dominated each Latin American nation and why urban modernity remained little more than a facade.

Urban culture with its capitalist-consumer imprint fostered a new mentality among city dwellers, an outlook different from the traditions of most rural inhabitants. In an urban environment, traditional relationships tended to bend under necessity or examples of newer ones. The more intimate living conditions of the city and the greater familiarity of the city dwellers with foreign cultures exposed them to alternative ideas and values. Many read newspapers and participated in public events and became aware of changes in the world. They knew of the opportunities open to the trained and the talented and were willing to strive for those possibilities. Consequently they laid plans for the future and worked to shape their own destinies. The educational opportunities, varied careers, and job possibilities afforded by the city encouraged them to aspire toward upward mobility.

The statistics on Latin America's urban boom are impressive. Argentina's urban population doubled between 1869 and 1914, when 53 percent of Argentines lived in cities, 25 percent of them in Buenos Aires alone. Brazil witnessed similar urban growth. Between 1890 and 1914, the government created approximately 500 new municipalities. From 1890 to 1920, Recife and Rio de Janeiro doubled in size; São Paulo increased eightfold, making it one of the fastest growing cities in the world. In 1910, British diplomat James Bryce described São Paulo, a city approaching half a million, as "the briskest and most progressive place in all Brazil.... The alert faces, and the air of stir and movement, as well as handsome public buildings, rising on all hands, with a large, well-planted public garden in the middle of the city, give the impression of energy and progress."

The reality of urban living, however, condemned many in the lower classes to grinding poverty. Brazilian novelist Aluísio Azevedo captured the urban reality in his novel, *The Slum* (*O Cortiço*, 1890). He described the growth of the slum as "a brutal and exuberant world." "And on the muddy ground covered with

Street cars and European-style architecture are among the signs of modernity in the Plaza de la Constitución, in Montevideo, Uruguay, photographed between 1880 and 1900.
(Library of Congress)

puddles, in the sultry humidity, a living world, a human community, began to wriggle, to seethe, to grow spontaneously in that quagmire, multiplying like larvae in a dung heap." It was an environment in which employment opportunities failed to keep up with the numbers of potential workers, and in which women labored as washerwomen, domestics, and prostitutes.

For the elites, however, the cities offered delights. By the beginning of the twentieth century, nearly all the capitals and many of the largest cities boasted electricity, telephones, streetcar service, covered sewers, paved streets, ornamental parks, and new buildings reflecting French architectural influence. Nurtured in an increasingly prosperous urban environment, intellectual activity flourished. Some of Latin America's most prestigious newspapers were founded: *El Mercurio* in Chile and the *Jornal do Commércio* in Brazil, both in 1827; *El Comercio* in Peru in 1839; and *La Prensa* in 1869 and *La Nacíon* in 1870 in Buenos Aires. Starting with the University of Chile in 1843, the major universities of Latin America began to publish reviews, journals, annuals, and books, an activity that further stimulated intellectual development.

Romanticism, with its individuality, emotional intensity, and glorification of nature, held sway in literary circles for much of the nineteenth century. José Mármol of Argentina, Jorge Isaacs of Colombia, and José de Alencar of Brazil were masters of the romantic novel. Excesses in romanticism prompted literary experiments by 1880 in modernism and realism, already in vogue in Europe. The brilliant Nicaraguan poet Rubén Darío helped to introduce modernism into Latin America, and by the end of the century, he dominated the field of poetry. Critics considered him one of the most original and influential poetic voices of his time. Under the sway of realism, urban writers depicted and denounced the injustices they observed in their society. Clorinda Matto de Turner's *Birds Without a Nest* protested the abysmal conditions under which the Quechua and Aymara of Peru lived. She saw the indigenous as victims of iniquitous institutions, not least of which in her opinion was the Church.

Latin American culture aped European trends, particularly those originating in Paris, with one notable exception. Argentine José Hernández's epic poem, *Martin Fierro* (1872), proclaimed the gaucho to be Argentina's true hero, one who was exploited and mistreated by the Liberal government. The poem, originally printed in installments in the newspaper, rescued the reputation of the formerly despised gaucho—once the gaucho had ceased to be a threat.

Education continued to be a privilege of the elite. Overwhelming numbers of the masses remained illiterate. In Brazil, the illiteracy rate never dropped below 85 percent in the nineteenth century. Argentina dedicated much of the nation's budget to improving education. As a result, literacy in Argentina rose from 22 percent in 1869 to 65 percent in 1914, an enviable record throughout most of Latin America, where the illiteracy rate ranged from 40 to 90 percent during the second half of the nineteenth century. It was always much higher in the countryside than in the city and among women and the indigenous. A marked contrast existed between the well-educated few and the ignorant many, most of whom could not attend the new schools because they had to work. The children of the elite, however, seldom set foot in a public school. Their parents hired tutors or sent them to private schools for a typically classical education. Such segregation further removed the future leaders from national realities.

The modernizing elites began to worry that they could not progress as long as mothers—at least elite mothers—lacked the skills to teach their sons to be good citizens. Mexican journalist Florencio del Castillo declared in 1856, "The most effective way to better the moral condition of the land is to educate women." Just as women served as transmitters of Spanish culture during the colonial era, educated women were now to pass on the republican, progressive values so important to Latin American modernization. In Argentina, Juan Bautista Alberdi and Domingo Faustino Sarmiento included women in their vision of an educated, progressive country.

Of course, Conservatives worried that education would encourage women to leave their traditional roles, and the idea of their education was not immediately accepted. It took more than just the enlightened attitudes of men like Sarmiento to win expanded education for women. Many women, particularly from the middle class, campaigned for educational opportunities. They founded journals and editorialized about the need for education. An article in the Brazilian journal *O Sexo Femenino* contended: "It is to you [men] that is owed our inadequacy; we have intelligence equal to yours, and if your pride has triumphed it is because our intelligence has been left unused. From this day we wish to improve our minds; and for better or worse we will transmit our ideas in the press, and to this end we have *O Sexo Femenino*; a journal absolutely dedicated to our sex and written only by us."

As educational opportunities for women improved, so did employment. Because girls and boys were usually taught in separate schools, women were trained to teach the girls. As schools later became coeducational, women were considered the most appropriate teachers of young children in primary schools. Normal schools were founded to train teachers for this role.

The bringing together of women in schools and producing journals gave impetus to an early feminist movement, which was concerned not only with basic education but also with the opening of higher education and the professions to women. Their actions were not taken lightly by the men at the top. Mexico's Minister of State Justo Sierra made the Porfirian attitude toward female education quite clear: "The educated woman will be truly one for the home: she will be the companion and collaborator of man in the formation of the family. You [women] are called upon to form souls, to sustain the soul of your husband; for this reason, we educate you … to continue the perpetual creation of the nation. *Niña querida*, do not turn feminist in our midst."

Chile pioneered in offering professional education to women. A special governmental decree in 1877 permitted women to receive professional degrees, and in 1886 Eloisa Díaz became the first woman in all of Latin America to receive a medical degree. In 1892, Matilde Throup graduated from the law school in Santiago and became the first female lawyer in Latin America. The Brazilian government opened professional schools to women in 1879.

Women also became activists in the antislavery movements in Cuba and Brazil. Much like women in the United States, their focus on lack of freedom for blacks led them to see the limits on their own freedom. As the slaves were freed in Brazil in 1889, one journal proclaimed, "once more we [women] ask equal rights, freedom of action, and autonomy in the home." The demands, however, remained unmet.

5.7: Superficial Modernization

During the second half of the nineteenth century, new forces appeared that challenged the social, economic, and political institutions deeply rooted in the colonial past. Urbanization, industrialization, and modernization formed a trinity menacing to tradition. Once introduced, these mutually supporting forces could not be arrested. The center of political, economic, and social life, once located on the plantations and haciendas, shifted gradually but irreversibly to the cities.

By the beginning of the twentieth century, some of the Latin American countries—certainly Argentina, Brazil, Chile, and Mexico—conveyed at least the outward appearance of having adopted the modes of the most progressive European nations and of the United States. Their constitutions embodied the noblest principles of Western political thought. Their governmental apparatus followed the most progressive models of the day. Political stability replaced chaos. Expanding transportation and communication infrastructures permitted the governments to control a larger area of their nations. New industries emerged. Larger banking networks facilitated and encouraged commerce. Society was more diversified than at any previous time. In the capitals and the largest cities, the architecture of the new buildings duplicated the latest styles of Paris: In how many Latin American cities do the local citizens proudly point to the opera house and claim it to be a replica of the Paris Opera? To the extent that some of the Latin American states formally resembled the leading nations of the Western world, which they consciously accepted as their models of modernity, it is possible to conclude that those nations qualified as modern. However, many would argue that such modernity was only a veneer. It added a cosmetic touch to tenacious institutions while failing to effect the changes implied by the concept. Modernization in Latin America lacked real substance.

The superficiality of modernization guaranteed the continued domination of the past. The rural aristocracy still enjoyed power, their estates remained huge and generally inefficient, and their control over their workers was complete. The latifundia actually grew rather than diminished in size during the nineteenth century, at the expense of indigenous communities and their traditional landholdings, the properties confiscated from the Church, and the public domain.

By early 1888, slavery had been abolished throughout the region. Still, former slaves and their descendants occupied one of the lowest rungs of the social and economic ladder. The doors to education, opportunity, and mobility remained tightly closed to them except in the rarest instances. The indigenous fared no better. Debt servitude of one or another variety characterized part of the labor market. It would have been foolish to have expected the law to intervene to restrict such abuses as the making, enforcing, and judging of the laws rested in the hands of the landowners and their sympathizers.

The wealthiest and most powerful class in Latin America in general was white or near white in complexion. Heirs of the creoles and mazombos, they enjoyed age-old economic advantages, to which, after independence, they added political power. The group that surrounded Porfirio Díaz spoke of themselves, symbolically enough, as the "New Creoles."

Rural tradition comfortably allied itself with the invigorated capitalism of nineteenth-century Latin America. Positivism even provided a handy ideological umbrella for the two. Like so much of the economic and political thought in Latin America then and since, it recognized no incompatibility in the imposition of capitalist industrialization on a traditional rural base. With its emphasis on order and hierarchy, positivism assured the elites their venerable privileges;

relative prosperity and selective progress held out promise of the same to the restless middle sectors. There was little promise, however, for the lower-class majority.

But social inequality, paternalistic rule, privilege, and dependency clearly were incompatible with the new trends started during the last half of the century. The Chilean intellectual Miguel Cruchaga Montt pointed out in his *Study of Chile's Economic Organization and Public Finance* (1878) that the colonial past was still dominant and frustrated development. But the past could no longer exist unchallenged once Latin Americans began to think in terms of national development.

At the opening of the twentieth century, Colombian Rafael Uribe stated that the basic economic question concerned the quality of life of the "people": "Are they able to satisfy their basic needs?" The reality was they could not. A majority of the Latin Americans were no better off at the dawn of the twentieth century than they had been a century earlier. In fact, a persuasive argument can be made that they were worse off. The negative response to Uribe's question foretold conflict.

5.8: The Popular Challenge

The Latin American nations marched toward progress to a tune played by the elites but not without discordant chords sounded by large numbers of the humble classes. For the majority of Latin Americans, progress was proving to mean increased concentration of lands in the hands of fewer owners; falling per capita food production with corollary rising food imports; greater impoverishment; less to eat; more vulnerability to the whims of an impersonal international market; uneven growth; increased unemployment and underemployment; social, economic, and political marginalization; and greater power held by the privileged few. The more the folk cultures were forced to integrate into world commerce, the fewer the material benefits they reaped. But poverty through progress must be understood in more than the material terms of declining wages, purchasing power, or nutritional levels. Traditional ways of life and the cultural arrangements they embodied were destroyed along with the indigenous villages. A tragic spiritual and cultural impoverishment accompanied the physical poverty.

The impoverished majority both bore the burden of the inequitable institutional structure and paid for the modernization enjoyed by the privileged. The deprivation, repression, and cultural attack of the majority by the minority created tensions that frequently gave rise to violence. The poor protested their increasing misfortunes as modernization increased. For their own part, the privileged were determined to modernize and maintain the order required to do so. They freely used whatever force was necessary to accomplish both. Consequently, the imposition of progress stirred social disorder.

Indigenous rebellions flared up from Mexico to Chile. The indigenous refused to surrender their remaining lands quietly as the large estates intensified their encroachment. The arrival of the railroads, which accelerated those encroachments, spawned greater violence. Doubtless the major indigenous rebellion in terms of length, carnage, and significance in the Americas of the nineteenth century was the Caste War of Yucatan between the Mayas and the peninsula's whites and mestizos.

In the years after Mexico declared its independence, the sugar and henequen plantations had expanded to threaten the corn cultures of the Maya by incorporating their lands into the latifundia and by impressing the Indians into service as debt peons. The Maya fought for their land and freedom, defending their world. On the other side, the Yucatan elite professed that they fought for "the holy cause of order, humanity, and civilization." Much of the bloodiest fighting occurred during the period from 1847 to 1855, but the war lingered on until the early twentieth century. During those decades, the Mayas of eastern and southern Yucatan governed themselves.

Free from white domination, the Mayan rebels took the name of Cruzob, turned their backs on the white world, and developed their own culture, a synthesis of their Mayan inheritance and Spanish influences. The Cruzob retained their knowledge of agriculture and village and family organization from the pre-Columbian past. Unique to the Cruzob was the development of their own religion, based largely on their interpretation of Christianity. Incorporating the Indigenous folkways, it strengthened the Cruzob and provided a spiritual base for independence that other indigenous people lacked. What was notable about the Cruzob was the emergence of a viable indigenous alternative to Europeanization. Although infused with Spanish contributions, it bore a strong resemblance to the pre-Columbian Mayan society. Reviving their indigenous culture by repudiating "foreign" domination and substituting their own values for "foreign" ones, the Cruzob revitalized their society and became masters of their own land again.

Powerful forces at work in the closing decades of the nineteenth century overwhelmed the Cruzob. The poor soil of Yucatan, exhausted under corn cultivation, no longer yielded sufficient food. Disease reduced the Mayan ranks faster than battle did. At the same time, Mexico, increasingly stable under Porfirio Díaz, showed less tolerance for the Cruzob and more determination to subdue them in order to exploit Yucatan. A treaty between Mexico and Great Britain closed British Honduras to the Cruzob, thus cutting off their single source of modern weapons and ammunition. Finally, the expanding wagon trails and railroads from northern Yucatan, which accompanied the spread of the prosperous henequen plantations, penetrated the Cruzob territory. A growing market for forest woods even sent the whites into the seemingly impenetrable forest redoubts of the Cruzob. Consequently a declining Cruzob population and relentless Mexican pressures brought to an end the Mayan independence of half a century. The long and tenacious resistance testified to the indigenous preference, a rejection of the modernization preferred by the elites.

The indigenous were not the only rebels. In Brazil, where slavery lingered after midcentury, the slaves vigorously protested their servitude. Sober members of the elites regarded the slave as "a volcano that constantly threatens society, a mine ready to explode," as one nineteenth-century intellectual phrased it in his study of Brazilian slavery. Foreign visitors also sensed the tensions created by slave society. Prince Adalbert of Prussia visited one large, well-run plantation, which he praised as a model. After noting the seemingly friendly relations between master and slaves, he revealed, "The loaded guns and pistols hanging up in his [the master's] bedroom, however, showed that he had not entire confidence in them [the slaves] and indeed, he had more than once been obliged to face them with his loaded guns." The decade of the 1880s, just prior to emancipation, witnessed mounting slave resistance. The slaves fled the plantations, killed the masters, and burned fields and buildings. One fiery Afro-Brazilian abolitionist leader, Luis Gonzaga de Pinto Gama, declared, "Every slave who kills his master, no matter what the circumstances might be, kills in self-defense." He also preached the "right of insurrection." Once freed, African Americans throughout the Americas protested their poverty and the institutions that they felt perpetuated their problems. For example, Panama City seethed with racial tensions during the decades from 1850 to 1880. The African American urban masses resented their depressed conditions and used violence—robberies, fires, and rioting—as a means of protest. Many referred to the situation as a "race war," exacerbated by an economic reality in which the poor were black and the rich were white.

Rural rebellions abounded, signifying still other challenges to the elite institutions and commitment to modernization. More often than not, the ideology behind those rebellions was vague and contradictory. Somehow, the rebels hoped to save their lands, improve their standards of living, and share in the exercise of power. Two popular revolts, one in Brazil, another in Argentina, illustrate the motives, the violence, and the repression.

The Quebra-Quilo Revolt, which took place from late 1874 to early 1875, ranked high in significance because the subsistence farmers of Brazil's interior northeast succeeded in checking the government's new modernization drive (which was underway in 1871 but ineffectual by 1875). The causes of that revolt were not unique: new taxes and the threat smallholders felt from the large landowners absorbing their farms, complicated by the imposition of the metric system with the requirement of fees for official alteration and authentication of weights. A journalist covering the revolt called it "the direct consequence of the suffering and deprivation … of the working classes of the interior," whereas a participant claimed, "The fruit of the soil belongs to the people and tax ought not be paid on it." As riots multiplied in the marketplaces, the municipal and provincial authorities feared the "forces of Barbarism" were poised to sweep across the northeast. The campesinos were unusually successful. They ignored the new taxes, destroyed the new weights and measures, and burned official records and archives (thus protecting their informal title to the land by reducing to ashes the legal documentation). The subsistence farmers in most cases had taken physical possession and worked the land over the generations without title. They faced possible eviction by anyone who could show the proper paper authenticating legal ownership. By destroying the records, they removed evidence—the local notarial registers of land, for example—from use in judicial proceedings, thereby putting themselves on equal legal footing with the local landed elite. Momentarily, then, the sporadic riots that constituted the revolt achieved their goals, while temporarily frustrating the penetration of the elites into their region.

In Argentina, revolts shook the province of Santa Fe in 1893. Small farmers there protested a tax on wheat to pay for the government's innovations, including railroads, which seemed to favor the large landowners. Furthermore, they resented the fact that immigrants received land and preferential treatment denied the locals. In the meantime, social disorder rose dramatically in the Argentine province of Tucumán between 1876 and 1895. During those two decades, the number of arrests, ones involving mostly illiterate workers, jumped from under 2,000 per year to over 17,000, while the total population merely doubled during the same period.

Popular protest also took the form of millenarian movements. Millenarianism has its roots in the Christian Bible's Book of Daniel, which predicts the apocalypse, a period of devastation ending with the forces of good triumphing over evil and heralding the second coming of Christ. Christ would then reign for a thousand years over a peaceful and just world before his final judgment. At times of great change, the seeming chaos might be interpreted as a sign of the coming apocalypse.

Brazil witnessed a remarkable array of millenarian movements. The best known took place in the dry, impoverished backlands of the state of Bahia where the mystic Antônio the Counselor gathered the faithful between 1893 and 1897. Thousands flocked to his settlement at Canudos to listen to the Counselor preach and stayed to establish a flourishing agrarian community. He alienated the government by advising his adherents not to pay taxes. Furthermore, his patriarchal ideas smacked of monarchism to the recently established republican government in Rio de Janeiro. The Church authorities denounced him, resenting his influence over the masses. The local landlords disliked him because he siphoned off the rural workers and stalemated the expansion of their fazendas. Those powerful enemies decided to arrest Antônio and scatter the settlers at Canudos. However, they failed to consider the strength and determination of his followers. It took four military campaigns, all the modern armaments the Brazilian army could muster, and countless lives to suppress the millenarian movement. The final campaign, directed by the minister of war himself, devastated the settlement at Canudos house by house. The people refused to surrender. The epic struggle inspired Euclydes da Cunha's 1902 masterpiece, *Rebellion in the Backlands*.

Messianic movements also flourished among the Andean people after the conquest. They yearned for a return to an order, basically the traditional Incan one destroyed by the Spanish conquest, which would benefit them rather than the outsider. Exemplary of such movements in the nineteenth century was one that occurred among the Bolivian people of Curuyaqui in 1891 and 1892. An individual called Tumpa, known as "the supreme being," appeared in the community announcing his mission "to liberate them from the whites." Under his prophetic new system, Tumpa promised that the whites would work for the indigenous. His followers took up arms as urged by the messianic leader; the whites fled to the cities, and the army arrived to brutally crush the uprising. The carnage disproved at least two of Tumpa's prophecies: first, that only water would issue forth from the soldier's guns, and second, that anyone who did die for the cause would return to life in three days.

Northwestern Mexico was the scene of the miracle cures of Teresa Urrea, referred to by hundreds of thousands of devotees as Teresita or the Saint of Cabora. In 1889, after a severe psychological shock, she lapsed into a comatose state. Considered dead, she regained life just prior to her burial. She reported having spoken to the Blessed Virgin, who conferred on her the power to cure. By 1891, pilgrims flooded Cabora seeking her help. Teresita's compassion for the poor earned her the devotion of the masses and the suspicion of the Díaz government. The Yaqui and Mayo confided in her and unburdened their sufferings before her. Believing she enjoyed influence with God, they pressed her for help and advice. In 1890, the Tarahumara mountain village of Tomochic adopted Teresita as their saint, placing a statue of her in their Church, to the disapproval of the region's priest. Local disputes led to the imposition of an outsider as the ranking village official. In 1892, Tomochic rebelled against the government and requested Teresita to interpret God's will to them. The government reacted immediately and harshly, but it still took several armed expeditions to quell the rebellion. The village was destroyed, and not a man or boy over thirteen years of age survived the slaughter.

Later that year, a group of approximately 200 Mayo attacked the town of Navojoa shouting, "Viva la Santa de Cabora!" Considering her a dangerous agitator of the masses, the Díaz government exiled Teresa Urrea to the United States. Teresita, herself opposed to violence, had served more as a figurehead, a catalyst, and a remarkable charismatic personality, whose compassion gave unity of expression to the miserable masses of northwestern Mexico.

Banditry was another form of rebellion, attracting the desperate, those who had lost out in the system whether they were the poor or members of the impoverished gentry. It was as much a means of protesting injustice or righting wrongs as it was of equalizing wealth or taking political revenge. Although unsympathetic to banditry, the Brazilian jurist of the mid-nineteenth century, Tavares Bastos, realized that the bandits often were victims of the State who, no longer confiding in its laws, made their own justice.

Bandits roamed the Brazilian interior in the nineteenth century, particularly in the impoverished northeast, where many won the admiration of the poor and the respect of the wealthy, who not infrequently co-opted them and utilized their services. Some scholarship correlates the rise of banditry in the late-nineteenth century with the breakdown of the patriarchal order in the countryside. Brazilian popular poetry abounds with tales of the bandit hero. A well-known poem sung at the beginning of the twentieth century related the history of Antônio Silvino, who became a bandit in 1896 to avenge an injustice: His father was slain by a police official who went unpunished by the government. Others relate the adventures of Josuíno Brilhante, also seemingly forced into banditry to avenge injustices against his family. He assaulted the rich and distributed their goods and money among the poor, boasting that he never robbed for himself.

Banditry characterized much of Spanish America as well. Mexican banditry flourished, and interestingly enough, regions that produced bandits, such as Chalco-Río Frío, eastern Morelos, and northwestern Puebla, spawned agrarian revolutionaries before the century ended, providing further evidence of the social dimension banditry could assume on occasion. Peru offers numerous examples of peasant bandits. In his study of them, Enrique López Albujar described banditry as "a protest, a rebellion, a deviation, or a simple means of subsistence." He concluded that nineteenth-century Peruvian banditry produced an array of folk heroes because those bandits corrected injustices, robbed to help the poor, and protested social and economic inequities. For their part, Chilean officials tended to lump together indigenous and bandits of the rugged Andes as "criminals." They also routinely complained that local populations supported the bandits, thus facilitating their antiestablishment activities.

The motives and activities of the bandits varied widely, but at least in part they could be explained as protests against the wrongs of society as they viewed it. Because of their strength and because they often opposed the elites and official institutions, they received the support, indeed the admiration, of large numbers of the humble classes, who often hid them, lied to the authorities to protect them, guided them through strange terrain, and fed them. To the poor, the bandits were caudillos who by default helped them to sustain their folk cultures and deflect modernization.

The folk by no means rejected change simply to preserve the past unaltered. Rather, they wished to mediate change over a longer period of time. They opposed those changes imposed by the elites that they judged harmful or potentially threatening. They perceived the threat that the export economy and the capitalist mentality promoting it posed to their remaining lands and to their control over their own labor. The elites showed little patience and less tolerance with folk preferences, caution, and concerns. They dealt severely with any protest and thus further raised the level of violence. In the last analysis, the elites triumphed. After all, they controlled the police, the militia, and the military. Furthermore, the popular protests tended to be local and uncoordinated. Thus, despite the frequency of such protests, the elites imposed their will and their brand of progress. The triumph of that progress set the course for twentieth-century history in Latin America. It bequeathed a legacy of mass poverty and continued conflict.

Questions for Discussion

1. Latin American elites chose to focus their economies on the export of primary products. What were the strengths and weaknesses of this strategy?
2. Who benefited from the economic changes of the late-nineteenth century? Who lost?
3. How did industrialization and urbanization impact Latin American societies?
4. How did the Latin American masses and folk communities react to the changes created by modernization?
5. The table "Export Commodity Concentration Ratios, 1913" shows the two most important export products in each Latin American country. What patterns emerge, and what do they suggest about Latin American development?

6 New Actors on an Old Stage

As Latin America approached its independence centennial, two trends, one of external and the other of internal origin, emerged with greater clarity. The first was the emergence of the United States as a major world power. Motivated by economic interests, U.S. leaders adopted a philosophy of Manifest Destiny, justifying aggression by claims of concern about security and by blatantly racist attitudes toward Latin America. The second trend was the emergence of two new significant population groups within Latin America: a small middle class, which aspired to the wealth of the elites via the path of progress shown by the United States, and the working class, which sought to improve its lot more through European examples of labor organizing, introduced by the flood of immigrants to such countries as Brazil and Argentina. Together, the United States and the Latin American middle sectors helped to shape Latin American history during the twentieth century.

6.1: The Presence of the United States

Foreign influence, especially Great Britain's, shaped much of the historical course of Latin America during the nineteenth century. At first, Great Britain had no serious rivals to its economic domination of the region. But as the United States grew stronger, its capitalists and politicians were determined to spread across the North American continent and to dominate the Caribbean. By the early twentieth century, the United States had succeeded in doing both and was well on its way to replacing Britain's century-long domination of Latin America. Concerns for security and desires for trade shaped U.S. attitudes and policies toward Latin America. Obviously the United States and Latin America shared the same geography, the Western Hemisphere, thus being neighbors, in a certain romantic sense, even though much of the United States was closer to Europe than it was to most of South America.

Despite the tough talk of the Monroe Doctrine in 1823, the United States was not at the time a world power, and Washington officials ignored the proclamation for decades. Much to the chagrin of Latin Americans, the United States did nothing while European powers intervened in Latin America at will. The British reoccupied the Falkland Islands despite vigorous Argentine protests, the French intervened in Mexico and the Plata area, and the French and British blockaded Buenos Aires, all without the United States reminding the European interlopers of the content or intention of Monroe's statement. Only when the British and French maneuvered to thwart the union of Texas with the United States did President John Tyler invoke the principles of the doctrine in 1842 to warn the Europeans to keep out of hemispheric affairs.

Indeed, as later used, the doctrine provided a handy shield for North American expansion, well under way by the mid-1840s. As President James Polk gazed westward toward California, he notified the Europeans that his country opposed any transfer of territory in the New World from one European state to another or from a nation of the Western Hemisphere to a European nation. However, by his interpretation the Monroe Doctrine did not prohibit territorial changes among the nations of this hemisphere. Such an interpretation complemented the annexation of Texas by the United States in 1845 and of Arizona, New Mexico, and California in 1848. Expansionist sentiment rose to a fever pitch as the stars and stripes fluttered across the continent toward the Pacific Ocean. An editorial in the influential *De Bow's Commercial Review* in 1848 expressed the ebullient mood of a confident nation:

> The North Americans will spread out far beyond their present bounds. They will encroach again and again upon their neighbors. New territories will be planted, declare their independence, and be annexed. We have New Mexico and California! We will have Old Mexico and Cuba!

The editorial reflected the era's dominant theme: *manifest destiny,* an expression coined in 1845 by John L. O'Sullivan, editor of the *Democratic Review*. The term meant that the United States, as a result of Anglo-Saxon superiority, was destined to absorb its neighbors. O'Sullivan railed against other nations, whose resistance to U.S. annexation of Texas was aimed at "thwarting our policy and hampering our power, limiting our greatness and checking the fulfillment of our manifest destiny to overspread the continent allotted by Providence for the free development of our yearly multiplying millions." That destiny was made more explicit in congressional debate about the war with Mexico: Maryland Rep. William Fell Giles said, "We must march from Texas straight to the Pacific ocean.... It is the destiny of the white race; it is the destiny of the Anglo-Saxon race."

Race was a key factor in manifest destiny, and race in the United States was closely tied to the issue of slavery. Slave owners in the U.S. south looked longingly at Cuba, Mexico, and Central America as potential slave states. While Mexicans resisted U.S. expansion efforts, slave owners in Cuba actually considered annexation to the United States. They feared that the British campaign against the slave trade would result in Spain outlawing the institution on which the Cuban sugar trade was built. Indeed, O'Sullivan's brother-in-law was

Cristobal Madán, a Cuban planter who served as leader of the annexationist lobby in New York in 1845.

Great Britain also served as the major check on U.S. expansion southward into Middle America and the Caribbean during the mid-nineteenth century. The best the United States could arrange at the time was an agreement in 1850, the Clayton–Bulwer Treaty, in which both nations promised not to occupy, fortify, colonize, or otherwise exercise domination over Central America. Specifically, neither country would seek to build an interoceanic canal. Significantly, no representatives of the targeted Latin American nations were a party to the treaty. At about the same time, the attention of the United States focused inward once again, as a divided nation girded itself for internal strife.

While civil war rent the United States, several European nations pursued their own adventures in the New World. Spain reannexed the Dominican Republic and fought Peru and Chile. France intervened in Mexico. Only after it became apparent that the North was winning the U.S. Civil War and was determined to oppose European ventures in the Western Hemisphere did Spain depart from the Dominican Republic and return the Chincha Islands to Peru. When Napoleon III hesitated to withdraw French forces from Mexico, the government in Washington dispatched a large army to the Mexican border to help the French emperor make up his mind. Once these European threats to Latin America had ended, the United States seemed content to ignore the region while the nation concentrated its energies on domestic reconstruction, railroad building, and industrialization.

Many Latin American leaders hoped the United States would intervene as needed to hold European powers at bay, but these leaders also wanted the United States to refrain from interfering with Latin American interests. Venezuelan President Antonio Guzmán Blanco asserted in 1865, "Venezuela had come to regard the United States as a Protector of her principles and institutions, and to look up to her as the senior and most powerful Republic on the Continent." In 1876, Guzmán appealed to the United States to invoke the Monroe Doctrine and aid Venezuela in its boundary dispute over the encroaching British Guiana, a request made repeatedly and unsuccessfully for nineteen years. Yet in 1880, Venezuelan statesman Simón Camacho complained in *La Opinión Nacional*, "To pretend that we would refuse French or any other capital we can get cheaply to bring out the natural resources of this country and that we could shut our door to progress because it happens not to suit your 'Monroe Doctrine' or any other idea of a protection you wish to exercise at a distance of two thousand miles would be the height of folly on our part."

These contradictory viewpoints toward the United States were not simply a function of Latin American desire to use the United States as a protector while avoiding it as an interloper. There were mixed feelings regarding the character of U.S. culture and society. Venezuelan consuls and business leaders reported approvingly from the United States about the use of science to improve agriculture and the educational system's focus on logic rather than rote memorization.

But Venezuelans also criticized U.S. culture as boorish and materialistic. An 1890 editorial in *La Opinión Nacional* warned those going to the United States: "You will see the God of money in all his grandeur; but don't expect to find an impressive civilization, nor good manners, nor even mediocre education. There one goes to work, not to live. There is no art nor anything that approaches it."

Latin American ambivalence about U.S. materialism came in the context of the country's increasing industrial power in the post–Civil War years. Business and government leaders looked outward once again in search of new markets—and none seemed more promising than Latin America, long the domain of European trade. The United States had long appreciated the strategic importance of Latin America but had been slow to develop trade relations. Now the region's economic potential beckoned.

In the eyes of many, Latin Americans appeared too slow in fulfilling the destiny that nature provided. A growing sector of the United States thought that Latin America needed a dose of "Protestant virtues and Yankee know-how" to turn potential into reality. The Reverend Josiah Strong summed up this opinion in his influential book *Our Country* (1885) when he wrote, "Having developed peculiarly aggressive traits calculated to impress its institutions upon mankind, [the United States] will spread itself over the earth. If I read not amiss, this powerful race will move down Central and South America, out upon the islands of the sea, over upon Africa and beyond. And can anyone doubt that the result of this competition of races will be the 'survival of the fittest'?" Imbued with the popular ideas of Spencer and Darwin, the good reverend evinced the enthusiasm, confidence, and arrogance of his age.

During the second half of the nineteenth century, North American trade and investment in Latin America rose gradually. Secretary of State James G. Blaine, understanding the need for cooperation among nations of the hemisphere, sought closer commercial relations to solidify the inter-American community. Blaine envisioned a fraternal hemispheric trade in which the United States supplied the manufactured goods and Latin America the raw products. With that idea in mind, he presided over the first Inter-American Conference, held in Washington in 1889–1890. The sessions appeared cordial, but it became increasingly obvious that the Latin Americans were less interested in placing orders for new industrial products than in containing the expansion of an ambitious neighbor by obtaining a promise of respect for national sovereignty. Cuban poet and independence leader José Martí, who attended the conference as consul for Uruguay, wrote in *La Nación* of Buenos Aires: "From Independence down to today, never was a subject more in need of examination than the invitation of the United States to the Pan American Conference. The truth is that the hour has come for Spanish America to declare its second independence."

Ironically, Martí's homeland of Cuba had not yet achieved its first independence, a goal Cubans first tried to reach during the Ten Years' War (1868–1878). In 1870, at the age of sixteen, Martí was imprisoned and exiled by Spanish colonial authorities for his vocal opposition to their rule. He spent most of his life in exile, where he earned a reputation as a major literary figure. His was one of the most significant voices cautioning Latin America against blindly adopting the ways of Europe and North America to govern Latin America. In Our America (*Nuestra America*, 1891), Martí cautioned: "The struggle is not between civilization and barbarity, but between false erudition and Nature. . . . How can the universities produce governors if not a single university in America teaches the rudiments of the art of government, the analysis of elements peculiar to the peoples of America? The young go out into the world wearing Yankee or French spectacles, hoping to govern a people they do not know. In the political race entrance should not go for the best ode, but for the best study of the political factors of one's country."

In 1892, Martí founded the Cuban Revolutionary Party, and in 1895, he sailed to Cuba to join the new independence uprising. In his last letter, he warned of the dangers of United States imperialism: "I am in daily danger of giving my life for my country and duty, for I understand that duty and have the courage to carry it out—the duty of preventing the United States from spreading through the Antilles as Cuba gains its independence, and from overpowering with that additional strength our lands of America. . . . I have lived in the monster and I know its entrails; my sling is David's." Martí died in battle the next day. His concerns about U.S. interests were prophetic.

Two months after Martí's death, Secretary of State Richard Olney broadly interpreted the Monroe Doctrine to allow the United States to settle Venezuela's long-simmering border dispute with British Guiana. Olney wrote, "Today the United States is practically sovereign on this continent and its fiat is law upon the subjects to which it confines its interposition. Why?. . . It is because. . . its infinite resources combined with its isolated position render it master of the situation and practically invulnerable as against any or all other powers."

The United States displayed that invulnerability in 1898 by intervening, unbidden, in Cuba's independence war. The United States had long been interested in Cuba, dating back to Thomas Jefferson's unsuccessful attempt to purchase it from the Spanish in 1808. In 1848, President James Polk's bid was rejected, and in 1854, Spain turned down the Franklin Pierce administration's offer, complicated by the notorious Ostend Manifesto, a document indicating that the United States would be justified in taking Cuba from Spain if the island could not be purchased. Despite these failures, by the end of the century, the United States dominated Cuba's economy. In 1893, Cuba's exports to the United States were worth $79 million, compared to only $6 million of exports to Spain. By 1894, Cuba sent 90 percent of its exports to the United States, and 40 percent of Cuban imports came from the United States.

In 1898, Cuba was close to winning its independence. Spanish commanders were refusing to fight; units were being withdrawn. The United States was aware of the change of fortunes. Assistant Secretary of State William Day warned,

". . . it is now evident that Spain's struggle in Cuba has become absolutely hopeless." But the United States did not find the idea of Cuban independence acceptable. U.S. Minister to Spain Stewart L. Woodford noted in March 1898: "I do not believe that the population is today fit for self-government. . . . I see nothing ahead except disorder, insecurity of persons and destruction of property. There is but one power and one flag that can secure peace and compel peace. That power is the United States and that flag is our flag."

The United States already had the perfect excuse to intervene. The government had sent the U.S. Maine to Havana in January, ostensibly to protect U.S. citizens. Spanish authorities did not seem to object to the arrival of the ship. The ship's captain even went to a bullfight with the Spanish commander. On February 15, 1898, the Maine exploded, killing 266 soldiers, most likely because of the gunpowder that it carried. The United States blamed the Spanish and entered the war in April. By December, it was over. Secretary of State John Hay labeled it a "splendid little war."

Results for the United States were splendid indeed, gaining control of Guam, the Philippines, and Puerto Rico. Convinced that Cubans could not govern themselves, the U.S. Marines occupied the country from 1899 to 1903. The condition for U.S. departure was that Cuba add to its constitution the Platt Amendment, written by U.S. Senator Orville Platt, giving the United States the right to intervene in Cuba whenever it saw fit.

The reason for United States interest was clear. In 1900, the United States had $50 million invested in Cuba. By 1914, the United States had $140 million in direct investment in Latin America, $79 million of that in Central America, and nearly $31 million in Latin American public debt. But these figures represented only 14 percent of the region's public debt and 18 percent of direct investment. In 1913, only 25 percent of Latin America's trade was with the United States. The United States was intent on supplanting Europe as the dominant economic force in the region. But U.S. economic interests collided with a new sense of nationalism in the region.

The U.S. role in turning the Cuban War of Independence into the Spanish–American War made many Latin American elites see the Colossus of the North in a different light. In 1900, José Enrique Rodó of Uruguay published

The sinking of the USS Maine was the ostensible reason for the United States to intervene in Cuba's war of independence from Spain. (American Photo Co/Library of Congress Prints and Photographs Division [PAN FOR GEOG - Cuba no. 2 (F size)])

Ariel, an essay in which, after Shakespeare's *The Tempest*, he characterized the United States as Caliban, the evil spirit of materialism and positivism, in contrast to Latin America's Ariel, the lover of beauty and truth. He cautioned Latin Americans against mimicking the efficient but soulless United States and to appreciate the moral and spiritual superiority of Latin America. The United States was similarly criticized by Rubén Darío, the Nicaraguan poet and father of literary modernism. In his poem "To Roosevelt," published in 1904, he warned, "O men with Saxon eyes and barbarous souls,/our America lives. And dreams. And loves./And it is the daughter of the Sun. Be careful./ Long live Spanish America!"

It was not just the literary figures who took aim at the United States. Diplomats tried to restrict the actions of U.S. and European actors, especially in 1902, when war threatened over British, German, and Italian attempts to collect claims against Venezuela. Luis M. Drago, Argentina's Minister to Washington, proposed a policy that prohibited use of force to collect foreign debts. Drago contended, "all states, whatever be the force at their disposal, are entities in law, perfectly equal to one another, and mutually entitled by virtue thereof to the same consideration and respect."

President Theodore Roosevelt (1901–1909) took a different view. In his 1904 message to Congress, Roosevelt recognized the right of a nation to use force to collect debts, but the nation doing so would be the United States. In what became known as the Roosevelt Corollary to the Monroe Doctrine, he asserted: "Chronic wrongdoing, or an impotence which results in a general loosening of the ties of civilized society, may in America, as elsewhere, ultimately require intervention by some civilized nation, and in the Western Hemisphere the adherence of the United States to the Monroe Doctrine may force the United States, however reluctantly, in flagrant cases of such wrongdoing or impotence, to the exercise of an international police power."

The exercise of police power, as well as commercial concerns, would be served by the long-sought interoceanic canal, a high priority for Roosevelt. The first step was to abrogate the old Clayton–Bulwer Treaty. London agreed in the Hay–Pauncefote Treaty in 1901 to permit the United States to build, operate, and fortify a canal across the isthmus. U.S. officials were torn between Nicaragua and Panama as potential canal sites. But Nicaraguan President José Santos Zelaya was not amenable to some of the U.S. demands, especially regarding sovereignty over a canal zone.

Washington then proceeded to negotiate with Colombia for rights across Panama, but the Senate in Bogotá balked at the terms suggested. At that point, Panamanians seceded from Colombia and declared their independence on November 3, 1903. Panamanians, far from Bogotá, had long sought independence; but by rebelling at this time, they were guaranteed to have powerful outside support. They were helped by the presence of two U.S. gunboats, the Nashville and the Dixie, which prevented Colombian troops from countering the bloodless Panamanian insurrection. The commander of the Nashville had orders to take over the railroad, preventing the Colombian troops from traveling from Colón to Panama City. Within a week of Panama's declaration of independence, the United States also sent the Atlanta, Maine, Mayflower, and Prairie to Colón and the Boston, Marblehead, Concord, and Wyoming to Panama City.

Panamanians found their new sovereignty heavily compromised by the treaty signed fifteen days later by U.S. Secretary of State John Hay and Frenchman Philippe Bunau-Varilla, an international adventurer who purported to represent Panama's interests. The treaty he signed in Panama's name granted the United States "in perpetuity" control of a ten-mile strip across the isthmus with power and jurisdiction "as if it were sovereign." It was negotiated without consulting the Panamanians. Work on the canal began in 1904 and terminated a decade later. Controversy over the canal and the treaty raged until the canal was finally handed back to the Panamanians on December 31, 1999. Roosevelt, however, was undisturbed by the controversy. "There was much accusation about my having acted in an 'unconstitutional' manner," Roosevelt said. "I took the isthmus, started the canal, and then left Congress—not to debate the canal, but to debate me. . . . While the debate goes on, the canal does too; and they are welcome to debate me as long as they wish, provided that we can go on with the canal."

Roosevelt insisted that the United States had brought peace to a region that suffered from endemic unrest. His attitude was typical in the United States. In his 1908 study of U.S.–Latin American relations, George W. Crichfield spoke pompously of the duty to impose "civilization" on Latin America: "The United States is in honor bound to maintain law and order in South America, and we may just as well take complete control of several of the countries, and establish decent governments while we are about it." More than half the nations, he huffed, had "sinned away their day of grace. They are semibarbarous centers of rapine. . . . They are a reproach to the civilization of the twentieth century."

Diplomacy dictated that such ideas be expressed more subtly, but there can be no doubt that the same sentiments governed Washington's twentieth-century behavior in Latin America in the twin pursuits of trade and security. Theoretically, the United States was ready to put aside the gunboat diplomacy of the turn of the century for the more genteel "dollar diplomacy." Roosevelt started the United States on this path with his settlement of a crisis in the Dominican Republic in 1904–1905 when the country was threatened with invasion by European creditors. The solution was a customs receivership, headed by the United States, which controlled Dominican customs receipts and distributed 55 percent of the revenues among the creditors.

Dollar diplomacy was more fully articulated by Roosevelt's successor, William Howard Taft (1909–1913). In his 1912 message to Congress, Taft said, "This policy has been characterized as substituting dollars for bullets. It is one

President Theodore Roosevelt posed at the controls of a steam shovel at the Culebra Cut of the Panama Canal in 1906, three years after helping Panama secede from Colombia, which had turned down the U.S. offer for a canal route. (Library of Congress)

that appeals alike to idealistic humanitarian sentiments, to the dictates of sound policy and strategy, and to legitimate commercial aims. It is an effort frankly directed to the increase of American trade upon the axiomatic principle that the government of the United States shall extend all proper support to every legitimate and beneficial American enterprise abroad." But Taft, who was influenced by his experiences as the first civil governor (1901–1904) of the U.S.-controlled Philippine Islands and provisional governor of Cuba (1906), did not hesitate to send troops to protect those dollars.

The United States intervened militarily in Latin America twenty-three times between 1890 and 1913. While U.S. troops were sent to Argentina in 1890 and Chile in 1891, the countries that bore the brunt of U.S. domination were in the circum-Caribbean, an area the United States came to see as its backyard. Troops went to Nicaragua in 1894, 1896, 1898, 1899, 1907, 1910, and 1912, which was the beginning of a twenty-year occupation. Neighboring Honduras also had its share of interventions: 1903, 1907, 1911, and 1912. Haiti, Panama, Puerto Rico, the Dominican Republic, and Mexico all felt the U.S. wrath. Although Woodrow Wilson (1913–1921) claimed to believe in nonintervention, the policy apparently

was not intended for Latin America. During the Wilson years, the United States intervened in Mexico, Haiti, Cuba, and Panama. In fact, the United States did not substantially change its policy of Latin American intervention until the Good Neighbor Policy in the 1930s.

The rationale for U.S. interventions varied. In the Dominican Republic and Haiti, the Marines landed ostensibly to forestall threatened European intervention to collect debts; in Nicaragua, the country's alleged chaotic finances partially explained the U.S. presence, but probably more significant was the rumor that the Nicaraguan government might sell exclusive canal rights through its territory to either Japan or Great Britain. Threats, real or imagined, against U.S. citizens or property occasioned repeated U.S. interventions.

It was no wonder that Mexico's Porfirio Díaz supposedly lamented, "Poor Mexico! So far from God, so close to the United States."

U.S. Relations with Latin America

1823	Monroe Doctrine
1835–1845	Anglo-American settlers in Texas revolt against Mexico, establish independent nation, and join United States.
1845	John L. O'Sullivan, editor of *Democratic Review*, coins phrase "Manifest Destiny."
1846–1848	War of North American Invasion (Mexican–American War).
1848	Treaty of Guadalupe Hidalgo cedes northern half of Mexico to United States.
1850	Clayton–Bulwer Treaty limits power of United States and Great Britain in Central America.
1853	Gadsden Purchase gives United States Arizona and New Mexico.
1854	Ostend Manifesto recommends United States take Cuba from Spain by force.
1855–1857	U.S. filibuster William Walker and mercenaries help Nicaraguan Liberals defeat Conservative rivals. Walker becomes president, with U.S. recognition, before being expelled by Central American forces.
1889	First Inter-American Conference held in Washington, D.C.
1895	United States forces Great Britain into arbitration over Venezuela boundary.
1898–1902	United States invades and occupies Cuba.
1899	United Fruit Company (UFCO) formed, buys seven independent companies in Honduras.
1901	Great Britain cedes Central American canal rights to United States in Hay-Pauncefote Treaty. United States forces Cubans to adopt Platt Amendment.
1903	Theodore Roosevelt sends gunboats to Panama to aid independence movement. United States is rewarded with sovereignty in Panama Canal Zone.
1904	Theodore Roosevelt issues his Corollary to the Monroe Doctrine. U.S. forces place Dominican Republic under a customs receivership.

1905	U.S. Marines land in Honduras.
1906–1909	U.S. forces occupy Cuba.
1909–1913	William Howard Taft promotes "Dollar Diplomacy."
1909–1910	United States helps Nicaraguan Conservatives overthrow José Santos Zelaya.
1912–1925	U.S. Marines intervene in Nicaragua.
1912	U.S. Army forces land in Cuba, Panama, and Honduras.
1914	Panama Canal opens.
1914	U.S. Navy fights rebels in Dominican Republic. U.S. Army invades Haiti. U.S. forces shell and then occupy Vera Cruz, Mexico.
1915–1934	U.S. Marines intervene in Haiti.
1916	Pancho Villa raids Columbus, New Mexico. U.S. forces invade Dominican Republic.
1916–1917	U.S. Expeditionary Force under Gen. John J. "Black Jack" Pershing unsuccessfully pursues Pancho Villa in northern Mexico.
1916–1924	U.S. Marines occupy Dominican Republic.
1917–1922	U.S. forces invade and occupy Cuba.
1918	U.S. Army lands in Panama to protect United Fruit plantations.
1918–1919	United States intervenes in Mexico.
1919	U.S. Marines land in Honduras during presidential campaign.
1920–1921	U.S. troops support a coup in Guatemala.
1921	United States intervenes in Costa Rica and Panama.
1924	U.S. Army intervenes in Honduras during elections.
1925	U.S. Army lands in Panama during a general strike.
1926–1933	U.S. Marines occupy Nicaragua and fight nationalistic forces of Augusto César Sandino.

6.2: The New Middle Class

Latin American elites were challenged from within their countries as well as by foreign powers. In the cities there emerged a small middle class, which grew in size and influence during the twentieth century. Members of the liberal professions, schoolteachers and professors, bureaucrats, military officers, businessmen, merchants, and those involved in the nascent industrialization composed the ranks of that group. The common denominator of the middle sectors rested on the fact that they were neither admitted to the ranks of the traditional elite nor associated with the lower and poorer ranks of society. The observant James

Bryce noted during his tour of South America at the end of the first decade of the twentieth century, "In the cities there exists, between the wealthy and the workingmen, a considerable body of professional men, shopkeepers, and clerks, who are rather less of a defined middle class than they might be in European countries." They possessed a strong urge to improve their lot and tended to imitate, as far as it was possible, the elite.

Initially, the heirs of the creoles and mazombos tended to predominate in the middle sectors, but increasing numbers of mulattos and mestizos entered their ranks as well. Education and military service provided two of the surest paths of upward mobility, but initially the climb was too steep for any but the exceptional or the favored. In many Latin American nations, mestizos or mulattos formed the largest part of the population. Mexico, Guatemala, Ecuador, Peru, Bolivia, and Paraguay, for example, had large mestizo populations, whereas the Dominican Republic, Venezuela, and Brazil had large mulatto populations. Representatives of the mestizos and mulattos entered the middle sectors in large numbers during the last decades of the nineteenth century and claimed their right to play a political and economic role in their nations' destinies. In some cases, the traditional social elites accommodated their ambitions; in others, their frustrations mounted as they were excluded from positions of control, prestige, or wealth.

Although few in number, the dominant presence of the middle sectors in the capital city of each nation allowed them to wield influence far out of proportion to their size. A high percentage of the intellectuals, authors, teachers, and journalists came from their ranks, and they had a powerful voice in expressing what passed for public opinion in the late nineteenth century. By the end of the century, they increasingly exerted influence on the course of events in some nations, particularly in Argentina, Brazil, Chile, Mexico, Uruguay, and Costa Rica. Only later were they large and articulate enough to wield a similar influence in other countries.

Only an educated guess permits some approximation of the size of the middle sectors. At the end of the nineteenth century, it is estimated that Mexico had an urban middle group numbering roughly three-quarters of a million, while another quarter of a million constituted a rural middle group. In contrast, there was an urban proletariat of more than one-third of a million and a huge peon class of 8 million working on the haciendas. In Mexico, Chile, Brazil, Argentina, and Uruguay, the middle sectors may have included as many as 10 percent of the population by the turn of the century. In many of the other countries, it fell far short of that.

The swelling tide of foreign immigration contributed to the growth of the middle sectors. Many of the new arrivals were from the lower class, but still a high percentage represented Europe's middle class, and there was a high incidence of upward mobility among the immigrants in the lands of their adoption. In Argentina in 1914, immigrants held 46 percent of the jobs associated with the middle sectors. Chile received only 100,000 European immigrants before

Thousands of immigrants flocked to Argentina, as illustrated by this crowd of men in the dining room of an immigrant hotel in Buenos Aires, photographed between 1890 and 1923. (Library of Congress)

World War I. They constituted at that time only 4 percent of the population, yet owned 32 percent of Chile's commercial establishments and 49 percent of the industries.

Certain characteristics of the middle sectors increasingly became evident. The majority lived in the cities and boasted an above-average education. Their income level placed them between the wealthy few and the impoverished many. Although the heterogeneous middle sector never unified, on occasion a majority might agree on specific goals, such as improved or expanded education, further industrialization, or more rapid modernization, and on certain methods to achieve them, such as the formation of political parties or the exaltation of nationalism. They consented to the use of the government to foment change, and with minimal dissension they welcomed the government's participation and even direction of the economy. Still, political preferences within their ranks varied from far right to far left.

Although the middle sectors expressed strong nationalistic sentiments, they also looked abroad for models, as did the elites. This contradiction was expressed most emphatically in their ambivalence toward the United States: on the one hand, a model of progress, and on the other, a too-powerful neighbor with aggressive tendencies. Their nationalism prompted frequent outcries against "Yankee imperialism." Yet the middle sectors regarded the United States as the example of a New World nation that had "succeeded," an example of the

"progress" preached but not always practiced in aristocratic Europe. True, the United States embraced impressive examples of poverty as well as wealth, but it seemed that large numbers, even a majority, lived somewhere between those extremes so characteristic of Latin American society. The aggressive strength of their counterparts in the United States inspired the Latin American middle sector.

They attributed part of the apparent success of the United States to industrialization and education, and so they prescribed industrialization as a panacea for their national ills. The high North American literacy rate seemed to provide the proper preparation for an industrial society, and the middle sectors appreciated the mobility that education afforded U.S. citizens. U.S. educator Horace Mann became a revered figure to many Latin American leaders, and they eagerly imported not only his doctrines but also Yankee books and schoolteachers to go with them. Domingo F. Sarmiento met Mann, imbibed his ideas, and as president of Argentina (1868–1874) hired New England teachers to direct the new normal schools he established. President Justo Rufino Barrios in Guatemala (1873–1885) encouraged North American missionaries to set up Protestant schools. Finally, the comfortable life style of the U.S. middle class, with its

Women's Suffrage in Latin America

Country	Year
Ecuador	1929
Brazil	1932
Uruguay	1932
Cuba	1934
El Salvador	1939
Dominican Republic	1942
Panama	1945
Guatemala	1945
Costa Rica	1945
Venezuela	1947
Argentina	1947
Chile	1949
Haiti	1950
Bolivia	1952
Mexico	1953
Honduras	1955
Nicaragua	1955
Peru	1955
Colombia	1957
Paraguay	1961

increasing arsenal of consumer goods, impressed Latin American aspirants to middle-class status.

In the early twentieth century, middle-class women began to organize to demand suffrage. As in the United States, women in Cuba and Brazil were often drawn into the feminist movement through their roles in the struggle for abolition of slavery. Middle-class reformers who sought suffrage had no radical reforms in mind. In fact, many did not expect women to hold public office and merely wanted the right to vote for male candidates. Their position was often a moral one: Women could bring their superior morality to bear on the political world through voting, without actually having to participate in the rough and tumble of political contests. Not all feminists agreed with the goal of suffrage. Some argued that it gave too much credence to an ineffective political system in which votes were generally meaningless anyway. Many were more concerned about material needs.

The middle sectors favored reform over revolution. They sought entrance into the national institutions, not necessarily destruction of them. In fact, they demonstrated a preference for economic improvement and less concern with altering political structures. Although elites at first distrusted the middle sectors, they eventually understood their potential as allies and not only incorporated them into their privileged institutions but also in due course let them administer them—a trust that the middle class did not betray.

6.3: The Working Class

The elites were less sure of the other significant element of the urban population: the working class. As urbanization and industrialization expanded, a larger, more cohesive, and militant proletariat appeared. Slowly becoming aware of their common problems and goals, these workers unionized despite relentless government opposition. The first unions appeared after 1850, evolving from mutual aid societies. They tended to be small, ephemeral, and local organizations.

Typographers, stevedores, railroad employees, artisans, miners, and textile workers were the first to organize, and most of the early union activity concentrated in Buenos Aires, Montevideo, Havana, Santiago-Valparaíso, and Mexico City as well as in the mining regions of northern Chile and central Mexico. By 1914, about half a dozen nations boasted well-organized unions, and at least some attempt had been made in the rest to institute them.

In Chile, union organizing was initiated by maritime, port, and rail workers, who quickly found support among the nitrate miners. Starting in 1900, the groups came together and formed the *mancomunal*, a regional organization designed to bring together skilled and unskilled workers. The mancomunal functioned as both mutual aid society and defender of the working class. Soon there was a mancomunal in every major port town and in the southern coal-mining region. In 1904, fifteen mancomunales with 20,000 members gathered for their first national convention. The newspaper *El Proletario* reported, "It is not just one mancomunal that is on strong footing today, but all of them; from Iquique to Valdivia one sees incredible movement of workers, a swelling of the ranks, a marvelous enthusiasm."

Labor conditions were generally poor throughout Latin America, but the situation of Chile's nitrate workers was among the worst. The deposits of sodium nitrate were located three to ten feet below the surface of the hot and hostile Atacama Desert. A worker called a *barretero* would locate the deposits, excavate holes, and place explosives in them. After the blast—which could well injure or kill the barretero—the other miners would break up the nitrate, load it into wheel barrows and take them to the mills, where the ore was crushed, dissolved in water, and then dried. Heat, lack of water, and poisonous fumes were a routine part of the job.

In 1907, several nitrate companies denied stevedores and boatmen the right, won by railway workers, to have their salaries paid according to a stable exchange rate. They also wanted raises in salary, which barely sufficed to buy food. When the companies ignored them, they asked to be transported back to their hometowns in the south. When that demand was also denied, the workers organized a strike, which was quickly joined by the miners. They presented the nitrate companies with petitions, which were not accepted.

Within a few days, some 8,000–10,000 workers gathered in the town of Iquique, where they set up a makeshift camp in the Plaza Manuel Montt and nearby Santa María School. Along with the workers were women and children. As more workers marched toward Iquique, the worried government sent two regiments of troops to back up the two based in the area. While strike leaders met with government officials, six people were killed and others injured when they were shot to prevent them from joining the strike. Immediately after the funerals, civil liberties were suspended, and the strikers were ordered to abandon the school and the plaza. When the strikers did not comply, the officials opened fire on the school; sources have estimated the number killed as anywhere from 130 to 1,000. The massacre had a chilling effect on the labor movement for several years, but poor conditions inevitably led to renewed organizing.

In Argentina, the elites faced a labor shortage as they tried to industrialize. Despite repeated attempts, there was no way to induce the gauchos of the pampas to accept work discipline. In search of an industrial labor force—and in the hopes that Europeans would help whiten the population—the government encouraged foreign immigration. Officials hoped to attract northern Europeans, particularly Germans, who they saw as very industrious. To their chagrin, the call was answered by Italians and Spaniards, and they came in droves. In 1914, at least 50 percent of the Buenos Aires population was foreign-born, compared with 30 percent in New York at the same time.

By the turn of the century, 60 percent of the Buenos Aires population consisted of manual workers, many of them foreign-born. By 1914, immigrants constituted 60 percent of the urban proletariat. The Italian and Spanish immigrants brought with them ideas about socialism, anarchism, and anarcho-syndicalism. In 1895, the Socialist Party was formed, and the 1910 centennial celebration was marred by anarchist demonstrations and government repression.

Brazil also welcomed large numbers of Europeans, particularly Italians, Portuguese, and Spaniards. Between 1891 and 1900, approximately 112,500 immigrants arrived annually. The trend continued and reached record yearly averages just before World War I. From 1911 through 1913, half a million immigrants entered. However, the proportion of immigrants to the total population in Brazil never surpassed 6.4 percent, a figure reached in 1900. Nonetheless, because of their concentration in the south and the southeast, and particularly because of their importance in the cities of these two regions, they exerted an influence far greater than their numbers might indicate. The traditional elite soon grew wary of the immigrants and blamed them for many of the ills the burgeoning urban centers began to experience.

The labor agenda was increasingly important for women as they joined the workforce in ever-greater numbers. In Colombia in 1870, some 70 percent of the artisans were women. In Mexico City in 1895, more than 275,000 women worked as domestic servants. Women led strikes in the tobacco and textile industries in Mexico as early as 1880, organized in such groups as the *Hijas de Anahuac* (Daughters of Anahuac). In Chile, working women were organizing around workplace issues long before middle-class women began to organize as feminists.

But women workers were not always welcomed as allies by men. Some anarchists urged women to stay home because they feared that by enlarging the labor pool, women depressed wages—especially because employers paid women less than men. However, women were usually relegated to the least technical and mechanized positions; in the Argentine textile industry of the turn of the century, they often still worked at home as weavers and seamstresses. As the industry slowly expanded, women's positions were limited because social reformers pushed for protective legislation. As late as 1914, as many women industrial workers were at work in their homes as in the factories.

Women in the cities who could not find work in industry found few options: domestic service, laundering, ironing, and prostitution. In Guatemala City starting in the 1880s, prostitution was legalized and state operated. Women over fifteen years old who were found guilty of "bad conduct" could be sentenced to a house of prostitution. Both those forced into the brothels and those who went voluntarily signed contracts that amounted to debt servitude. Such bordellos existed until 1920. While Guatemala may have been alone in its sentencing of women to brothels, most Latin American countries tried to regulate the occupation and often forced women into other kinds of employment. For example, in Mexico vagrancy laws were used to force women to work in textile factories and bakeries, in which women were virtual prisoners.

In Argentina, poor women were often placed into domestic service under the theory that they would be protected by the family. Domestic service, however, has often offered the worst possible conditions of employment. Live-in servants were often given miserable lodging and food, expected to work day and night, and subjected to sexual abuse by men in the home. If they became pregnant, they were usually fired. Complaints were dismissed as signs of ungratefulness from "girls" who were supposedly regarded as members of the family.

When working women became politically active, they were more likely to join the socialist and anarchist movements than the new middle-class parties being formed. The Argentine Socialist Party supported women's rights from its founding in 1896. Alfredo Palacios, the first Socialist elected to the Chamber of Deputies (1804–1808, 1912–1915), gave testimony before the chamber showing statistics on the deplorable conditions of women workers. He was a supporter of the Feminist Center, which demanded shorter working hours, daycare centers, and safer equipment, measures that were not enacted.

In the face of such worker unrest and anticapitalist sentiment, elites found the middle class to be far less threatening.

6.4: The Middle Class in Politics

As the heterogeneous middle class emerged, they sought admission to the power structure to have a political voice in the administration of the State. Their role was most dramatically demonstrated in Brazil, Uruguay, and Argentina.

In Brazil, the middle sectors had become strong enough by the 1880s to be a key ally of the disgruntled military who overthrew the monarchy in 1889. The composition of Brazilian society had altered considerably in the nineteenth century. At the time of independence the new empire counted barely 4 million inhabitants, probably half of whom were slaves of African birth or descent. Sixty-five years later, there were 14 million Brazilians, roughly 600,000 of them slaves. At the other end of the social scale were 300,000 plantation owners and their families. The majority were impoverished, illiterate rural folk who fell somewhere between the two extremes. But there was an important, growing body of urban dwellers, many of whom qualified for the ranks of the middle class.

The gulf between the countryside, with its many vestiges of the colonial past, and the city, with its increasingly progressive outlook, widened during the last decades of the nineteenth century. Urban dwellers were less favorably disposed to two basic institutions inherited from the past, slavery and monarchy. They viewed those institutions as the means by which the traditional rural elite retained most of what was colonial in Brazilian society and economy while rejecting colonial status in the stricter legal sense. The military, hostile to slavery, ignored by the emperor, and restless, shared the view of the urban middle sectors, to whom officers were closely related by both family ties and philosophy. Together, they brought an end to slavery and monarchy.

Not surprisingly the new republican government established by the military in 1889 reflected the aspirations of Brazil's middle sector. The new Chief of

State, Deodoro da Fonseca, was the son of an army officer of modest means, and his cabinet consisted of two other military officers, an engineer, and four lawyers. They were sons of the city with university degrees, a contrast to the aristocratic scions who had formed previous governments. The new leaders hoped to transform the nation through industrialization. The government raised the tariff on items that competed with national goods and lowered the duty on primary goods used in national manufacturing. To augment the number of technicians, four new engineering schools were opened in the 1890s. A high income from coffee exports, generous credit from the banks, and the government's issue of larger amounts of currency animated economic activity to a fever pitch. Speculation became the order of the day. Bogus companies abounded, and unfortunately for Brazil, the speculation resulted in little real industrial progress. In 1893, a political crisis complicated the economic distress. The navy revolted and the southern state of Rio Grande do Sul rose in rebellion; together they threatened the existence of the republic.

The powerful coffee planters, with their wealth and control of the state governments of São Paulo, Minas Gerais, and Rio de Janeiro, held the balance of power between the government and the rebels. They promised aid to the government in return for a guarantee of an open presidential election in 1894. Both kept their sides of the bargain, and in the elections, the coffee interests pushed their candidate into the presidential palace. The political victory of the coffee interests reflected the predominant role coffee had come to play in the Brazilian economy. Cheap suitable land, high profits, large numbers of immigrant workers, and a rising world demand made coffee a popular and lucrative crop. By the end of the nineteenth century, it composed half of the nation's exports.

The alliance of the coffee planters and the federal government in 1894 superseded all previous political arrangements. Thereafter, the political dominance of the coffee interests characterized the First Republic (1889–1930). The new oligarchy, principally from São Paulo but secondarily from Minas Gerais and Rio de Janeiro, ruled Brazil for its own benefit for thirty-six years. The coffee interests arranged the elections of presidents friendly to their needs and dictated at will the policies of the governments. Sound finances, political stability, and decentralization were the goals pursued by the coffee presidents. The urban middle groups, whose unreliable ally, the military, was torn by disunion and bickering, lost the power they had exercised for so brief and unsettled a period.

In neighboring Uruguay, the middle sectors had their greatest success, providing one of the best examples of peaceful change in Latin America. Independent Uruguay emerged in 1828 as a result of the stalemate between Argentina and Brazil, which had continued the centuries-old Luso–Spanish rivalry over the left bank of the Río de la Plata. Uruguayans divided into two political camps, the Conservatives (Blancos) and the Liberals (Colorados). From independence until 1872 they fought each other almost incessantly for power. When the Liberals won power in 1872, they managed to hang on to it until 1959, despite challenges from the Conservatives and the military. During the last decades of the nineteenth century, relative peace settled over the small republic, by then in the process of an economic metamorphosis. Prosperity helped to pacify the nation. Exports of wool, mutton, hides, and beef rose. New methods of stock breeding, fencing, the refrigerated ship, and railroad construction (the mileage jumped from 200 to 1,000 miles between 1875 and 1895) modernized the economy.

During the same period, Uruguay constructed the foundation of its enviable educational system. New teacher-training institutes and public schools multiplied. Uruguay was on its way to becoming South America's most literate nation. Expanded and improved education was among the foremost concerns of the middle sectors, and the attention given to education in Uruguay reflected their increasing influence.

The outstanding political representative of the middle class at that time, not only in Uruguay but also in all of Latin America, was José Batlle. He first exerted influence as the articulate editor of a prominent newspaper in Montevideo that spoke for the interests of the middle sectors and, by providing them with a voice, helped to organize that always amorphous group. By the end of the nineteenth century, he led the Colorado Party. He served twice as president (1903–1907, 1911–1915), and his influence over the government lasted until his death in 1929. During those decades, he sought to expand education, restrict foreign control, enact a broad welfare program, and unify the republic.

At the turn of the century, the Conservatives controlled some of the departments (local territorial units) to the extent that they were virtually free of the control of the central government. Batlle extended the power of his government over them by assuring the Conservatives proportional representation in the central government. He managed to balance the budget, repay foreign creditors, and strengthen the national currency. National banks grew in confidence and were able to lend to Uruguayans so that they no longer had to look abroad for much of their capital. To protect national industry, congress raised the tariffs. The government began to enter business, taking over light, power, insurance, and many other formerly private enterprises. The government entered the meatpacking business to offer competition to foreign companies that had long been engaged in the industry, so vital to a nation dependent on stock raising.

Batlle's policies were aimed at the working class as well as the middle sectors. The enactment of advanced social-welfare legislation guaranteed workers their right to unionize, a minimum wage, an eight-hour day, pensions, accident insurance, and paid holidays. Batlle said the government should play a positive role in improving the living conditions of the less-favored citizens. On one occasion he announced, "There is great injustice in the enormous gap between the rich and the poor. The gap must be narrowed—and it is the duty of the state to attempt that task."

These reforms, like others taking place in Latin America at the time, affected only the urban areas and never extended into the countryside. Batlle never directly challenged the landowners or the rural socioeconomic structures. In fact, he saw no reason to, as he stated in 1910: "There is no pressing agrarian problem requiring the attention of the government. The division of the landed estates will take place in response to natural forces operating in our rural industries." It was a point of view shared by the middle-sector leaders of the period. Thus they permitted the continuation of the oldest and most fundamental land and labor institutions. The neglect of rural reforms reflected the middle sectors' fear of the power of the landowners, their preoccupation with the city, their own intermarriage and connections with the landowning families, and a desire to acquire estates of their own.

The climax of the Batlle reforms came in the new constitution, written in 1917 but promulgated in 1919. It provided a model of the type of government the middle class of the period wanted, which was one, of course, that guaranteed them power. It authorized direct elections, reduced the powers of the president and created a National Council of Administration to share the presidential powers (with the hope of eliminating any future threat of dictatorship), established a bicameral legislature elected by means of a proportional representative system, reduced the military to a minor institution, separated the Church and State, and provided a comprehensive program of social welfare. The creation of a welfare state was designed to provide for working-class needs and convince them that reform, not revolution, could answer their demands. As Batlle wrote in 1918, ". . . the class struggle, the class war that it proclaims, is a monstrous plan that will never give good results."

In Argentina, it was elite fears of the working class that led them to share power with the previously excluded middle class. The political leaders, the Generation of 1880, were members of or allied to the landowning class. They monopolized the instruments of state power—the army and the electoral system—and used fraud when necessary. They controlled the ruling Partido Autonomista Nacional (PAN) and made decisions by *acuerdo*, informal agreement. But by the beginning of the twentieth century, these elites faced challenges from newly prosperous landowners, the old aristocracy from the interior who did not profit from the export boom, and from the rising middle class.

An unsuccessful armed revolt in 1890 challenged the power of the ruling elites and paved the way for recognition in 1892 of a new political party dominated by the middle class, the Radical Civic Union. The new political party was initially frustrated at the polls and continued to attempt armed revolts in 1893 and 1905. But the elites were even more worried about demonstrations led by foreign-born anarchists and socialists. In 1902, the Law of Residence provided for expulsion of immigrants "who compromise national security or disturb public order." The 1910 Law of Social Security, which provided for expulsion of anarchists, coincided with anarchist demonstrations that were planned to mar the centenary of Argentine independence. Worker mobilization convinced conservative elites that their interests would be best served by an alliance with the middle classes against the masses. In 1912, President Roque Saenz Peña supported passage of the Saenz Peña law, giving universal male suffrage to Argentine citizens, with compulsory voting in elections by secret ballot. Excluded from the new law were the many foreign-born workers. The law paved the way for election of the first president from the Radical Civic Union, Hipólito Yrigoyen, in 1916.

Throughout Latin America, elites generally came to the conclusion that the middle class could be an important ally against the interests of the poorer masses. While the middle classes advocated a wide variety of reforms, in practice they proved to be essentially conservative, fearful that too much reform might harm rather than benefit them. After all, the middle class aspired to move up and join the elites, not to slide down into the precarious position of the working class. Much of the middle class provided goods and services to the elites while others in the middle sector managed their workers. If the elites doubted that the admission of the middle class to political power was a wise move, all they had to do was look at the violent results of exclusion in Mexico.

Questions for Discussion

1. What were the reasons for U.S. interest in Latin America?
2. What role was played by the Monroe Doctrine, Manifest Destiny, and the Roosevelt Corollary?
3. How did Latin Americans react to U.S. involvement in the region?
4. What were the economic changes that led to the rise of the middle class and urban labor?
5. What were the political interests of the middle class? Of labor?
6. How did immigration affect the emergence of new middle groups?
7. What ideologies did immigrants bring with them? What impact did they have on Latin America?
8. How did feminist movements emerge? How did they differ by class?

7
The Mexican Explosion

The struggles for political voice and economic opportunity that characterized Latin America at the turn of the century were not resolved at the electoral polls in Mexico. Mexico continued to represent the extremes of Latin American patterns, from its exceptional chaos in the early nineteenth century to the amazing endurance of its Liberal, modernizing caudillo, Porfirio Díaz, at century's end. The dictator refused to open the political system to the new economic elites, and their desperation for a role in the system ignited long-simmering disputes with the masses. The result was the Mexican Revolution, or perhaps more accurately, revolutions (the "long" revolution ran from 1910 to 1940).

By revolution, we mean the sudden, forceful, and violent overturn of a previously relatively stable society and the substitution of other institutions for those discredited. Change by revolution thus denotes sweeping change, the destruction of old social, political, and economic patterns in favor of newer ones. Use of this definition divides genuine revolutions from the innumerable palace coups, military takeovers, civil wars, and the wars of independence, which were nothing more than shifts in the holding of power within the same or similar groups unaccompanied by fundamental economic, social, or political changes. These changes are also accompanied by new ideology and national mythologies reflecting a change in the balance of power.

Octavio Paz wrote one of the most famous and moving descriptions of the Mexican Revolution: "Like our popular fiestas, the Revolution was an excess and a squandering, a going to extremes, an explosion of joy and hopelessness, a shout of orphanhood and jubilation, of suicide and life, all of them mingled together." Historians debate what was achieved in that explosion and whether the results were worth the loss of more than 1 million lives.

7.1: Cracks in the Regime

By 1910, Porfirio Díaz and the "New Creoles" had ruled for thirty-four years, without a popular mandate, for the benefit of privileged native elites and foreign investors. The economy still depended upon foreign whims and direction, a neocolonialism clearly seen in the statistics: 75 percent of all dividend-paying mines in Mexico were owned by U.S. interests. Foreign capital represented 97 percent of investment in mining, 98 percent of rubber, and 90 percent of oil. Some Mexicans resented the high level of foreign investment. Others were concerned that the Porfirian prosperity was narrowly based, relying mostly on mining, utilities, commerce, and large-scale agriculture, with relatively little industry.

The majority of the population did not share the wealth. Real wealth actually declined for the majority: Hacienda peons earned an average daily wage of thirty-five cents, which remained almost steady throughout the nineteenth century, while corn and chile prices more than doubled, and beans cost six times more than they had at the beginning of the century. Life was little better for urban workers, who worked eleven- to twelve-hour days, seven days a week, and lived in squalid housing with one bathhouse per 15,000 residents.

Land, a principal source of wealth, remained in the hands of a few. Foreigners owned between 14 and 20 percent of it. Ninety-five percent of the rural population owned none. Not even 10 percent of the indigenous communities held land. Fewer than 1,000 families owned most of Mexico. In fact, fewer than 200 families owned one-quarter of the land. Private estates reached princely proportions. The de la Garza hacienda in the state of Coahuila totaled 11,115,000 acres; the Huller estate in Baja, California, sprawled over 13,325,650 acres. Productivity was low, and absentee landlords were common. The fact that a majority of Mexicans lived in the countryside and worked in agriculture made the inequity of land distribution all the more unjust.

At the same time, the growing mestizo urban classes were dissatisfied with the inequitable institutions inherited from the past. The mestizos had grown rapidly in number. By the end of the nineteenth century, they surpassed the indigenous in number and overshadowed the tiny "creole" class. It was obvious from their size, skill, and ambitions that the mestizos held the key to Mexico's future. The urban mestizo working and middle sectors voiced discontent with their inferior and static position in Porfirian Mexico.

As early as 1900, intense criticism of the Porfiriato was launched by the anarchists Jesus, Enrique, and Ricardo Flores Magón in their weekly newspaper, *Regeneración*. The brothers were jailed in 1901 for criticizing the political chief of Oaxaca. From their subsequent exile in St. Louis, Missouri, they issued the Liberal Plan, calling for an eight-hour workday and a six-day week; abolition of the notorious plantation store (*tienda de raya*); abolition of payment in scrip, which could only be used at the tienda de raya; restoration of the ejidos; and redistribution of uncultivated land.

However, it was actions within Mexico that revealed the intensifying contradictions of the Porfiriato. In 1906, workers went on strike at Colonel William Greene's Cananea Consolidated Copper Company, one of the ten largest mines in the world. A chief complaint was that Mexicans were paid less than U.S. workers, who held all the technical and managerial posts. Furthermore, the foreigners were paid in gold dollars, whereas Mexicans received silver pesos, worth far less.

When 3,000 Mexican workers went on strike, Greene refused to negotiate. When unarmed workers tried to force their way through a locked gate at the company lumberyard, high-pressure water hoses were turned on them. The gate collapsed, and workers streamed into the yard. Company guards fired into the crowd, killing dozens of workers. The chaos spilled into town, where Greene's guards shot indiscriminately into crowds. When Díaz could not get his rurales to the scene quickly enough, Greene turned to the United States, and 275 Arizona Rangers were brought in, a direct challenge to Mexican sovereignty that angered Mexican elites.

Six months later, new violence broke out at the Rio Blanco textile mill in Orizaba, Veracruz, where poorly paid workers put in twelve-hour days, and the workforce included children as young as eight years old. The 1907 conflict began when workers' wives were denied credit for food at the company store. What began with pushing ended with shots, as rurales fired into the crowd at point-blank range, killing women, children, and workers. When survivors returned later to claim the bodies, they were again assaulted.

The initial target of the revolution was Porfirio Díaz, who by the time of his overthrow in 1911 had ruled for thirty-six years. (Library of Congress)

Despite these harbingers of problems, it was not the oppressed workers who led the uprising. It was the upper classes, who had their own complaints about the regime. The theoretical justification for the Porfiriato was "Order & Progress"—but order had become rigidity, and progress had slowed. These dissatisfied elites wanted a share of political power, and they questioned the dictatorship. Northern elites in particular chafed under Mexico City's control, made possible by the telegraph, the railroad, and Díaz's grip on centralized power. Furthermore, his support of foreign commercial interests brought unwanted competition to the north, hurting their economic fortunes. It is not surprising, then, that it was a northern son who led the uprising.

7.2: Effective Suffrage and No Reelection

Francisco I. Madero was the son of a wealthy rancher from the border state of Coahuila, Evaristo Madero, who had supported Díaz in the early days of his leadership. Francisco was educated at the University of California, Berkeley, and the Sorbonne. Among the family's vast holdings was the Compañía de Tierras de Sonora, comprising 1,450,000 acres of land, as well as iron and coal mines.

Despite their wealth, northern elites such as the Maderos were too far from Mexico City to compete with foreign interests that gathered around Díaz. His concessions to foreign companies translated into direct losses for Mexican businesses. For example, the Maderos were among cotton planters mired for years in lawsuits to reduce the unlimited water rights of the British-owned Tlahualilo Company. The Maderos were also the only serious competitor to the Rockefellers' Continental Rubber Company, which tried to eliminate competition by merging with the U.S. Rubber Company and glutting the market, driving the price of rubber down from $1.00 to 25 cents per pound. In addition, the Maderos were among only a few Mexicans who owned smelters that could compete with foreign mining firms and the only ones to survive after Díaz gave generous concessions to the Guggenheims' American Smelting and Refining Company.

In 1908, the elites were taken by surprise when Díaz announced in an interview with journalist James Creelman in *Pearson's Magazine*, a U.S. publication, that he would not seek reelection in 1910. "I have waited patiently for the day when the people of the Mexican Republic would be prepared to choose and change their government at every election without danger of armed revolutions and without injury to the national credit or interference with national progress. I believe that day has come. . . . No matter what my friends and supporters say, I retire when my present term of office ends, and I shall not serve again."

His opponents immediately began to express their views. In 1909, Andrés Molina Enriquez published *The Great National Problems*. Molina was no revolutionary. In fact, he was a positivist who hoped that reform, particularly agrarian reform, could avert revolution. The same year, Francisco I. Madero published *The Presidential Succession of 1910*, calling for political change. Challenges were felt on the local level as well: In Morelos, Patricio Leyva challenged Díaz candidate Pablo Escandón. Leyva's candidacy drew support from such local community leaders as Emiliano Zapata.

Despite his claims, Díaz chose to run again in 1910. Madero, under the banner of the Antireelectionist Party, ran as well. His platform called for political reform and fair elections. His motto was the same one that Díaz himself had raised in the 1876 Revolution of Tuxtepec against President Sebastián Lerdo de Tejada: Effective suffrage and no reelection. When asked about economic issues, however, Madero replied that the Mexican people wanted liberty, not bread. Madero was more than willing to compromise with Díaz, even offering to run as his vice presidential candidate. But Díaz refused and instead threw Madero in jail the night before the election. To no one's surprise, Díaz declared his victory at the polls—with more than 1 million votes to less than 200 for Madero—and took office for the eighth time.

Madero escaped to the United States. When he saw the popular response that his political opposition to the old dictator had aroused, he chose to launch a revolution. In San Antonio, Texas, he wrote his revolutionary plan and then crossed the border in October 1910 to announce the Plan de San Luis Potosí on Mexican soil. His plan showed the limited goals of his movement: the forced resignation of Díaz and electoral reforms. Repeatedly, his followers voiced the slogan "Effective suffrage and no reelection," a clue to the exclusively political, urban, and middle-sector origin of the revolution.

7.3: Patrias Chicas

Mexico's vast territory was a country of regions, each with its own set of local traditions and problems. The reaction to Madero's call for revolution depended as much on local conditions as on resentment toward the Díaz regime. Indeed, the impact of Díaz policies differed by region. The two main geographic divisions were north and south. The north included the contiguous northwestern states of Chihuahua, Sonora, Sinaloa, Durango, and Coahuila. The south comprised the five adjoining central states of Guerrero, Morelos, Puebla, Tlaxcala, and Veracruz.

In general, the north was characterized by its distance from the political power center of Mexico City. There was no sedentary indigenous population, and settlers on the frontier had been united in their struggle against the Apaches. Because it was a border region, restless workers could cross to the United States in search of better jobs. It was a region where the Catholic Church had less impact, indeed, where Protestants and Mormons from the United States had made inroads. If there was any tradition in the northern states, it was one of independence and mobility. That independence was threatened during the Porfiriato by increasing state intrusion without a corresponding inclusion of local elites in the national power structure. Within this region, states had their own peculiarities. For example, residents of Chihuahua chafed against the monopoly over politics and economics exerted by the Terrazas-Creel family, whereas in Durango, Díaz had alienated the middle class by his generous support to foreign mining companies.

The south and central states, on the other hand, were places where the Catholic Church and the large hacienda had formed the main structures of society. Small villages of campesinos struggled to maintain their farms against the encroachment of modernizing and expanding hacendados, eager for their land and labor. Nowhere was this more apparent than in Morelos, where the sugar plantations threatened the continued existence of traditional haciendas, the small rural settlements known as ranchos, and of entire villages. In contrast, in Veracruz, industrial workers labored in foreign-owned textiles mills under miserable conditions.

In isolated areas of Mexico, the revolution initially had little impact. It was not until 1913 and 1914, well after the Maderista phase had ended, that people in Chiapas and Tabasco were aware of the revolution. Isolation was only one factor keeping Yucatán removed from revolutionary upheaval: There was also the iron-fisted power of the henequen industry. As a result, the revolution did not come to the peninsula until 1915. Oaxaca, on the other hand, was quiet because most of the land was still in traditional indigenous villages, relatively untouched by the capitalist displacement and political intrusion of the Porfiriato.

7.4: The Maderista Revolt

Madero called for Mexicans to rise in revolution on November 20, 1910. The only response to his call came in Chihuahua, where Abraham Gonzalez, head of the state's Antireelectionist Party, called for an uprising. When the conspirators found that their plans for November 20 had been discovered by the authorities, they rose instead on November 14. They were led by Toribio Ortega, a campesino leader who had led his native village in a 1903 attempt to recover lost lands. Ortega began with sixty men, and they were soon joined by the residents of nearby villages who had also struggled over lost lands. In Parral, a mining town, it was a wealthy merchant, Guillermo Baca, who led forty men in an attack on the *jefe* politico and was soon joined by 300 men. The revolt spread quickly throughout Chihuahua's mining towns and old military colonies. While they were responding to Madero's call, the local rebellions were aimed primarily at unpopular local authorities and were fueled by local grievances. When the revolutionaries failed in attacks on local police, they retreated to the mountains to regroup for guerrilla warfare.

There were many local leaders like Ortega and Baca, but two men emerged to coordinate the local uprisings in the north: Pascual Orozco and Pancho Villa. Orozco was a member of the new middle class, the minimally educated son of a store owner. He made his money as an enterprising muleteer leading convoys of precious metals through the mountains. His knowledge of the region and the danger of his occupation made him a natural revolutionary leader. He was at first much better known than Pancho Villa, who entered the revolution as the leader of only twenty-eight men.

Villa was born Doroteo Arango into a family of hacienda peons in the state of Durango. Legend has it that he shot the hacendado, or perhaps his son or an administrator, after the man attacked Villa's sister. He fled to Chihuahua and began a career as a small-time bandit; he was arrested for minor robberies and sent to the army, where he served one year before deserting. He fled to Chihuahua and changed his name to Francisco "Pancho" Villa. Much has been made of Villa's life as a bandit with several gangs, eventually switching from holdups and robberies to cattle rustling. But he also worked at a number of legal jobs, mostly for foreign companies; he was a muleteer for a silver mine and a contractor for a railway line, as well as the organizer of cockfights. In all of these legal activities, he proved to be honest and reliable as well as an effective leader of men. Perhaps it was for those reasons that Abraham Gonzalez recruited Villa to the revolution. As the first to defeat regular government troops, Villa's fame as a revolutionary drew hundreds of men to serve under him.

Revolutionary leader Pancho Villa rides among his troops on a dusty trail in 1916 during raids in the United States and northern Mexico. (Library of Congress)

Initially, there was less action in the south. Would-be revolutionaries, including Zapata, met in Morelos in November, but they took no action until a delegation sent to Madero in mid-December returned in February with formal appointments to leadership positions. On March 11, the leaders marched to the Villa de Ayala and read Madero's Plan de San Luis Potosí, gathered seventy men, and headed for the mountains south of Puebla.

Despite his October call to action, Madero had remained out of the country until February, when he arrived in Chihuahua to assume control. He ordered an attack on Ciudad Juárez, a border city that would have given him control over customs duties. The attack failed, but Madero's presence in Mexico inspired more uprisings. By March 21, the federal army was in retreat, and on April 1, Díaz promised reforms, including land reform and an end to reelection. But the promises came too late and served only to show the revolutionaries the weakness of the regime.

By mid-April, the revolutionary guerrilla bands had become organized armies, and Madero again ordered them to march on Ciudad Juárez. Orozco and Villa each headed a column of 500 riders, and Madero led another 1,500. The Díaz government frantically proposed negotiations, and Madero implemented a ceasefire in order to carry out the talks. On April 22, Madero and government representatives agreed to a treaty that did not demand the resignation of Díaz, though Madero was told confidentially that the president would step down. Villa and other leaders were outraged, leading Madero to change his position and insist that Díaz resign. At that point, negotiations broke down, but Madero continued the ceasefire. He had been convinced by the government that an attack would provoke the Taft administration, which in March had sent 20,000 U.S. troops to the border and ships to patrol Mexico's coastline.

Orozco and Villa chose disobedience at this point and fired on the federal troops. They took the city easily, and on May 21, the Treaty of Ciudad Juárez ended the war. The revolutionaries had no more respect for the treaty than for the ceasefire, for it promised little change. Díaz and his Vice President, Ramón Corral, would resign; Foreign Minister Francisco de León de la Barra would become provisional president until new elections were held in the coming months. Madero would approve the interim cabinet and name fourteen provisional governors, but nothing else would change: judges, mayors, state legislators, and police—all would stay in office. The revolutionary army, however, was required to disband, a sure signal of Madero's limited goals.

On May 26, 1911, Porfirio Díaz boarded the ship *Ypiranga* and sailed to exile in Paris, where he died in 1915. As he left, Díaz commented, "Madero has unleashed a tiger; let us see if he can control him."

Map 7.1 The Overthrow of Díaz

SOURCE: http://users.erols.com/mwhite28/mexico.htm

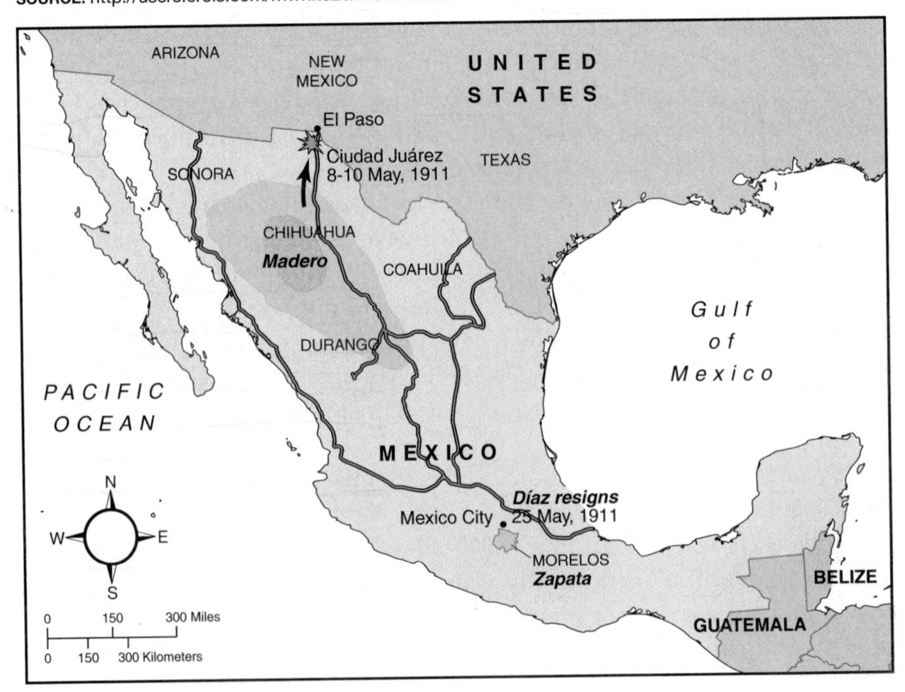

7.5: Madero in Power

On the same day that Díaz left the country, two other key events occurred. A victorious Emiliano Zapata rode into Cuernavaca at the head of 4,000 troops waving the image of the Virgin de Guadalupe. Zapata and his followers had been motivated by the third clause of Madero's Plan de San Luis Potosí, a vague plank calling for illegally obtained land to be returned to the rightful owners. But that same day, Madero issued his first manifesto since the end of the war. He noted, "The aspirations contained in the third clause of the Plan of San Luis Potosí cannot be satisfied in all their amplitude."

In Morelos, land had been the reason for fighting. Zapata had given orders that village lands be restored, and haciendas were occupied during the struggle. The terms of the treaty, however, essentially restored the Díaz regime in Morelos as the governor, state legislature, jefes politicos, and municipal presidents were all reinstated. Once subject to arrest for resisting the revolution, the old leadership now expected the revolutionaries to respect their authority.

On June 7, Madero arrived in Mexico City. Among those greeting him at the train station was Zapata. When they met the next day, Zapata tried to convince Madero of the importance of land reform. Madero dismissively told Zapata that it was a complicated issue, and that it was far more important for Zapata to disband his rebel troops. Stunned, Zapata wondered how Madero could trust the army to be loyal to an unarmed revolutionary government. The unspoken answer was that there was nothing revolutionary about Madero's aims.

Zapata convinced Madero to visit Morelos, but the visit backfired. Madero was influenced by the Mexico City press, which called Zapata "the Attila of the South," and by the Morelos "revolutionary" elites, who insisted that Zapata could not control his "barbaric" troops.

Madero stood by ineffectually as de la Barra, the interim president, ordered troops south to Morelos to disarm the

Emiliano Zapata was the revolutionary leader who embodied the goals of the poor campesinos. He is shown in what forever became the image of the Mexican revolutionary: wearing a large sombrero, chest crossed with bandoleras, and holding a Winchester single-action rifle and a sheathed saber. (Library of Congress)

Zapata forces. Zapata was at his wedding celebration on August 9 when he was informed that more than 1,000 troops were entering the state under the leadership of Brigadier General Victoriano Huerta. By August 29, Zapata had been declared an outlaw. Huerta's forces rampaged through Morelos, and the campesinos became Zapatistas.

During those summer months, there was little peace in Mexico. Despite conciliatory statements, Madero was criticized by both the right and the left. Madero turned his forces on the anarchist *Partido Liberal Mexicano* (PLM, Mexican Liberal Party) in Tijuana and then began to demobilize his troops. In July, miners formed unions and began a series of strikes. In August, right-wing leader Bernardo Reyes began to campaign against Madero, but he fled to the United States in September after he was physically attacked by Maderistas and congress refused to postpone the election.

Madero won a massive victory at the polls in October and took office in November. But he was quite unprepared for the task he faced. His political platform contained some vague planks on political reform and almost nothing solid on social or economic change. He represented the traditional liberalism of the nineteenth century, which conflicted with the newer demands being made. Although he restored some ejidos, villagers bore the burden of proof to reclaim lands. He used troops to break up strikes, and his education budget was a mere 7.8 percent, little more than Díaz's 7.2 percent. Instead of changing the old Porfirian regime, he worked with it to end the social instability that threatened elite wealth. Coahuila Governor Venustiano Carranza complained that Madero was "delivering to the reactionaries a dead revolution which will have to be fought over again." His words were prophetic.

7.6: ¡Viva Zapata!

Emiliano Zapata was not a typical campesino. He had estate land to sharecrop and apparently a personal relationship with hacendado and Díaz's son-in-law Ignacio de la Torre y Mier, who had used his influence to help Zapata leave the army after a scrape with the law. Zapata was a renowned horseman, had some education, and sometimes hired laborers, making him a *patrón* and a member of the rural middle class. He was well respected and had been elected to head his village of Anenecuilco. In that role, he once had armed eighty men to retake the land that the government had allowed big landowners to take from the village.

Zapata went on to earn admiration during his fighting for Madero. But when Madero abandoned the cause of land reform, Zapata abandoned Madero. In his Plan of Ayala, issued in November 1911, Zapata called for the overthrow of Madero and the return of land to the people. While Madero's slogan had been Díaz's "Effective Suffrage and No Re-election," Zapata's was "Land and Liberty!" Campesinos rallied to his cause. A new force had been unleashed, and it represented what distinguished the Mexican Revolution from previous

Map 7.2 The Madero Regime

SOURCE: http://users.erols.com/mwhite28/mexico.htm

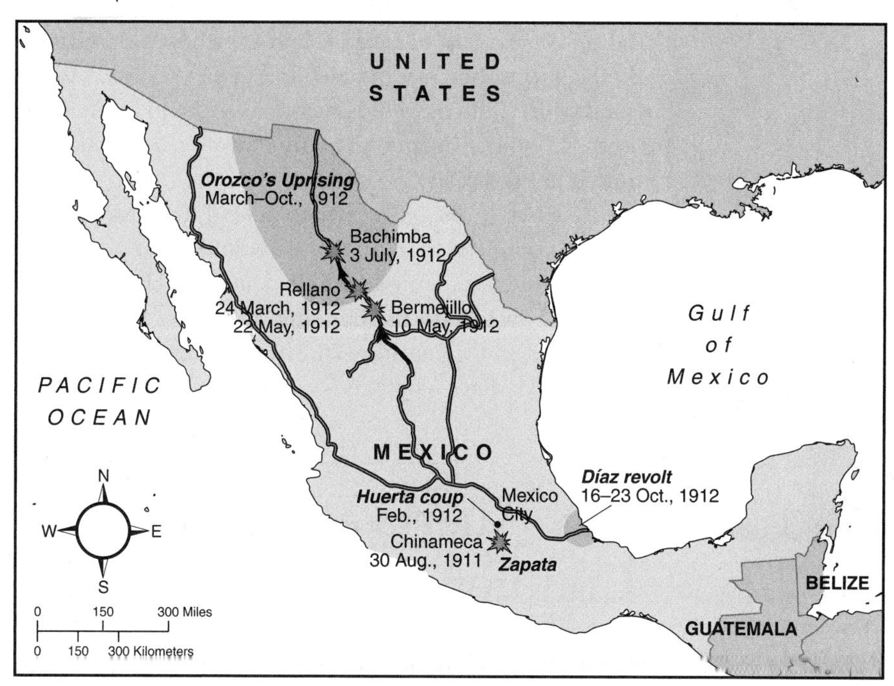

movements in Latin America: the stirring of the masses. It became clear that a social revolution had begun.

By January 1912, the Zapatista example had inspired movements in Tlaxcala, Puebla, Mexico state, Michoacán, Guerrero, and Oaxaca, creating a crisis in the south just as Pascual Orozco was becoming increasingly estranged from Madero in the north. Zapata called on Orozco to lead the new revolution. At first, Orozco remained loyal to Madero, but by March, Orozco, too, had turned on the government, won over by the Terrazas-Creel clan. Despite the support from wealthy elites, his Plan Orozquista called for the end of the ten-hour day and child labor, higher wages, better working conditions, the end of the tienda de raya, and, like Zapata, agrarian reform. Villa, however, remained loyal to Madero, who asked Villa to merge his forces with Huerta's; together they defeated Orozco in May. The Huerta–Villa relationship was, naturally, conflictive, given that Huerta had originally defended the Díaz regime against Villa. On trumped up charges, Huerta had Villa arrested, and he spent seven months in jail before escaping to El Paso, Texas, in December 1912.

As before, the attacks on Madero came from the right as well as the left. Bernardo Reyes had returned to Mexico and was imprisoned after leading an unsuccessful revolt against Madero. In prison, he and Felix Díaz, Porfirio's

The initial revolution was led by Francisco I. Madero, a northern rancher who wanted political democracy but worried less about economic change. (Library of Congress)

nephew, bribed a general to release them and launched another coup in February 1913. In what became known as the *Decena Tragica*, the Tragic Ten Days, Mexico City became a maze of barricades and trenches. Cars burned in the streets, horses ran wild, and gunfire claimed thousands of civilian lives. Businesses closed and food shortages drove the desperate to live on rats, while bodies rotted in the streets.

Madero relied on Huerta to defend the government. Huerta promised Madero that he would bring peace, but it was a peace brokered by Felix Díaz and Huerta at the U.S. Embassy, with the blessing of Ambassador Henry Lane Wilson. On February 20, Huerta switched sides and arrested Madero, who was soon killed.

7.7: Huerta and the Counterrevolution

Huerta became the symbol of reaction, of the counterrevolution. A native of Jalisco, Huerta was the son of a poor Huichol farmer who worked an unirrigated plot of land. He joined the military in order to receive schooling, and a general took him on as a personal secretary and aide. He was sent on to military college, and his career coincided with that of Porfirio Díaz; both fought the Yaqui and Mayo peoples during their uprisings. His coup was supported by the upper classes, business interests, the Church, and the federal army. Although Orozco initially opposed Huerta, by March he had joined him.

Huerta's coup was deplored by both Villa and Zapata, but neither leader had national standing. Only one Madero government official called for resistance to Huerta, and that was Coahuila governor Venustiano Carranza. Carranza was an unlikely revolutionary. A rich hacendado from an old colonial family, he had held positions in the Díaz regime and been a supporter of Bernardo Reyes. When Díaz blocked Carranza from becoming governor, Carranza turned against the regime; but he only joined Madero after Reyes went into exile. Carranza was even more conservative than Madero on all but one issue: nationalism. As governor of Coahuila, he had supported strikers at foreign-owned companies.

Carranza tried to rally the other northern governors, but he soon found himself alone as the governors were murdered and imprisoned. He then tried to negotiate with Huerta, and when that failed, he took up arms. Villa returned from the United States to organize resistance in Chihuahua, while in Sonora, forces were led by Alvaro Obregón, the owner of a medium-sized ranch who had also worked as a mechanic, schoolteacher, and tenant farmer. Although he was not involved in the Madero revolution, Obregón had fought against the Orozco uprising. In the struggle against Huerta, Obregón quickly emerged as one of the most talented military leaders of the revolution. In March, Carranza issued the Plan of Guadalupe, calling for the overthrow of Huerta and restoration of constitutional government. Carranza declared himself "First Chief of the Constitutionalist Army" and claimed to be Madero's rightful successor. Zapata, meanwhile, continued to fight on his own in the south, trusting neither Huerta nor the Constitutionalists to restore the ejidos.

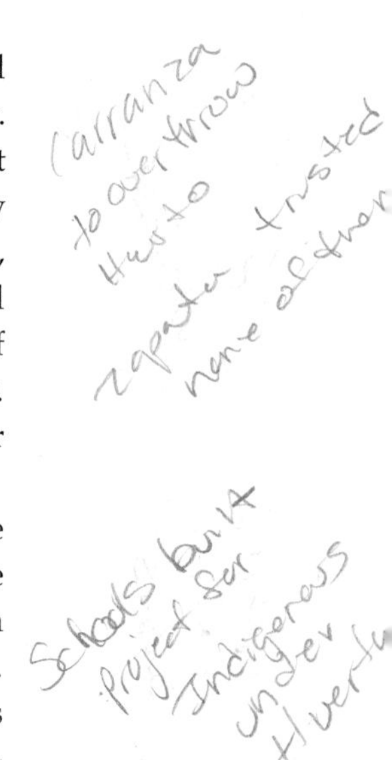

From 1913 to 1914, Mexico devolved into a bloody civil war. At the same time, however, Huerta's government initiated some domestic reforms: The education budget rose to 9.9 percent of the national budget, and Education Minister Nemesio Garcia Naranjo overturned Gabino Barreda's positivism. One hundred thirty-one rural schools were built, and community projects were initiated in many indigenous villages, including the restoration of seventy-eight ejidos to Yaqui and Mayo Indians of Sonora.

However, any achievements were outweighed by his increasingly dictatorial measures. Huerta closed congress, ordered the assassination of his rivals,

Map 7.3 Huerta vs. the Constitutionalists

SOURCE: http://users.erols.com/mwhite28/mexico2.htm

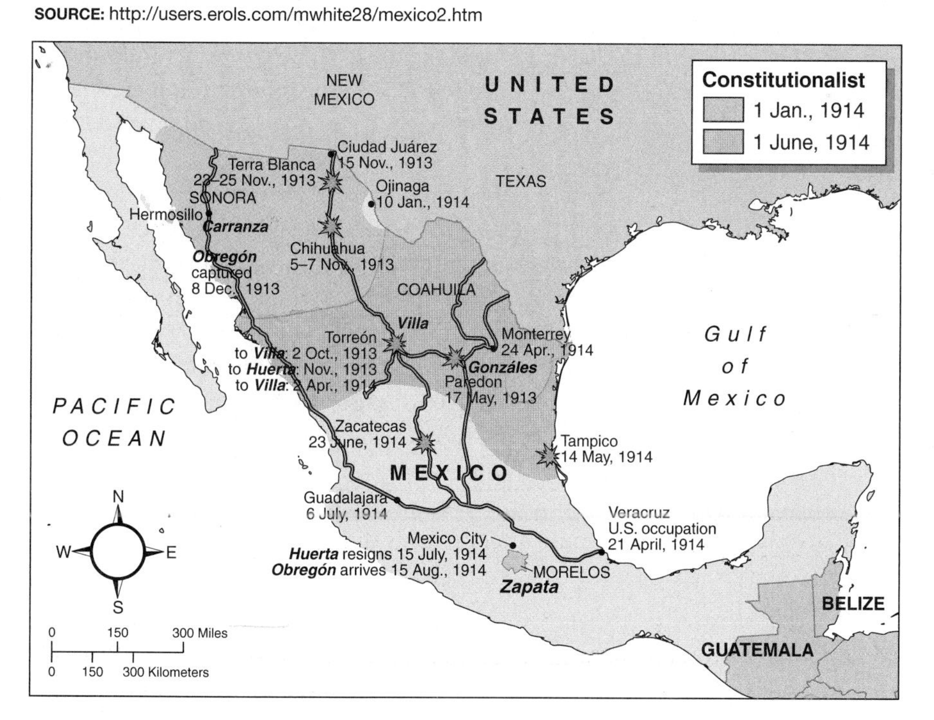

and dragged poor conscripts from the streets in the detested *leva* (military draft). But perhaps his biggest problem was U.S. President Woodrow Wilson's refusal to recognize the Huerta regime, despite the urging of the ambassador. And it was the United States that Huerta eventually blamed for his downfall.

In April 1914, the USS *Dolphin* was stationed off the shores of Tampico. The ship's captain sent a small group of sailors to get fuel, and they wandered into a restricted area. They were immediately arrested, but released within a few hours, along with an official apology. But the United States refused to let the small incident go so easily, instead demanding that the Mexicans hoist the U.S. flag and give a twenty-one-gun salute. The Mexican government reluctantly agreed, but said the U.S. officials should then salute the Mexican flag, which Wilson saw as tantamount to recognition of Huerta's government. Before they could resolve the matter, the United States received word that a German ship was bringing weapons to Huerta. U.S. naval forces were ordered to occupy Veracruz, where they prevented the Germans from landing, controlled the customs house, and killed hundreds of civilians. Villa was not particularly concerned about U.S. occupation, if it would help topple Huerta. Zapata said the U.S. actions made his "blood boil," but so did the idea of uniting with Huerta against the United States. The Constitutionalist forces raged against the insult of U.S. invasion, but obviously, Huerta could not ally with the Constitutionalists. Spontaneous demonstrations erupted in Mexico against U.S. actions, and Huerta concentrated his troops in the east to counter the occupation.

The northern revolutionaries capitalized on the U.S. distraction. In June, in the bloodiest battle of the war against Huerta, Villa captured Zacatecas, a railway junction essential to the route to Mexico City. The battle resulted in the deaths of 6,000 federal troops and 1,000 rebels, and the injured numbered 3,000 federales and 2,000 constitutionalists. On July 15, Huerta resigned; like Díaz, he fled to Europe on the *Ypiranga.*

But Huerta's fall was no more of a unifying force than the fall of Díaz had been. The revolutionaries were able to agree when focused on the overthrow of a particular leader. But their interests conflicted, and there was no peace to be gained. Instead, each revolutionary force tried to be the first to reach Mexico City and take control of the federal government. The northern forces split into factions led by Carranza and Villa, with Obregón eventually swinging his support to Carranza. Villa's and Zapata's forces discussed a possible alliance, while the other revolutionaries tried to keep both of them from the capital.

On August 20, Carranza arrived in Mexico City and declared himself the new chief executive. He called a convention for October 1 and tried to make sure that only Carrancistas would attend. To his dismay, the convention voted to move to Aguascalientes, where Villistas could join them. The delegates then declared their sovereignty and invited Zapata to attend. The convention was split, with the forces of Carranza and Obregón on one side and Villa and Zapata on the other. Carranza and Obregón represented the elites and middle class, those whose interests focused more on political than economic and social change.

Zapata and Villa represented the lower classes, the Mexican masses that hungered for redistribution of land and economic opportunity. The convention sought a compromise in Eulalio Gutierrez, who was installed with the backing of Villa's troops. A furious Carranza fled and established his own government at Veracruz.

As 1914 drew to a close, Zapata and Villa met, trying to ally their more radical Conventionist forces against the elite Constitutionalists. Unfortunately, neither Zapata nor Villa was able to articulate a message that went beyond their regional and social bases. Neither had an overarching ideology, and some would argue that Zapata's concerns were anachronistic, envisioning a return to a past that was long gone.

Villa and Zapata fought on, but by late 1915, the tide had turned; Gutierrez abandoned Mexico City to rule from Nuevo León, and Carranza was able to return to Mexico City and consolidate his position. Zapata was driven out of Mexico City, although his forces kept pressure on the capitol. Villa controlled Chihuahua, but he struggled in the rest of the north against Obregón's forces.

The beginning of the end for Villa came with his defeat in April 1915 at the town of Celaya. Obregón, using the new tactics of the European war, dug trenches, enclosed his defensive positions with barbed wire, and used machine guns to mow down 14,000 of Villa's men. With a subsequent defeat at León, Villismo was destroyed as a national force. Villa's army disintegrated, and even

Map 7.4 Carranza vs. the Conventionists

SOURCE: http://users.erols.com/mwhite28/mexico2.htm

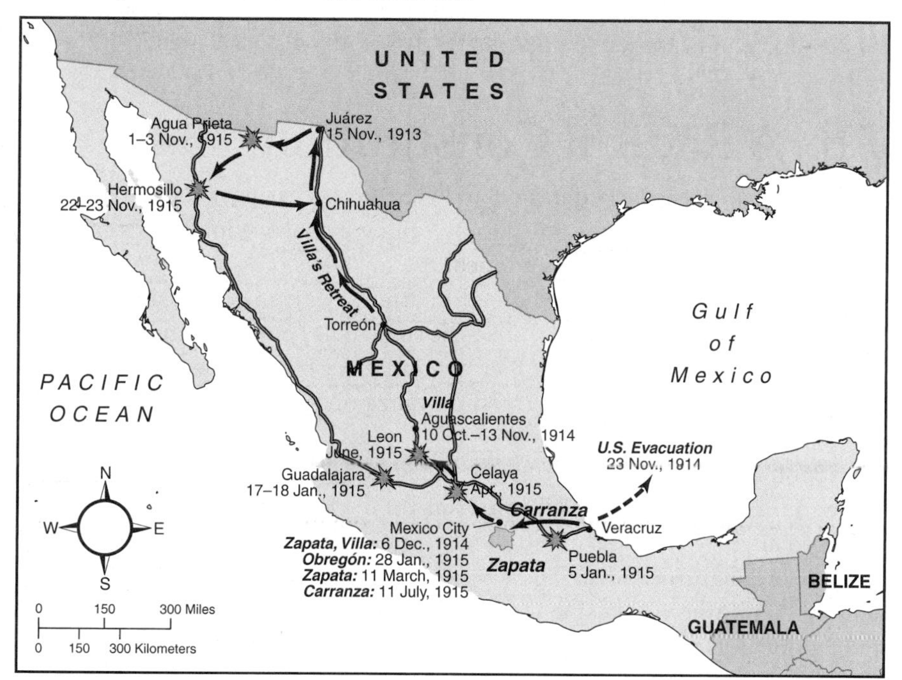

many of his closest collaborators turned against him. In his heyday, the Division of the North had boasted 50,000 men. By the time he headed to Sonora, he was down to 12,000 demoralized troops, short on food and supplies. The coup de grace came with an ambush at Agua Prieta on November 1, 1915, made possible by U.S. President Wilson giving permission for Carrancista troops to pass through the United States. Villa disbanded his army and disappeared into the mountains with a few hundred men. No one expected to hear from him again.

But Villa blamed Wilson for his defeat. He wanted revenge, and he hoped to provoke a U.S. invasion that would cause a backlash against the Carranza regime. First, Villista forces stopped a train carrying U.S. mining engineers and technicians from the Cusi Mining Company, hoping to reopen their mine. Instead, they were dragged from the train, and fifteen were killed. Villa took his revenge again in March on the town of Columbus, New Mexico, where seventeen U.S. citizens were killed. Villa failed to get money, supplies, or arms and lost 100 men in the raid. But he succeeded in luring the United States into Mexico.

President Wilson sent General John J. Pershing and 5,000 U.S. troops on what was called the Punitive Expedition, which included cavalry, infantry, artillery, and eight planes. In April and May 1916, Villa was injured and in hiding, with 10,000 U.S. troops and 10,000 federales occupying Chihuahua. He seemed defeated. Yet, by the end of the year, his forces numbered 6,000–10,000, and they controlled a substantial part of Chihuahua, where he had the support of the majority of the people.

The year was not as successful for the Zapatistas. Carranza's forces invaded Morelos, wreaking such havoc that the campesinos adopted the word *carrancear*, meaning to loot. By the fall of 1916, Zapata had disbanded his 20,000-member regular army and reverted to guerrilla warfare with only 5,000 men.

7.8: A Radical Constitution

With the Zapatistas nearly defeated and U.S. forces chasing Villa, Carranza decided to institutionalize his position by calling a constitutional convention at Querétaro. This time he did not repeat the mistakes of Aguascalientes: Only Constitutionalists would be allowed to attend. Carranza envisioned minor changes to the 1857 constitution. But as he soon found out, the Constitutionalists were no more unified than the other revolutionary bands. The desire for more radical social changes dominated, and Carranza was presented with a document far from the one he had envisioned.

Ideological differences split the delegates. The radicals, supported by Obregón, gained control and imposed their views. The constitution that emerged after two months of bitter debate at Querétaro contained many of the traditional enlightened ideas characteristic of the former constitution. In the customary Latin American fashion, the constitution conferred strong authority on the president. However, it went on to alter significantly some fundamental and traditional concepts.

The new constitution exalted the state and society above the individual and conferred on the government the authority to reshape society. The key articles dealt with religion, labor, and land. Article 130 placed restrictions on the Church and clergy. Churches were denied juridical personalities and could not own property. States could limit the number of clerics by law, and priests were not allowed to vote, hold office, or criticize the government. The Church was barred from participating in primary education. These provisions would give the State the supremacy over the Church that Mexican Liberals had sought since the mid-nineteenth century.

Article 123 protected the Mexican workers from exploitation by authorizing the passage of a labor code to set minimum salaries and maximum hours. Workers were to receive accident insurance, pensions, and social benefits. The right to unionize and strike was guaranteed. Because foreign investment in Mexican industrialization was significant, this article could potentially be used as one means of bridling the operations of the foreign capitalists.

The most significant change was Article 27, which laid the foundation for land reform and for restrictions on foreign economic control. It declared government ownership of mineral and water resources, subordinated private property to public welfare, and gave the government the right to expropriate land. This clause annulled all alienations of ejidos since 1857 and recognized communal ownership of land.

Map 7.5 Carranza in Charge

SOURCE: http://users.erols.com/mwhite28/mexico2.htm

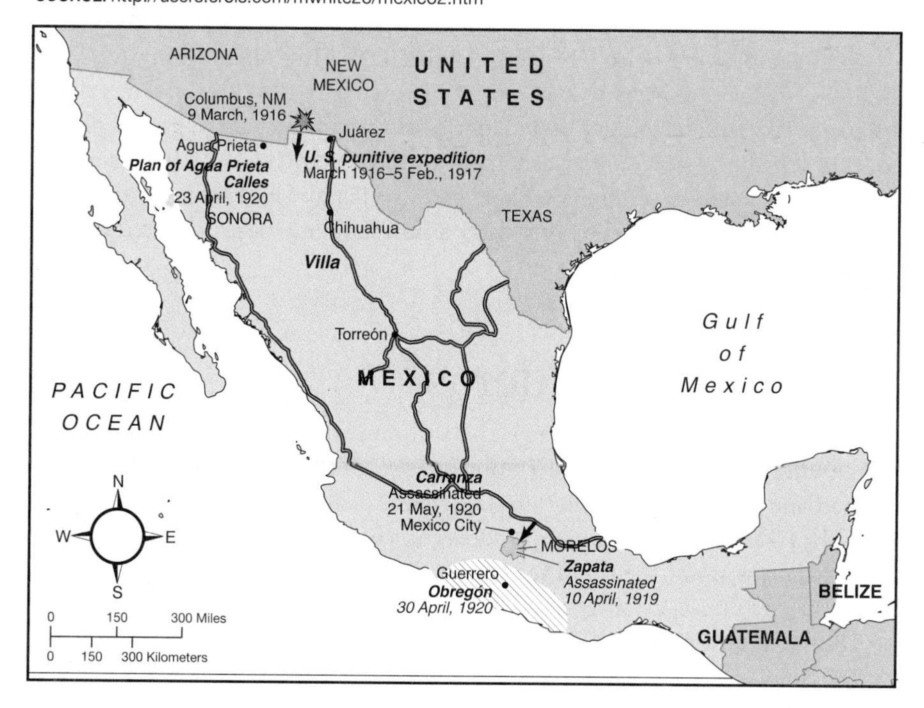

Carranza was unhappy with the constitution and had little intention of following its tenets when he was sworn in as president in May 1917, after an election in which he was unopposed. Nonetheless, the constitutional reforms stole some of Zapata's thunder.

7.9: The Radicals Lose

By the summer of 1917, Zapata was struggling to hold his revolution together. In March 1918, he issued new manifestos, for the first time recognizing the different needs of different regions, but no new support was forthcoming. In November, Spanish flu swept Morelos, where the campesinos had already suffered the effects of typhus, malaria, and dysentery. The population dropped 25 percent in 1918 alone, and Zapata was left with only 2,000 soldiers, each armed with only 304 cartridges. The final blow came on April 10, 1919, when Zapata walked into an ambush. As he walked into a meeting with supposed Carrancista defectors, a bugle sounded three times and soldiers who appeared ready to shoot in his honor instead lowered their guns and fired twice at point-blank range.

In the north, it became clear that Villa would never be caught. Growing more concerned about events in Europe, the United States began withdrawing the Pershing Expedition in early February 1917. Villa continued to fight Carranza, but there seemed to be little justification. By the end of the year, Villa had lost popular support. With his limited forces, he failed in an attack on Parral in 1918 and in an ill-conceived plan to kidnap Carranza in 1919.

Confident that the wars were over, Carranza attempted to name his own successor to the presidency. His arrogance was greeted with another revolt, led by three Sonorans: Alvaro Obregón, Adolfo de la Huerta, and Plutarco Elias Calles. Carranza fled and was killed by his own guard. With Carranza dead, Villa was ready to negotiate. De la Huerta, as interim president, gave amnesty to Villa's men and gave Villa a ranch, Canutillo, where he could have security and a means to live in retirement. Nonetheless, Villa met the same end as Zapata when he was assassinated in 1923 after a newspaper article indicated that he might be planning a comeback.

7.10: Las Soldaderas

As in all of Mexico's wars, women played important roles in the revolution. When villages rose up in revolt, women were among the throngs. As armies formed and moved on, whether on foot, horseback, or the trains, women went with them. Most served as camp followers, cooking for and tending to their men. Many women were kidnapped and raped, forced to accompany the soldiers.

But a significant group fought in the wars alongside the men, strapping on the *bandoleras*, the crisscrossed cartridge holders that became emblematic of the Mexican revolutionary. The *soldaderas* (women soldiers) came from all classes, though most were lower-class mestizas and indigenous. They apparently were welcomed by Zapata, but resisted by Villa. Stories are told of Villa casually executing enemy women and turning a blind eye to his soldiers' depredations. Other stories tell of Villa ordering his men to escort women to safety when they tried to fight with his troops, only to find that the women rushed back to battle.

It is impossible to know how many women picked up arms. A few names of particularly strong leaders are remembered: Genovevo de la O served as one of the leaders of Zapata's insurrection; Rosa Bobadilla served as a Zapatista colonel and fought in some 168 battles. Carmen Parra de Alanís fought with Villistas at the battle of Ciudad Juárez and later became a Zapatista. Carmen Vélez led 300 men into battles in Tlaxcala. These soldaderas are among the *generalas* and *coronelas* whose names are known and the many whom history has not recorded.

When the wars ended, however, the women found that they did not receive the same rewards as the men. They were denied military pensions and were not included in the new armed forces. And while the radical 1917 constitution considered workers and campesinos, women were largely left out. The 1927 law to implement Article 27 specified that those eligible for ejidos were "Mexican, males over eighteen years of age, or single women or widows who are supporting a family." Single women without children were not included unless as part of a man's family.

Furthermore, the constitution did not grant women the right to vote. Many male revolutionary leaders feared that Mexican women, who tended to be more religious than men, would listen to priests and become a conservative drag on revolutionary reforms. Women were granted local voting rights in the Yucatán in 1922 and Tabasco in 1934, but the vote was not extended to all women and in national elections until 1953.

7.11: From Destruction to Construction

The periodization of the Mexican Revolution typically casts 1910–1920 as the violent phase and 1920–1940 as the constructive phase. In some ways, the periodization is a bit misleading. Full-scale war had largely ended by 1917, when Villa's and Zapata's movements were in disarray and the constitution promised a new political order. Violence, however, continued to plague the countryside and to play a role in politics well into the 1920s.

The period 1920–1934 was dominated by two veterans of the revolution who became president: Alvaro Obregón (1920–1924) and Plutarco Elías Calles (1924–1928). Obregón was the undisputed leader of the revolution, the architect of Villa's military defeats and leader of the coup that prevented Carranza from maintaining control of the presidency. Both were from a lower middle-class background in Sonora. But Obregón was the epitome of the successful middle-class entrepreneur; he raised chickpeas on a 150-acre farm and became an expert mechanic, even patenting a harvester. Calles was less fortunate. The illegitimate son of a prominent hacendado, he had little success in his business endeavors before the revolution. As a schoolteacher, he was fired because parents complained about him. He was fired from a position as city treasurer because funds were missing from his department. He tried hotel management—the hotel burned down; farming—the farm went bankrupt; and milling—a business that failed. His true success came with the revolution: He was an early supporter of Madero, rose to be a divisional general and governor of Sonora. Calles served as Obregón's government secretary before becoming president himself.

These leaders from Sonora declared themselves the rightful heirs of Madero and proclaimed an era of peace and reconstruction. A practical man, Obregón tried to reach compromises among the various conflicting sectors in order to rebuild the country. The destruction of the war had indeed been considerable: The railroad was bankrupt and in ruins; more than 1,000 miles of telegraph lines were destroyed, and agricultural and mining production had fallen by half. The foreign debt topped $1 billion, interest payments were overdue, and foreign governments demanded compensation for destruction of the property of their citizens. And the military consumed 60 percent of the national budget.

At the same time, veterans of the wars expected their goals to be met. As Saturnino Cedillo of rural Morelos told the new government: "I want land. I want ammunition so that I can protect my land after I get it in case somebody tries to take it away from me. And I want plows, and I want schools for my children, and I want teachers, and I want books and pencils and blackboards and roads. And I want moving pictures for my people, too. And I don't want any Church or any saloon."

Obregón wisely named Zapatistas to be minister of agriculture and to head the new agrarian reform commission. But it took campesinos five to eight years to get their land grants. Obregón distributed 3 million acres of land benefiting 140,000 people. But Luis Terrazas, who Obregón allowed to return from exile to Chihuahua, owned as much as Obregón distributed. More than half of Mexican territory in 1923 was still owned by 2,700 wealthy families, and 25 percent of the land belonged to a mere 114 families. Furthermore, the majority of Mexico's rural population consisted of resident farmhands, and they were not eligible to get land. Calles went further than Obregón, distributing 8 million acres of land, mostly in the form of ejidos from 1924 to 1928. He accompanied the land grants with credit and agricultural schools. But from 1928 until 1934, agrarian reform became a low priority, and little land was distributed.

Table 7.1 Land Distribution in the Mexican Revolution

Year	Ejidos (%)	0.1–1,000 Hectares	1,000 Hectares
1910	1.6%	26.6	71.8
1923	2.6	19.6	77.9
1940	22.5	15.9	61.6

SOURCE: Based on data in Adolfo Gilly, *The Mexican Revolution* (London: Verso Editions, 1983), 334.

The record on labor reform was equally mixed. During the war, Obregón had cultivated the Mexico City workers' organization *Casa del Obrero Mundial* (House of the World Worker). He supported their strike against the Mexican Telegraph and Telephone Company, and in turn 5,000 union members joined his army, forming the Red Battalions. When he ran for president in 1919, Obregón again sought out the workers, forging an alliance with the *Confederación Regional Obrera Mexicana* (CROM; Regional Confederation of Mexico Workers). Nonetheless, real wages remained well below the three-peso-a-day minimum that, according to the national labor commission, was necessary for subsistence. Calles took the alliance even further, naming CROM boss Luis Morones to his cabinet as minister of industry, commerce, and labor. A massive union organizing drive brought even Mexico City's prostitutes into unions.

One of the most significant aspects of building the new Mexico was the construction of a new culture. Obregón named as his education minister José Vasconcelos, one of Mexico's leading intellectuals. Vasconcelos was determined to bring education to the masses, partly through a commitment to rural education and partly by making art available to the masses by fostering the painting of murals in public places.

The beginning of the mural movement generally is attributed to the paintings in the National Preparatory School, which prepared students to enter the National University. It was there that the great artists Diego Rivera, José Clemente Orozco, and David Alfaro Siquieros made their first contributions to the movement. In 1922, Rivera initiated the formation of the Syndicate of Revolutionary Technical Workers, Painters, and Sculptors of Mexico, which urged collective art creation that would play a significant role in the struggle for justice for the masses. The syndicate rejected easel painting and championed "monumental art in all its forms, because it is public property." However, the largely conservative art community denounced the murals, which were even attacked by the students, who threw mud and stones at the murals by Orozco and Siquieros.

All three went on to paint other murals, which reflected their radical political views. But it was Rivera's work that would become emblematic in its adulation of traditional folk elements of Mexican culture. Critics charged that he represented an idealized indigenous past, whereas the real Mexican masses would gladly trade their colorful pottery and woven clothing for modern, mass-produced goods. The veneration of the indigenous campesino went hand in hand with Vasconcelos's desire to bring education to the countryside. But while the revolution supposedly celebrated rather than denigrated the campesino, it shared the old liberal doctrine of wanting to assimilate the indigenous into a society that was largely not of their making. Vasconcelos set up an "Indian Department" focused on teaching Spanish to the indigenous so that they could enter the school system and be incorporated into mestizo society.

A tremendous effort was made to educate teachers, many of them women, and send them to the countryside, where more than 1,000 rural schools were established from 1920 to 1924. The curriculum was basic—reading, writing, arithmetic, geography, and history. The tone was secular, and teachers often encountered hostility both from the Church, which was losing its educational role, and from rural families themselves, who often viewed the teachers as atheist outsiders. Nonetheless, the government continued to emphasize education, adding another 2,000 rural schools from 1924 to 1928.

In the 1920s, education in Mexico took a socialist turn. Math lessons calculated excess profits of factory owners, and geography classes examined imperialist exploitation. New history texts focused on the struggle of the oppressed masses against capitalists, imperialists, and importantly, the Church. A textbook published by the state of Tabasco's Redemption Press in 1929 cautioned, "The worker's ignorance is very dangerous, for it allows him to be victimized by the exploiters, priests, and alcohol."

The attitude toward the Church displayed in the schools was one of the factors leading to the Cristero Revolt, an armed conflict between militant Catholics and the government. The catalyst for the uprising was Calles's determination to enforce the anticlerical provisions of the 1917 constitution. The Church had been hostile to the revolution almost from its beginning. In 1911, the clergy formed the *Partido Católico Nacional* (PCN; National Catholic Party), whose candidates for the 1912 Congressional elections swept Jalisco and Zacatecas, making it a major political party. The Maderistas annulled the results, turning the PCN against Madero and the revolution. Huerta courted the Church hierarchy, and both Zapata and Villa had cordial relations with the Church. The Zapatistas in particular tended to be religious, riding into battle with banners of the Virgin de Guadelupe. Carranza was hostile to the Church and welcomed the anticlerical aspects of the 1917 constitution, but he did not try to enforce them.

Obregón was committed to reconciling all elements of Mexico, so he did not enforce the constitution either. But during the heated days of the war, in 1914 and 1915, Obregón had been unsparing in his pursuit of antirevolutionary clergy. Church leaders did not forgive him for having priests and nuns imprisoned, seizing churches and convents, and closing Catholic schools. During Obregón's presidency, Church leaders became more openly opposed to the revolution. The Church also challenged the State and its connection to labor in 1920 by creating the *Confederación Nacional Católica de Trabajadores* (National Catholic Confederation of Workers); by 1922, the organization boasted 80,000 members, as many as the CROM, which responded to the competition by sending thugs to attack priests and churches.

Calles was determined to demonstrate the supremacy of the State and to end the Church's challenge. He took aim at the Church within weeks of taking office by reminding the state governments that they needed to control the activities of the clergy. The states immediately began to take action; for example, the state of Tabasco limited the number of priests to six, one for each 30,000 residents. In 1926, the archbishop of Mexico, José Mora y del Río, commented in a newspaper interview that, given such incidents in the various states, Roman Catholics could not accept the Constitution. His comments gave Calles an excuse to attack the Church by disbanding religious processions; deporting foreign-born priests and nuns; and closing monasteries, convents, and Catholic schools. The Church, in turn, went on strike, refusing to conduct masses or administer the sacraments. The strike lasted three years, leaving babies unbaptized and sending the old to the grave without last rites.

Loyal Catholics were incensed and ready to fight. Most priests—some 3,390 out of 3,600—fled to the cities. Of those who remained, only about forty openly supported what became known as the Cristero Rebellion, named after their cry, ¡*Viva Cristo Rey*! (Long live Christ the King!). The rebellion was essentially a series of uncoordinated local protests. There was little activity in the generally secular northern states or in the south, where strong indigenous cultures in such states as Oaxaca and Chiapas competed with the Church. But the central states of Michoacán, Jalisco, Guanajuato, and Colima were in open revolt.

The violence was as severe as the darkest days of the revolution. Catholic militants murdered teachers and burned down government schools. Government troops retaliated by trying to kill one priest for every teacher. By the time the war ended, it had cost the lives of 90,000 combatants and thousands of people caught between the government and militant Catholics. The war was finally ended when Calles's successor, Emilio Portes Gil, agreed to enforce the laws in a less provocative way.

Portes Gil came to the presidency after the Cristeros claimed their ultimate victim—Obregón, who shortly after winning another term of office was assassinated by a Cristero militant. Portes Gil was appointed to serve from 1928 to 1930, and Calles protégé Pascual Ortiz Rubio won election to the 1930–1934 term. Ortiz ran as the first candidate of the new *Partido Nacional Revolucionario* (PNR, National Revolutionary Party) founded by Calles in 1929 to consolidate the revolution. Ortiz Rubio lasted two years before resigning and being replaced by Abelardo Rodriguez.

It was clear to everyone, however, that none of the three presidents who served from 1928 to 1934 had any power. It was Calles who continued to rule as *Jefe Máximo*, lending the period the name *Maximato*. During these years, Calles moved revolutionary action to the right, while creating a public image and ideology that spouted left-wing rhetoric. For example, Zapata was hailed as an official national hero, conveniently overlooking the fact that the winners of the revolution actively fought Zapata and were responsible for his death. The co optation of Zapata's image occurred as the actual distribution of agricultural land dwindled.

7.12: The Apex of the Revolution

Calles expected to continue ruling Mexico under the next president, Lázaro Cárdenas. Cárdenas, who joined the revolution when he was fifteen years old, had been a protégé of Calles, who called him "the kid." He moved up along with his mentor. In 1930, he was named president of the PNR, and in 1934 he was tapped as the party's presidential candidate.

Although his victory was all but assured, Cárdenas campaigned across the country, building a network of supporters. He developed a reputation for modesty, largely by refusing to live in Chapultapec castle, the traditional presidential palace. His immediate goal in the presidency was to wrest control from Calles. He purged his cabinet of Callistas and gave generous retirement benefits to generals so that he could replace them with his own men. He favored the new labor organization, the *Confederación de Trabajadores de México* (CTM; Mexican Workers' Confederation), rather than the CROM. The final confrontation took place in 1936, when demonstrating workers demanded that Calles be removed from the political scene. Cárdenas ordered Calles into exile and emerged as his own man.

In 1938, Cárdenas reorganized the revolutionary party into the Mexican Revolutionary Party (*Partido Revolucionario Mexicano*, PRM), based on four pillars of society: the military, labor, agrarian, and popular sectors. Mexicans were mobilized into groups that communicated directly with the government, encouraging a vertical hierarchy of base and state, rather than a horizontal linking of mass organizations. This corporate structure minimized social conflict, and during the Cárdenas years, it largely worked in favor of the masses. For example, labor unions dealt directly with the government, rather than industry; the government intervened on labor's behalf. However, the corporatist organization also created the basis for future state domination of society.

The president also turned his attention to agrarian reform, intent upon rebuilding the traditional ejido. During his administration, which was Mexico's first *sexenio*, or six-year term, the Cárdenas government distributed nearly 50 million acres of land, representing 66 percent of the total land distributed from 1917 to 1940. In addition to land distribution, the government created the *Banco de Crédito Ejidal* (Ejido Credit Bank) to provide financing. Massive distribution of land dramatically changed society in some parts of Mexico, destroying the old hacendado class. But that class was replaced by a welter of new elites: agrarian reform officials and local political bosses. The process to receive land was often long and torturous, the amount received frequently insufficient. And in many parts of Mexico, no redistribution of land took place at all.

Although Cárdenas has been portrayed as a beloved figure among the campesinos, many rose up against him over his hostility toward the Church and a renewed socialist emphasis in school curriculum. Even in his home state of

Michoacán, campesinos who were the intended beneficiaries of his policies boycotted schools and said they would prefer not to get land unless local churches were reopened.

The most dramatic event in the Cárdenas presidency was the expropriation of the foreign-owned oil companies. The conflict began in 1936, when workers went on strike to demand higher wages and improved working conditions. The dispute was sent to an industrial arbitration board, which ruled in favor of the workers. The oil companies appealed to the Mexican Supreme Court, which upheld the board's ruling. At that point, the companies simply refused to abide by the decision. Cárdenas insisted that foreign companies must follow national laws and on March 18, 1938, ordered the nationalizing of seventeen oil companies. The president wrote of his decision in his diary: "I believe that there are few opportunities so special as this for Mexico to achieve independence from imperialist capital, and because of this my government will comply with the responsibility conferred by the revolution." The confiscated companies were organized into *Petroleos Mexicanos* (PEMEX, Mexican Petroleum) and became a symbol of economic sovereignty.

However, the outcry from the oil companies was matched by the U.S. and British governments. Cárdenas negotiated a settlement of $24 million to the companies—a far cry from the $200 million they claimed and far above the $10 million he deemed to be fair. The United States and Great Britain responded by boycotting Mexican oil and silver, with oil output falling by nearly 60 percent and silver production falling 50 percent.

Nonetheless, the expropriation was wildly popular in Mexico. Even the Catholic Church supported the decision by raising the Mexican flag above the national cathedral. Throughout the country, people gathered to contribute their pesos to pay the indemnity to the companies. U.S. ambassador Josephus Daniels observed women lining up in the Zócalo, Mexico City's main plaza, to contribute: "They took off wedding rings, bracelets, earrings, and put them, as it seemed to them, on a national altar. All day long, until the receptacles were full and running over, these Mexican women gave and gave. When night came crowds still waited to deposit their offerings, which comprised everything from gold and silver to animals and corn."

7.13: A Revolutionary Balance Sheet

By 1940, Mexico was a far different country than it had been in 1910. There was progress in many areas, but the country was still beset with problems. Although nearly 23 percent of the nation's land had been redistributed into ejidos, 62 percent of the land remained in farms of 1,000 or more hectares (down from 72 percent in 1910). School attendance for children six to ten years old increased from 30 to 70 percent. But rural poverty was still endemic. In 1940, 27 percent of the population was too poor to afford shoes—a figure that rose to 75 percent in the mostly indigenous states of Chiapas and Tabasco. Eighty percent of the population outside Mexico City lived without indoor plumbing or sewage disposal.

By 1940, critics wondered whether the Mexican Revolution was dead or whether it had been a revolution at all. Some dismissed it as a bourgeois revolution, won by the middle classes, who now took the place of the former Porfirian elites. Others complained that the country was still run by caudillos and that the new political party was as corrupt as its predecessors.

Perhaps the best analysis is offered by historian Michael J. Gonzales: "The popular and agrarian character of the revolution makes it a social revolution. The conflict pitted landless peasants, elements of the working classes, and discontented provincial gentry against the dictator Díaz, his elite supporters, and the federal army. The revolution threw out the old guard, reinvented the state, and made possible historic social and economic reforms. The revolutionary state gave landless peasants hundreds of thousands of hectares of land, nationalized foreign-owned petroleum companies, and significantly expanded public education. If the final outcome failed to eradicate poverty, create democracy, or achieve economic independence, the event still remains revolutionary."

Questions for Discussion

1. Who opposed Porfirio Díaz and why? How did class position and geographic location contribute to their reasons?
2. Why did Francisco I. Madero's ascension to the presidency not end revolutionary demands in Mexico?
3. Who were the key revolutionary leaders, and how did they represent different sectors of Mexican society?
4. How did the U.S. role impact the course of the revolution?
5. The Constitution of 1917 was a landmark document of the twentieth century. What were its key provisions, and how did they challenge the old nineteenth-century order?
6. What was the significance of the cultural changes signaled by Alvaro Obregón, José Vasconcelos, and Plutarco Elias Calles?
7. Why is the administration of Lázaro Cárdenas often hailed as the high point of the revolution?
8. In the last analysis, did the Mexican Revolution truly bring revolutionary change? Who were the winners and losers?

8

From World Wars to Cold War

While Mexico dealt with its domestic divisions, the entire region was buffeted by a series of international shocks: World War I (1914–1918), the Great Depression (1929–1940), and World War II (1939–1945). The crises were particularly problematic because by the early twentieth century, the export economy was well entrenched throughout Latin America, making the region extremely dependent on the well-being of foreign economies. Latin American food sales depended on working-class buying power in Europe and the United States. The expansion of tin, henequen, and other raw material production depended on technological advances abroad. National governments had become dependent on export revenues to balance their budgets, keep the armed forces paid and equipped, and meet domestic demands for services.

The international crises disrupted Latin American foreign trade: Economic downturns led to political instability, and throughout the region dictators came to power. These strongmen increased government's role in the economy, and new institutions were created to help stem the crisis. Import substitution industrialization and agriculture made the Latin American region less dependent on the international economy and more interdependent.

In the aftermath of World War II, the United States replaced Europe as the model for Latin Americans. They looked north and considered two issues: democracy and development. The U.S. rhetoric of the war years in the fight against fascism inspired Latin Americans, who saw their dictators in a new light. And the unprecedented wealth of the United States in the postwar years prompted Latin Americans to reconsider how to modernize their economies. By the 1950s, however, Latin Americans would discover that their own attempts at democracy and development would become entangled with U.S. economic expansion and new Cold War realities.

8.1: Economic Crises

From 1914 to 1945, Latin America was buffeted by international affairs, paying a price for its extensive openness to the world economy. As Europe became immersed in World War I, investment in Latin America plummeted. Public loans to

Brazil fell from $19.1 million in 1913 to $4.2 million in 1914 and to zero in 1915. Commodity exports suffered from both shortages of shipping and decline in European demand for goods. Because most Latin American governments relied on import tariffs, there was a precipitous decline in government revenue. Chile, for example, saw government revenues drop by two-thirds from 1911 to 1915.

World War I also provided opportunities for a few Latin American countries. Venezuela began exporting oil, and Mexican oil exports increased. Peruvian copper, Bolivian tin, and Chilean nitrates were in demand. Despite the success of these products, economies suffered as import prices rose along with trade surpluses and budget deficits, fueling inflation. Inflation eroded urban real wages, leading to political upheavals.

The curtailment of trade with Europe benefited the United States, which became the main supplier not just of Mexico and Central America, but of the entire region. The United States proved to be a limited market, however, for those whose products competed with U.S. products, such as grain and beef. At the end of the war, Great Britain's slow recovery led the United States to consolidate its new prominence in Latin America by becoming the dominant lender. U.S. capital, however, was usually tied to foreign-policy concerns. This was the era of dollar diplomacy, and the United States gave substantial loans and then took over customs houses to guarantee repayment.

Europe's recovery in the 1920s did little to help Latin America. A drop in the birth rate in Europe led to a slowdown in demand for Latin American primary products. European investors turned their attention to their own continent. New synthetic substitutes dried up the market for Latin American cotton, rubber, forest dyes, timber, and nitrates. The instability of commodity prices led Latin Americans to increase production, which in turn led to a decline in prices. Furthermore, the increase in worldwide production of strategic war products, such as oil, copper, and tin, led to market gluts and a drop in prices.

To rebuild their economies, Latin Americans sought loans. The United States responded with so much money—$1 billion from 1926 to 1928—that it became known as the dance of the millions. Some of the money went into the still-small industrial sector, which by 1914 was well established in Argentina, Brazil, Chile, Mexico, Peru, and Uruguay. However, most expansion of industrial production involved a more intensive use of existing facilities. By the end of the 1930s, Argentina was the only country in the region where industry's share of gross domestic product exceeded 20 percent. Most of the region continued to rely on exports. By the end of the 1920s, almost all export earnings were still earned from production of primary products, and as few as three products accounted for at least half of all foreign exchange earnings. Seventy percent of foreign trade was conducted with the United States, Great Britain, France, and Germany.

When the Great Depression hit in 1929, the market for Latin American products shriveled. As industry contracted in the industrialized countries, they stopped importing minerals from the developing world. Demand and prices for primary products fell, with the unit value of exports dropping more than 50 percent from 1928 to 1932. In Argentina, export values dropped from $1,537 million in 1929 to $561 million in 1932. But while prices and volume of exports fell, the interest on the sizable foreign debts of the 1920s did not. Governments defaulted on their loans, and new credits were not forthcoming.

The economic turmoil caused Latin American attention to shift to yet another form of nationalism: economic. The difficult years emphasized once again to the Latin Americans the dependency and vulnerability of their economies. Their monocultural export economies collapsed. Cuba's economy broke down: Foreign trade in the early 1930s was 10 percent of the 1929 figure. Uruguay's exports dropped 80 percent in the early 1930s. Brazil's exports plummeted from $445.9 million in 1929 to $180.6 million in 1932. In short, by 1932, Latin America exported 65 percent less than it had in 1929, proving once again that foreign trade contributed mightily to the cyclical fluctuations of the Latin American economy.

Nationalists demanded that steps be taken to increase the viability of the national economies and to reduce their dependency on fluctuations of the international market. Plans were made to increase economic diversification and promote industrialization, which appealed to both common sense and pride. For one thing, it promised to diversify the economy; for another, it kept foreign exchange from being spent to import what could be manufactured at home. At the same time, an acute shortage of foreign currencies meant that either the nations manufactured their own goods or they did without.

The economic crisis motivated governments to play an increasingly active role in national economies. They introduced long-range economic planning, exerted new controls, and offered incentives, stimulating national industry with devalued currency, import controls, and higher tariffs. Wider government participation in the economies and mounting demands for faster development shifted leadership of the nationalist movement from intellectuals to the governments themselves. Government leaders began to understand the potential power of nationalist thought and sometimes mobilized urban working classes as their base of support for nationalism.

Latin America responded to the Great Depression with domestic production of formerly imported goods, both manufactured products and food crops, which had been imported by countries that had turned their fields over to export crops. The two strategies—import-substituting industrialization (ISI) and import-substituting agriculture (ISA)—coupled with export promotion helped foster domestic demand for goods. As a result, much of the region was able to recover from the worst effects of the Great Depression by the late 1930s. Faster recoveries were made in Brazil, Mexico, Chile, Cuba, Peru, Venezuela, Costa Rica, and Guatemala. Recovery came much slower for the rest of the region, and particularly lagged in Honduras, Nicaragua, Uruguay, Paraguay, and Panama. Some of the recovery can be attributed to new trade partners. From 1932 to 1938, Germany, Italy, and Japan became important importers of Latin American goods. The European market was purchasing 55 percent of exports and supplying

45 percent of imports by 1938. Those new markets, however, were cut off with the outbreak of World War II.

But World War II also presented new opportunities for Latin America, which became the only raw material producer in the world that was not a site of hostilities. Trade increased between Latin America and the United States, particularly for tin and oil, and trade increased among Latin American countries, with the larger, industrializing regions exporting to the rest. Argentina, Chile, Brazil, and Mexico even developed a limited capital goods industry. Brazil's President Getulio Vargas (1930–1945, 1951–1954) played Germany against the United States, convincing the United States to replace the German steel manufacturer Krupp's planned investment in the Volta Redonda steel mill.

Nationalism was an amorphous sentiment, but it could ignite passions over such inflammable issues as control over oil, which represented Latin American longing to control its own natural resources. Nationalists argued that discovery and exploitation of their own petroleum was not only economically desirable but was also the guarantee of real national independence and, in the case of several of the larger countries, of world-power status. "Whoever hands over petroleum to foreigners threatens our own independence," one Latin American nationalist leader warned. No acts in recent memory received more popular acclaim than the nationalization of foreign oil industries in Bolivia in 1937 and Mexico in 1938.

Brazil's Vargas ably used the oil issue to his advantage. He created the National Petroleum Council in the 1930s to coordinate and intensify the search for oil. In 1939, the first successful well was drilled. The excited nationalists at once called for the creation of a national oil industry. In a bid for wider popular support, Vargas urged the creation of a state oil monopoly to oversee exploration and promote development of petroleum resources. Its creation in 1953 followed a passionate national campaign and marked a major victory for the nationalists. They triumphed over those who argued that it would be more economical for experienced foreign companies to drill for oil and pay Brazil a royalty on whatever was pumped out. The nationalists denounced that argument. After all, the issue was an emotional, not an economic, one. They wanted Brazil to retain control over one of its most precious and important resources. Their arguments convinced the masses that a national oil industry represented sovereignty, power, independence, and well-being. The masses responded with enthusiastic support, a demonstration of the power that economic nationalism can muster.

8.2: Dictators and Populists

As the prosperity of the 1920s gave way to the depression of the 1930s, unrest rippled through Latin America, and governments fell in rapid succession. In 1930, the armed forces overthrew the governments in the Dominican Republic,

Bolivia, Peru, Argentina, Brazil, and Guatemala. In 1931, the pattern continued in Panama, Chile, Ecuador, and El Salvador.

The men who came to power in the 1930s tended to be personalist dictators who put their own stamps on government. They were known for their brutality and corruption, and they amassed fortunes during their years of absolute control. Some came to power through the institutions founded during the years of U.S. occupation: Rafael Léonidas Trujillo, who dominated the Dominican Republic from 1930 to 1961, was commander of the Army that formed out of the constabulary created by the United States before withdrawing its occupation forces in 1924. Anastasio Somoza García came to power in Nicaragua as head of the National Guard, which the United States created before withdrawing the Marines in 1933. Somoza took power in 1936 and ruled until his assassination in 1956. His elder son, Luís, inherited the mantle until his death in 1967, when he was succeeded by his younger brother, Anastasio.

Anastasio Somoza García was one of the dictators to come to power in Latin America during the 1930s. Somoza's rule, however, endured far beyond that of most of his contemporaries. He was followed in office by his sons, Luis and Anastasio Somoza Debayle. (Library of Congress)

The new dictators served U.S. interests by guaranteeing order. Their authoritarianism was ignored under the guise of Franklin Delano Roosevelt's (FDR's) new policy of nonintervention, known as the Good Neighbor Policy. The brutality of the regimes was no secret to the United States, as evidenced by the oft-told, but apparently apocryphal, tale that FDR once said of Somoza, "He's a son of a bitch, but he's our son of a bitch." The words were likely uttered by FDR's Secretary of State Cordell Hull, but FDR certainly shared his sentiments. Interestingly, some Dominicans insist that the words were spoken in relation to Trujillo, not Somoza.

Both Trujillo and Somoza censored the press, suppressed or co-opted labor movements, and used the police to punish enemies and dissenters. Trujillo was known for his megalomania; he held at least forty official titles, including "Genius of Peace" and "The First and Greatest of the Dominican Chiefs of State." His brutality was most clearly demonstrated in the 1937 military massacre of Haitian workers, killing as many as 25,000 people. Trujillo was determined to make the Dominican Republic a "white" country, although most Dominicans are black or mulatto. The inability to distinguish Dominicans from black Haitians was demonstrated during the massacre by the need to ask people to say the word *perejil* (parsley), a word that the French patois-speaking Haitians could not pronounce as the Spanish-speaking Dominicans.

Somoza García was not shy about using the National Guard to intimidate or eliminate his enemies, as evidenced by the assassination of Augusto César Sandino, the Liberal general whose ragtag army fought against U.S. occupation from 1927 to 1933. But he also saw the need to make alliances. He wooed labor during the 1930s and made deals with the political opposition in the 1940s, guaranteeing them representation but preserving his own power. Somoza's regime was characterized by his greed—he became perhaps the wealthiest man in Nicaragua—and by his lack of concern for the people. Reportedly, when it was suggested that Nicaragua might benefit from a more educated population, Somoza replied, "I don't want an educated population; I want oxen."

Somoza and Trujillo's brutality paled, however, next to Maximiliano Hernández Martínez, who took power in El Salvador in 1932 after overthrowing the constitutionally elected President Alberto Araujo. Araujo had adopted as his campaign the ideas of Salvadoran intellectual Alberto Masferrer, who called for redistribution of wealth, providing the "vital minimum." The countryside grew restive as the coffee economy collapsed, the price of coffee dropping from $15.75

Table 8.1 Establishment of Depression Dictatorships, 1930s

Country	Regime
Argentina	Gen. José Uriburu, 1930–1932
	Gen. Agustín P. Justo "elected" president (1932–1938)
Bolivia	Self-coup and "elections" with oversight by military junta, 1930; army "arrests" President Daniel Salamanca, 1934; Cols. David Toro, Germán Busch; "military socialism" (1936–1939)
Brazil	Military junta, then regime of Getúlio Vargas with army support (1930–1945)
Chile	"Socialist republic" (1932)
Cuba	Fulgencio Batista; "sergeants' coup" (1933)
Dominican Republic	Gen. Rafael Trujillo (1930–1938; 1942–1952)
Ecuador	Nineteen presidents from 1931 to 1948; none complete term; 1932, four-day civil war, Quito garrison drives congress from capital, provincial regiments restore congress; coup in 1935, army imposes Federico Páez (1935–1937); coup by Minister of Defense Gen. Alberto Enriquez (1937–1938)
El Salvador	Gen. Maximiliano Hernández Martínez (1931–1944)
Guatemala	Gen. Jorge Ubico (1931–1944)
Honduras	Gen. Tiburcio Carías Andino (1932–1947)
Nicaragua	Gen. Anastasio Somoza García (1936–1956)
Paraguay	Col. Rafael Franco (1936); Gen. José Félix Estigarribia (1937–1940)
Peru	Col. Luis Sánchez Cerro (1931–1933); Gen. Oscar Benavides (1933–1939)
Uruguay	President Gabriel Terra (self-coup) (1933–1938)
Venezuela	Gen. Eleazar López Contreras (1935–1941)

SOURCE: Brian Loveman, *For la Patria: Politics and the Armed Forces in Latin America.* Copyright 1999. Reprinted by permission of SR Books, now an imprint of Rowman & Littlefield Publishers, Inc.

per hundredweight in 1928 to $5.97 in 1932. As Araujo's vice president, Hernández Martínez worried as fledgling communist organizations under Faribundo Martí conducted hunger strikes and protests. Hernández Martinez overthrew Araujo, and then responded to an easily defeated uprising in Sonsonate with *la matanza* (the massacre), a brutal attack on the rural, mostly indigenous, population, leaving an estimated 30,000 dead. After the *matanza*, Hernández Martínez became known as El Brujo, the warlock, because of his devotion to the occult. He held seances in the presidential palace, encouraged children to go barefoot so they could "better receive the beneficial effluvia of the planet, the vibrations of the earth," and hung colored lights across the streets in San Salvador to stem a smallpox epidemic.

In Cuba, a democratically elected president became a dictator. Gerardo Machado was elected in 1924 on the "Platform of Regeneration," a reform program including diversification of exports, encouragement of new industry, and tariff reform. He was reelected in 1928, but his reforms depended on prosperity, which ended with the Great Depression. Sugar sold for 22.5 cents per pound at the height of 1920s' prosperity; but by 1930, it dropped to 2.5 cents. Sugar production dropped by 60 percent, sending shock waves through the economy. Machado greeted the inevitable social and political unrest with repression. By 1931, there was open warfare as moderate leaders were jailed. His brutal attempts to hold power, however, led the army to overthrow him in 1933 rather than face the possibility of U.S. intervention.

A longer-lasting rule was demonstrated in Brazil by Getúlio Vargas, a defeated presidential candidate who took power in a military coup in 1930. Vargas suspended Congress, the state legislatures, and local governments and ruled by decrees. In 1934, he staged elections and instituted a new constitution, which nationalized banking and insurance. He also emphasized industrialization and oversaw the formation of the national oil industry, Petrobras, and the founding of the steel mill Volta Redonda. The resulting increase in the working class led to worker mobilizations and fears of social revolution. Vargas responded in 1937 with the Estado Nôvo, or New State, an idea he took from Mussolini's Italy. His government became increasingly authoritarian, and he relied on press censorship and secret police to prevent dissent.

Vargas discovered, however, that he could win the masses to his side via populism, a new political tool. Populism was the political movement that characterized much of Latin America from 1930 to 1965. It was labeled as a people's movement because it was based on mass electoral participation and pledged to address popular concerns. But populism was also hierarchical and was directed from above by a charismatic leader. These urban movements relied largely on a coalition of the working class, labor unions, middle classes, and industrial elites. Populist leaders co-opted the more radical ideas of the masses and redirected them in a nonrevolutionary direction. Growing poverty, first because of the Depression and then because of the nature of industrial development and urban growth, led to fears of revolution. Populism was the answer, and it seemed to please most sectors of the increasingly diverse Latin American society—all except the oligarchic elites and their revolutionary opposition. The oligarchy objected to their loss of power and the transfer of wealth to the working class. The revolutionary left complained that populism depended on the largesse of the populist leader rather than on structural changes won by organized masses.

But the charismatic populist leaders spoke an intoxicating nationalist vocabulary. Rhetorically convincing, ideologically weak, they offered immediate benefits—better salaries, health services, and the nationalization of resources—rather than institutional reforms. With the notable exception of Mexico's Lázaro Cárdenas, they focused their attention on the cities.

Vargas clearly understood the importance of the growing proletariat in Brazil. After taking power in 1930, he created the Ministry of Labor to deal with the workers. He used the urban workers to check the formerly overwhelming power of the traditional elite. The workers pledged their support to him in return for the benefits he granted them. With a highly paternalistic—and some say demagogic—flourish, Vargas conceded to the workers more benefits than they had previously obtained through their own organizations and strikes. A decree ordered the Ministry of Labor to organize the workers into new unions under governmental supervision. By 1944, there were about 800 unions, with membership exceeding one-half million. The government prohibited strikes but established special courts and codes to protect the workers and to provide redress for their grievances. Under the government's watchful eye, the unions bargained with management. Further, Vargas promulgated a wide variety of social legislation favoring the workers. He decreed retirement and pension plans, a minimum wage, a work week limited to forty-eight hours, paid annual vacations, maternal benefits and child care, educational opportunities and literacy campaigns, safety and health standards for work, and job security. In short, Vargas offered to labor in less than one decade the advances and benefits that the proletariat of the industrialized nations had agitated for during the previous century. Little wonder, then, that the urban working class rallied to support the president. In 1945, Vargas created the Brazilian Labor Party; small at its inception, the party grew in size and strength during the following two decades, while the other two major parties declined in strength.

One of the icons of Latin American populism was Juan Domingo Perón, who came to power in Argentina as part of the string of military governments that took power beginning with a coup in 1930. Previous Argentine governments had done little to favor the workers, despite the industrial surge and expansion of labor's ranks. Perón perceived the potential of the working class and utilized it as a power base after the military coup d'etat of 1943. As secretary of labor in the new government, he lavished attention in the form of wage increases and social legislation on the hitherto neglected workers, who responded with enthusiastic endorsement of their patron. During the two years that Perón held the labor portfolio, the trade unions nearly quadrupled in size. Perón adroitly manipulated the labor movement so that only leaders and unions beholden to him were officially recognized. When military leaders, suspicious and resentful of Perón's growing power, imprisoned him in October 1945, workers from around the country angrily descended on the center of Buenos Aires and paralyzed the capital. The military, devoid of any visible popular support, immediately backed down and freed Perón. With the full backing of labor, he easily won the presidential elections of 1946 with 56 percent of the vote, and his followers dominated the new Congress. During his decade of government, Perón's base of support was the mass movement of urban workers.

Perón's popularity with the working masses was rivaled only by that of his wife, Eva Duarte de Perón (1919–1952), known affectionately as Evita. A former radio actress, Evita's dramatic flare made her an effective speaker in support of her husband. Furthermore, she drew on her own impoverished background, making her a sympathetic figure for the masses. She established the Eva Perón Foundation to provide aid to Argentina's poor, with funding coming from the private sector, induced sometimes by such strong-arm tactics as threats of government inspections, fines, and company closures. It has been suggested that the Peróns divided their political work along gendered lines, with Evita serving the maternal role, nurturing the poor and the needy. Although she scorned feminists,

Argentine President Juan Domingo Perón and his wife, the adored Evita, greet a throng of supporters from the balcony of the Casa Rosada, the presidential mansion, in 1950. The Peróns claimed to represent *"los descamisados,"* the shirtless ones. (Library of Congress)

Evita devoted much time to strengthening the Peronist Women's Party and was instrumental in achieving the enfranchisement of Argentine women in 1947. In the 1951 elections, more than 2 million women voted for the first time, and six female senators and twenty-four deputies, all Peronistas, were sent to Congress.

Critics have noted that these populist governments gave a great deal to workers, but the price was government control of labor movements. It would be too simplistic, however, to lament the demise of labor's freedom under the Peróns. In truth, labor never had enjoyed much liberty—at best it was tolerated—under previous governments, and certainly prior to Perón the unions had gained few victories. Unions did compromise their liberty, but they did so in return for undisputed advantages and for a greater degree of participation in government than workers enjoyed under the elitist leaders who had governed Argentina. Perón's nationalist rhetoric cheered the workers, who identified closely with his programs. They rallied behind him to taunt foreign and native capitalists who they believed had exploited them.

Perón, like the other populist leaders of the time, exuded a charismatic charm. He never lost the support of the working class. His fall from power in 1955 resulted from economic problems, loss of Church approval, the firm and increasingly effective opposition of the traditional oligarchy, and—most importantly—the withdrawal by the military of its previous support. The middle class and the elite rejoiced in his fall, but the event stunned great multitudes of the masses who had given their leader enthusiastic support in return for more benefits and dignity than they had received from all the previous governments combined.

Obviously such populist governments found little favor among Latin America's traditional elite and those who, thanks to greater social mobility in the twentieth century, had recently joined their ranks. They resented any erosion of their power from below. Increasingly the middle class seemed frightened by the prospects of populist government and consequently tended to align with the elite. The previous accords between urban labor and the middle class, noticeable in some instances during the early twentieth century, disintegrated as the middle class became apprehensive of a threat, real or imagined, to their status and ambitions from labor. By the mid-twentieth century, identifying more with the elite against whom they had once struggled but whose lifestyle they incessantly aped, the middle class when forced to choose between the masses advocating change and the elite-backed status quo tended to opt for the latter.

8.3: Latin America Turns Inward

As World War II ended, Latin America faced a new world order, dominated by the United States and dedicated to free trade. The new order was established by the allied powers meeting in 1944 at Bretton Woods, New Hampshire, where they determined how to rebuild the postwar economy. The United States guaranteed a more international economy by providing a market, through open borders and unilaterally reduced tariffs; a stable currency, by pegging the dollar to gold at $35 an ounce; and capital, by investing in the rest of the world. Institutionally, the new order rested on three pillars: the International Monetary Fund, which would lend money to governments to avert balance of payments and balance of trade crises; the International Bank for Reconstruction and Development (IBRD, World Bank), which was to lend money at long-term rates to developing countries to build infrastructure that would attract private investment; and the International Trade Organization, which led to formation in 1947 of the General Agreement on Trade and Tariffs (GATT), designed to negotiate tariff reductions and to free world trade.

At the Eighth Inter-American Conference held in Chapúltepec, Mexico, in 1945, U.S. representatives announced their commitment to free trade, ending the wartime agreements that had provided guarantees to Latin America. The United States then turned its attention to rebuilding the destroyed European economies via the Marshall Plan. Now that the United States could once again access Asian raw material supplies that had been cut off by the war, Latin America became a low priority.

Latin America tried to turn back to its former European markets, but Europe was struggling to rebuild and could buy little of Latin America's products. Although export volume grew slowly, export prices increased dramatically. Foreign exchange reserves grew, but they were quickly depleted by repayment of debts and the high demand for imports that once again flooded the region. Pessimistic about exports and buoyed by economic nationalism, Latin American leaders chose to look inward.

Their view was bolstered by the theories of the United Nations Economic Commission for Latin America (ECLA), founded in 1948 and headed by Argentine economist Raúl Prebisch. Prebisch contended that over time the terms of trade went against Latin America because it exchanged low-price primary products for higher-valued manufactured goods. Furthermore, international markets were dangerously unreliable, and commodity prices were subject to boom-and-bust cycles. ECLA's recommendation was to take the unofficial import substitution industrialization policy of the war and Depression years and make it a conscious policy decision. The policy of ISI gained many adherents: nationalists who believed it was important to have local industries to meet national needs; urban industrialists and workers, who wanted to see an expansion in industry; and the middle class, who anticipated the growth of the service sector associated with industrialization.

Argentina, Brazil, Chile, and Uruguay enthusiastically adopted the ISI model, erecting tariff barriers to limit imports that would compete with fledgling industries. They subsidized the inputs needed for industrial production, provided low-rate credit to create and expand an industrial base, and passed legislation requiring particular levels of domestic content for goods on the market. Brazil amply illustrated the growing importance of industry within the economy. In 1939, the industrial sector provided 17.9 percent of the national income; by 1963, it furnished 35.3 percent.

All nations of Latin America did not participate equally in the industrial surge. Industrialization concentrated in a few favored geographic areas. At the end of the 1960s, three nations, Argentina, Brazil, and Mexico, accounted for 80 percent of Latin America's industrial production. In fact, more than 30 percent of the total factory production was squeezed into the metropolitan areas of Buenos Aires, Mexico City, and São Paulo. Five other nations—Chile, Colombia, Peru, Uruguay, and Venezuela—produced 17 percent of Latin America's industrial goods, leaving the remaining 3 percent of the manufacturing to the twelve other republics.

Industrialization sowed the seeds of new problems for Latin America. For one thing, it was creating a new type of dependency in which the region relied ever more heavily on foreign investment, technology, technicians, and markets. For another, it funneled wealth increasingly into the hands of a few industrial elites. Furthermore, the ties between industrialists and large landowners were closer than most people realized or cared to admit. While at times there were separate landowning and industrialist classes, in many instances the two groups overlapped. Thus, one could observe a growing industrial concentration in the hands of a few alongside a great concentration of land ownership, and often the same persons played the dual role of landowner and industrialist. Such an interrelationship of interests complicated reform efforts.

One significant consequence of industrialization was the growth of a better-defined urban proletariat class conscious of its goals and powers. As the labor movement expanded between 1914 and 1933, its leadership spoke increasingly in terms of major social changes. The ideological content of the labor programs, the increasingly efficient organization of the unions, and the new power that labor wielded worried elites and the middle class. Governments yielded to some basic labor demands to limit working hours, set minimum wages, provide vacations, ensure sick and maternal leave, and legislate other social-welfare laws. But at the same time they moved to dominate, control, and finally co-opt the labor movement.

One of the things that labor demanded, especially when it could not secure higher wages, was cheap food. Yet, though agriculture still formed the basis for the area's economy, few nations tried to reform the centuries-old agrarian structure. Mexico was the sole exception until the 1950s. The landowning system in most of Latin America remained flagrantly unjust and was based, at least in part, on the accumulation of huge tracts of land often by means of force, deceit, and dubious measures approved by the passage of time and the connivance of bureaucrats.

In no other area of comparable size in the world did there exist a higher concentration of land owned by the few than in Latin America. In the 1950s, 80 percent of the land in the region was owned by 5 percent of the population. On the opposite end of the scale lay another problem: the minifundium, a property so small that it often failed to sustain its owner, much less to contribute to regional or national economies. These small farms constituted 80 percent of all farm units but only 5 percent of the land. Statistics for the early 1960s showed that 63 percent of the rural population, or 18 million adult farmers, owned no land; another 5.5 million owned insufficient amounts of land; 1.9 million possessed sufficient land; and 100,000 owned too much land for the social and economic good of the area.

In general terms, and with notable exceptions, a lack of efficiency characterized the large estates. They included much land their owners either did not cultivate or left undercultivated. Experts estimated that Latin Americans in the mid-twentieth century farmed only about 10 percent of their agricultural holdings. In Brazil, Venezuela, and Colombia, approximately 80 percent of the farmland was unused or unproductively used for cattle-raising at the end of the 1950s. A 1960 study of Colombia revealed that the largest farmers, in control of 70 percent of the agricultural lands, cultivated only 6 percent of it. Farmers of fewer than thirteen acres, on the other hand, cultivated 66 percent of it.

The landowners continued to hold their property, not to farm but for purposes of prestige, investment, and speculation. As in the past, control of the land ensured control of the workforce as well because the resulting landlessness forced people to work on the great estates. With such large supplies of cheap labor, estate owners saw no need to modernize their farming techniques. The slash and burn method remained the most popular means of clearing the land. Farmers rarely spread fertilizer or did so sparingly. Consequently, the land eroded, became easily exhausted, and depleted quickly. Productivity, always low, fell. The farmer used the hoe, unmodified for centuries. The plow was rare, the tractor even rarer.

Increasingly Latin America needed to import food. Chile, for example, shifted in the 1940s from being a net agricultural exporter to becoming a net importer. By the mid-1950s, agricultural products accounted for 25 percent of Chile's total imports. In short, that nation was spending about 18 percent of its hard-earned foreign currency on food that Chile could grow itself, a tragedy by no means unique to Chile. By 1965, foodstuffs constituted 20 percent of Latin America's purchases abroad. Even with imports supplementing careless local production, the poor masses could not afford import prices. Starvation was not unknown, and malnutrition was common in wide regions of Latin America.

Agriculture in the twentieth century lost none of the speculative, reflexive nature so characteristic of the mercantilist past. For its prosperity, agriculture continued to rely heavily on a few export commodities, which were always vulnerable on the international market. Prices depended on the demands of a few industrialized nations. Further, new producers and substitutes appeared to challenge and undersell them, thus increasing Latin America's economic vulnerability. Africa, in particular, emerged as a formidable competitor for international markets. After World War II, prices of agricultural products gradually declined, while prices of imported capital goods spiraled upward.

Still, the latifundia followed their usual practice of offering monoculture. The well-being of an alarming number of Latin Americans at the end of the 1950s depended on a fair price for coffee; coffee composed 67 percent of Colombia's exports, 42 percent of El Salvador's, 41 percent of Brazil's, 38 percent of Haiti's, 34 percent of Guatemala's, and 31 percent of Costa Rica's. Other Latin American economies depended on export of sugar, bananas, cacao, wheat, beef, wool, and mutton.

8.4: Guatemala

A Flirtation with Democracy

The economy was not the only aspect of Latin America to change in the postwar years. The rhetoric that accompanied the United States as it marched into World War II was heard throughout the hemisphere. The Allies' victory marked democracy's triumph over dictatorship, and the consequences shook Latin America. Questioning why they should support the struggle for democracy in Europe and yet suffer the constraints of dictatorship at home, many Latin Americans rallied to democratize their own political structures. A group of prominent middle-class Brazilians opposed to the continuation of the Vargas dictatorship mused publicly, "If we fight against fascism at the side of the United Nations so that liberty and democracy may be restored to all people, certainly we are not asking too much in demanding for ourselves such rights and guarantees."

In the surge toward democratic goals, the ideals of nineteenth-century liberals revived. The times favored the democratic concepts professed by the middle class. Governments out of step with them toppled. A wave of freedom of speech, press, and assembly was supported by the middle class, and new political parties emerged to represent broader segments of the population. Democracy, always a fragile plant anywhere, seemed ready to blossom throughout Latin America.

Nowhere was this change more amply illustrated than in Guatemala, where Jorge Ubico ruled as dictator from 1931 until 1944. Ubico, a former minister of war, carried out unprecedented centralization of the state and repression of his opponents. Although he technically ended debt peonage, the 1934 vagrancy law required the carrying of identification cards and improved the landlords' position in disputes with workers. The landlords, in turn, supported his regime because he prohibited independent labor organizing. He massacred the indigenous who rebelled, killed labor leaders and intellectuals, and made his friends rich.

In May and June of 1944, there were a series of anti-Ubico protests, led primarily by schoolteachers, shopkeepers, skilled workers, and students. They were influenced by FDR's four freedoms: freedom of speech and religion and freedom from want and fear. They were also influenced by Mexican President Lázaro Cárdenas's nationalization of the oil industry and agrarian reform program.

The series of nonviolent protests that led to Ubico's overthrow began when teachers refused to march in the annual Teachers' Day Parade. During one protest, the Guatemalan cavalry charged, killing and injuring some 200 people. Ubico then declared a state of siege. As the crisis mounted, a group of prominent teachers, lawyers, doctors, and businessmen drew up the Petition of the 311, calling on Ubico to resign. The petition was presented to Ubico by several prominent citizens who he considered to be friends. Defeated, he resigned on July 1, 1944, and turned power over to General Federico Ponce. Ponce stood for election that fall, and the opposition recruited as their victorious candidate Dr. Juan José Arévalo, a prominent educator and author of several textbooks on history, geography, and civics, who was in exile in Argentina.

Arévalo considered himself a "spiritual socialist," an inspiration he credited to FDR. In his inaugural address, Arévalo said of the U.S. president: "He taught us that there is no need to cancel the concept of freedom in the democratic system in order to breathe into it a socialist spirit." Arévalo dissolved the secret police, removed all generals from their positions, dissolved the National Assembly, and repealed the constitution. An elected constitutional assembly wrote a new constitution and elected a new national assembly. The four priorities of the new government were to carry out agrarian reform, protect labor, improve education, and consolidate political democracy.

The challenge was formidable: An experienced bank clerk in Guatemala earned about $90 a month. A farmworker earned about five to twenty cents a day. Only 2 percent of all landowners owned 72 percent of the land. Ninety percent of all landowners held only 15 percent of productive acreage. Indigenous workers were still held by debt labor for 150 days of the year. Illiteracy was 75 percent in the general population and 95 percent among the indigenous, who constituted more than half the population. Life expectancy for ladinos was fifty years and only forty for the indigenous.

Coffee production was dominated by Guatemalan elites, and banana production was dominated by the U.S. giant, the United Fruit Company (UFCO). Arévalo addressed these problems by adopting a labor code, setting a minimum wage, and creating a National Production Institute to distribute credit, expertise, and supplies to small farmers. He distributed the confiscated German landholdings, ensured that smallholder titles were registered, and passed the Law of Forced Rental to guarantee that land would be rented to anyone with less than one hectare (2.5 acres). He also established a national bank and national planning office.

In 1950, after historic free elections, Jacobo Arbenz took office, having won 65 percent of the vote in an election that all observers agreed was fair and honest. Although democracy had been consolidated by Arévalo, it fell to Arbenz to work for economic change. In 1950, the annual per capita income of agricultural workers was $87. Of 4 million acres in plantations, less than 25 percent of the land was cultivated. In his inaugural address, Arbenz pledged to convert the country from a dependent nation to economic independence; to convert from a "feudal" to a modern, capitalist economy; and to raise the standard of living for the masses.

His blueprint was a report issued in 1950 by the International Bank for Reconstruction and Development. The report recommended government regulation of the energy companies and creation of an autonomous National Power Authority, setting wages that took prices into account, regulation of foreign business, industrialization to decrease reliance on foreign trade, and institution of a capital gains tax. The report criticized Guatemalan elites for raking in exorbitant profits and then investing them abroad.

Arbenz knew that the place to begin was the countryside, which he claimed was feudal and needed to be restructured for successful capitalism. The tool was to be the Agrarian Reform Law of 1952. The law declared that uncultivated land on estates over 220 acres where less than two-thirds of the estate was under cultivation was subject to expropriation and redistribution. It left untouched farms smaller than 220 acres, and there was no upper limit on farm size as long as the land was fully cultivated. Confiscated land was to be compensated in twenty-five-year bonds at 3 percent interest, based on the value of the property as declared by the owners for tax purposes as of May 1952. Land was redistributed in maximum parcels of 42.5 acres, and 100,000 people were given 1.5 million acres of land. Among the property confiscated was 1,700 acres taken from Arbenz himself. Arbenz had a firm legal base for the reform he intended. The Constitution of 1945 declared large estates to be illegal and conferred on the government the power to expropriate and redistribute land. Nonetheless, the landowning elite immediately and vociferously charged that the reform was communistic.

The charge of communism was guaranteed to get the attention of U.S. officials. The United States had never been friendly to socialism and communism and had been wary since the founding of the Soviet Union, with whom the United States reluctantly allied in World War II. After the "hot" war of fighting the Axis powers was over, the United States became immersed in a new war: the Cold War. This time there would be no fighting—at least not directly between the United States and the U.S.S.R. But the United States was determined to stop the spread of communism—or anything perceived as communism—around the world. Communism is based on a planned rather than a market economy, and the United States was committed to markets, now more than ever.

Guatemala's agrarian reform was far less severe than the one carried out in neighboring Mexico. But it came at the wrong time and in the wrong place. In another decade, the United States would encourage similar agrarian reform programs. But in the 1950s, the attack on the UFCO was seen as an attack on U.S. capital. And in the zero-sum game of the Cold War, anything that limited capitalist development must be socialist. And socialism equaled the supposedly monolithic communism of the Soviet Union.

In Washington, the State Department watched apprehensively but said nothing until 1953, when the Guatemalan government seized 233,973 acres of unused land claimed by the UFCO, a figure later raised to 413,573 acres. The company exercised formidable economic powers. Not only was it the largest single agricultural enterprise in the country, but it also owned and controlled the principal railroad and the facilities in the major port, Puerto Barrios. It was extremely unpopular among the nationalists and vulnerable because it was foreign and exercised too powerful an influence over the economy. As a matter of record, the company paid the Guatemalan government in duties and taxes about 10 percent of its annual profits, a sum regarded by indignant nationalists as far too low.

The U.S. State Department rushed to support UFCO claims against the Guatemalan government on the basis that the compensation offered was insufficient. Yet, the sum offered equaled the value of the lands declared by the company for tax purposes. At that point, the State Department began to charge that the Guatemalan government was infiltrated with, if not controlled by, communists. The UFCO enjoyed unusually close connections to the U.S. government: The law office of Secretary of State John Foster Dulles had written the drafts of the UFCO's 1930 and 1936 agreements with the Guatemalan government. CIA director Allen Dulles, brother of the secretary of state, had served on the board of the UFCO. The family of John Moors Cabot, then assistant secretary of state for inter-American affairs, owned stock in the banana company.

Although UFCO may have had unusually close links to the U.S. government, it was certainly not unusual for the government to intervene on behalf of U.S. corporations. That intervention had, in fact, been U.S. policy for many years. Furthermore, U.S. corporations were not the only ones deeply opposed to Arbenz's economic policies. Guatemalan businessmen, particularly in the coffee industry, were equally incensed. They applauded U.S. opposition to Arbenz and would have asked for such help had it not already been forthcoming.

Despite the mounting pressure, Arbenz pushed ahead with his program to decrease Guatemala's dependency. He announced the government's intention to build a highway from Guatemala City to the Atlantic coast and thereby end the transportation monopoly of UFCO's International Railways of Central America. Further, he decided to construct a national hydroelectric plant. Until that time, foreigners produced Guatemala's electrical power, with rates among the highest in Latin America.

By late 1953, the Arbenz government feared intervention. Arbenz repeatedly requested arms, arguing that his more conservative neighbors in Honduras and El Salvador were encouraging exiles to attack. The State Department responded with an arms embargo. Unable to equip the army with U.S. materiel, Arbenz turned to another source. On May 17, 1954, a shipment of arms arrived from Czechoslovakia. To the State Department, its arrival served as final proof that Guatemala had fallen under communist control.

The U.S. Air Force at once ferried military supplies to Tegucigalpa in order to equip a small army under the command of a Guatemalan army exile, Colonel Carlos Castillo Armas. On June 18, 1954, Castillo Armas and approximately 150 men crossed the border into Guatemala. They penetrated about twenty-five miles but engaged in no significant action. They did not need to. The Arbenz government fell because the army refused to act and the workers were not armed. A series of air attacks on the unarmed capital terrorized the population and broke the morale of the people. The attacks caused more psychological than physical damage. The planes were furnished by the CIA and flown by U.S. pilots, as former President Dwight D. Eisenhower revealed later in a publicly recorded interview. U.S. Ambassador John Peurifoy handled the changing of the government and, with the enthusiastic endorsement of Washington, installed Castillo Armas in the presidency. In violation of United Nations and Organization of American States treaties and international law, the United States blatantly intervened in the small country. In a radio address on June 30, 1954, John Foster Dulles informed the American people of the changes in the Guatemalan government, which prompted him to declare, "The events of recent months and days add a new and glorious chapter to the already great tradition of the American states."

A year after the United States placed Castillo Armas in power, Vice President Richard Nixon wrote, "President Castillo Armas' objective, 'to do more for the people in two years than the Communists were able to do in ten years,' is important. This is the first instance in history where a communist government has been replaced by a free one. The whole world is watching to see which does the better job."

A small plane drops leaflets in Guatemala City during the 1954 U.S.-sponsored coup that overthrew President Jacobo Arbenz. (Library of Congress)

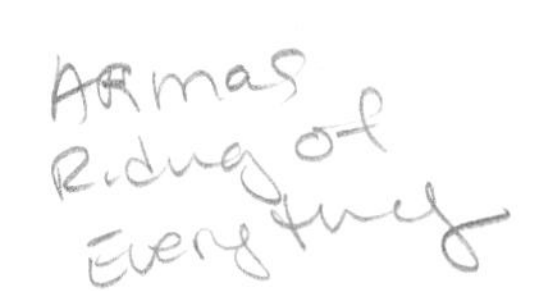

In a plebiscite echoing the Ubico practice of the 1931–1944 period, Castillo Armas confirmed himself in power. Deriving his major support from local elites, the UFCO, and the U.S. State Department, he ruled for three years without ever offering to hold elections, free or otherwise. Nonetheless, he took the precaution of abolishing all political parties that did not please him and disenfranchising all illiterates, thus canceling the voting rights of more than half the adult population. The police jailed, tortured, exiled, and executed political opponents, excusing their crimes with the accusation that the victims were communists. Castillo Armas eradicated whatever traces of communism might be found or fabricated, but in the process he also eliminated democracy and reform.

Guatemala after 1954 offered the first Latin American example of the reversal of a land reform. Castillo Armas returned to the UFCO the lands his predecessor had nationalized. What is more, he signed a new contract that facilitated the company's exploitation of Guatemala until 1981, limiting its taxes to a maximum of 30 percent of the profits. (That figure contrasted with the 69 percent oil companies had to pay Venezuela.) Approximately 1.5 million acres, which under the 1952 Agrarian Reform Law had been confiscated as idle and distributed to 100,000 landless rural families, reverted to the original owners. The newly created campesino class suffered severely. Much of the land returned to the original owners lay fallow, while Guatemala imported food that the country was perfectly capable of growing.

Castillo Armas was assassinated by one of his own presidential guards in 1957, but that did not put an end to the country's political repression. A series of military governments during the second half of the twentieth century, always under the banner of anticommunism, ignited the fires of a holocaust for the indigenous population. Reflecting on the brutality and arrogance of military rule in Guatemala, the *Los Angeles Times* (July 5, 1984) termed the CIA intervention in 1954 and its dismal aftermath "one of the most shortsighted 'successes' in the history of U.S.–Latin American relations."

Questions for Discussion

1. How did the world wars and the Great Depression impact the economies of Latin America?
2. What were the reasons for the rise of dictatorships in Latin America in the 1930s? Who supported these regimes?
3. Discuss the origins, nature, and impact of economic nationalism in Latin America. How did economic nationalism change after the end of World War II?
4. Who were the Latin American populists, and what policies did they pursue?
5. How did ideas about democracy change during the first half of the twentieth century?
6. What does the case of Guatemala show us about change and continuity in Latin America? Why was the role of the United States so significant?

9 The Revolutionary Option

By the late 1950s, much of the promise of development and democracy seemed hollow. In most nations and for major segments of the populations, reforms came too slowly and too ineffectively, if they came at all. On five occasions in the twentieth century, Latin Americans despairing of evolutionary change and reform opted for revolution as a quicker and surer path to change. The Mexican Revolution marked the first half of the twentieth century. Four more countries erupted in revolution during the second half of the century: Bolivia, Cuba, Chile, and Nicaragua.

All four revolutions had some common patterns: They were led mainly by young men and women who were the children of the middle class and sometimes even the elites, but who had opted to reject their class privileges and work for social change. The leaders created a coalition of middle-class, urban working class, and poor rural groups, joined by disaffected elites. They wanted revolution—a sweeping change of economic, political, social, and cultural structures—not just reform. And they wanted to create programs to benefit the poor masses. But specific circumstances of each country, and international circumstances at the time of each revolution, led to differences in the paths taken and in the outcomes.

Revolution is an extreme option that people choose when legal and peaceful paths to change fail. By revolution, we mean the sudden, forceful, and usually violent overturn of a previously stable society and the substitution of other institutions for those discredited. This is sweeping and fundamental change, not just in political organization but in social and economic structures and the predominant ideology. These are not just personnel changes, like the numerous coups d'etat (or *golpes de estado*) that changed the individual in the presidential palace but changed none of the social structures.

Bolshevik leader Leon Trotsky famously opined that if all one needed to make revolution was poverty and oppression, there would be revolutions every day all around the world. The causes of revolution do not develop overnight but are the result of long-term problems. Foreign control of the economy or political processes can cause local resentments. Another cause of revolution is the public's perception of economic conditions, which can include rising expectations because of economic growth and development programs, the social dislocations caused by rapid changes in the economy, or the hardship of economic crises.

Political disputes can also lead to revolution, particularly when new economic leaders want political power. Political openings can create increased hopes for democratization and a violent reaction when expected freedoms are not granted. A government's inability to meet the needs or desires of various constituencies can weaken its bases of support. When the government becomes the target of protest, threatened officials often respond with brutality, which turns more people against the government.

For revolutions to occur, there must be both leaders and followers. Eloquent leaders focus the discontent of the masses and plan actions, frequently violent ones, to ignite the revolution. But they also need the support of the masses, or their exhortations to action will fall on deaf ears; no leaders can make a revolution by themselves. Importantly, the leadership offers guidance and justification for revolution, shattering old myths and creating new ones. They articulate the goals of the revolution, frequently focusing on nationalism and some form of socialism or redistribution of wealth.

9.1: Bolivia

Although Bolivia led the new wave of revolutions, it turned out to be the least influential of the four. The Bolivian revolution fell apart because of internal problems. Furthermore, it was atypical because it was not actively opposed by the United States, which became the nemesis of the revolutions in Cuba, Chile, and Nicaragua.

Like Guatemala, Bolivia is a mostly indigenous country, and in the 1950s, its population was characterized by illiteracy, undernourishment, sickness, low per capita income, and short life expectancy. Bolivia depended primarily on exports of tin, and the mines were owned by three Bolivian families: Patiño, Hochschild, and Aramayo. Patiño's annual income exceeded the national government's, and the annual allowance of one of his sons was greater than the national education budget.

In 1941, urban intellectuals under the leadership of Victor Paz Estenssoro organized the *Movimiento Nacional Revolucionario* (MNR; National Revolutionary Movement). Its two primary goals were to nationalize the tin mines and combat international imperialism. In the presidential election of 1951, as was customary, the government controlled the electoral machinery, and suffrage was restricted to literate males, about 7 percent of the population. Despite these handicaps, the MNR won a resounding victory. The government and army refused to let Paz Estenssoro take office, leading to a bloody struggle in April 1952 in which the MNR seized power by force. The revolution represented a broad-based alliance of the progressive elements of the middle class, intellectuals and students, organized labor, and the rural landless. The urban and rural workers' militias provided the backbone of armed strength and emerged as a key power group.

President Paz Estenssoro nationalized the tin mines without compensating the owners, a move nearly the entire population supported. More controversial was the unexpected rural agitation that forced the urban movement to undertake agrarian reform. The indigenous rose up and took land, and Paz responded to the reality: He formalized the reforms in a decree-law of August 2, 1953.

After serving his four-year term, Paz Estenssoro turned over the presidency to MNR stalwart Hernán Siles, who won the 1956 elections. A term out of office permitted Paz Estenssoro to run again in 1960, and the voters enthusiastically endorsed his new bid for office. But when Paz tampered with the constitution in 1964 to allow himself another term, the army took the reins of power. Revolutionary activity declined and then disappeared.

The United States had few investments in Bolivia and did not consider the isolated country to be strategically important. Instead of aggressive opposition, Washington lavished funds, technical aid, and support on the MNR governments to support the moderate, middle-class leaders of the revolution, Paz Estenssoro and Siles, against the radical, working-class factions. George Jackson Eder, who the Bolivian government was required to accept in 1956–1957 as director of its stabilization program, reminisced in his book *Inflation and Development in Latin America* (1968) that he helped to break the power of the radical labor unions over the government. His stabilization program "meant the repudiation, at least tacitly, of virtually everything that the Revolutionary Government had done over the previous four years."

For the rest of Latin America, Bolivia was an example of a revolution that had imploded. It was co-opted by the middle classes and destroyed by its leadership's greater concern for power than for revolutionary goals. As a result, its impact was minimal. It did little to inspire other revolutionary movements. In fact, by 1967, Bolivia's situation was so dismal that Cuban revolutionary hero Ernesto "Che" Guevara tried to start a new revolutionary movement there, ending in his death. The Cuban, Chilean, and Nicaraguan revolutions would have a much greater impact on the region, and the role of the United States would become crucial in all three.

9.2: Cuba

Perhaps no other event in Latin American history has had the impact of the Cuban Revolution of 1959. It became the model for revolutionary change throughout Latin America and beyond. It also became a model for U.S. Cold War policy. The Cuban Revolution particularly rankled Washington leaders, partly because of the historically close relationship between the two countries and partly because it occurred a mere ninety miles from U.S. shores.

At first glance, Cuba in the 1950s seemed an unlikely place for revolution. By Latin American standards, Cubans enjoyed a high literacy rate and high per capita income. It was an urbanized nation with a relatively large middle class. However, aggregate statistics can be misleading. Much of the population lived in the countryside, where 43 percent of adults were illiterate, 60 percent lived in homes with dirt floors and palm roofs, 66 percent had no toilets or latrines, and only one in fourteen families had electricity. The largest group in Cuba comprised 600,000 rural wage workers, mostly sugarcane cutters, fully employed only during the harvest. Agricultural wage workers averaged only 123 days of work a year. Cuba was dependent on one crop, sugar, which brought in 75–85 percent of all export earnings. And it was further dependent primarily on one market, the United States.

The United States has loomed overly large in Cuba's history ever since 1898 and the military occupation that followed. In the process, the United States consolidated its hold on the Cuban economy. In 1953, U.S. owners produced 40 percent of Cuban sugar and owned 50 percent of the railways and 90 percent of the utilities. In 1958, U.S. investment totaled over $1 billion, more than in any other Latin American country except Venezuela, and in per capita terms, more than three times U.S. investment in any other Latin American country. Furthermore, the large-scale and attendant capital requirements of sugar production kept Cubans from expanding their share of ownership, and the development of secondary industries occurred in the United States rather than in Cuba.

The United States was Cuba's most important market, and the amount of sugar that Cuba could sell depended on quotas set by the U.S. Congress, with an eye toward the needs of sugar growers and refiners in the United States. This dependence on congressional action goes well beyond the disadvantageous position in which most agricultural-exporting peripheral countries find themselves in a world market controlled by the industrialized countries. In addition, in exchange for the quota, Cuba gave preference to more than 400 U.S. products, in essence trading away the island's industrialization opportunities.

All that was left to Cubans was real estate speculation and the government, which became synonymous with theft and corruption—clearly the practice of the dictator Gerardo Machado (1924–1933). Resistance to Machado mobilized masses of Cuban workers and students, who succeeded in driving him from office. However, the real power shifted to the military when the Sergeants' Revolt, an army coup led by Sergeant Fulgencio Batista, toppled the provisional government that replaced Machado. This paved the way for a new president, Ramón Grau San Martín, a doctor and professor who proclaimed a socialist revolution. Washington officials immediately became worried and urged Batista to step in again. Many Cubans felt that yet another chance at democracy and sovereignty had been taken from them by U.S. intervention.

From 1934 to 1940, puppets held the presidency while Batista held real power. He was popularly elected in 1940, and then retired to Miami in 1944 an estimated $20 million richer. In his absence, a now conservative and corrupt Grau finally won his turn in the presidency. Batista returned in 1948 and served in the Cuban Senate during the presidency of Carlos Prio Socorrás, which reached new lows of political corruption. Batista responded with a coup in 1952 and ruled with dictatorial powers until the 1959 revolution. It seemed there had been no change from 1924, when Machado took power, until 1959, when Batista was toppled. Indeed, some argued that there really had been no change since independence. The system was corrupt, elections were fraudulent, and the island was dominated by the United States. It was this long, tortured history that set the stage for the revolution.

Batista was toppled in a popular movement led by Fidel Castro, a brash young representative of middle-class students and intellectuals. Castro, the illegitimate son of a Spanish-born landowner and his maid, was an attorney who represented the poor. He was an Orthodox Party candidate for Congress, but elections were aborted by Batista's coup. Fidel used the creative approach of filing a lawsuit charging Batista with violating the 1940 constitution, but, of course, Batista's courts would not hear the suit. At that point Fidel followed Cuba's long historical tradition of armed uprising. In 1953, he and a ragtag band of 162 men set out to attack the Moncada Barracks, and they failed miserably. After Fidel's release from prison, he escaped to Mexico, where he organized an invasion.

Exultant Cuban revolutionaries pose with rifles on a monument in Matanzas in February 1959. The sculptures are of Cuban independence hero José Martí and a female allegorical figure of liberty brandishing broken chains. (Library of Congress)

Among the trainees in Mexico was the young Argentine medical doctor Ernesto "Che" Guevara. Guevara had traveled the length of Latin America and come to the conclusion that many of the medical problems he had hoped to treat were related to poverty and could only be solved by systemic change. He arrived in Guatemala in 1953 and was impressed by Arbenz's attempted reforms, only to see the government toppled in 1954 by the U.S. coup. He gladly joined Fidel's invasion force of eighty-two men who sailed to Cuba on a fifty-eight-foot yacht, the Granma, in 1956. Most of the men were killed when attacked by Batista's troops upon landing. The handful of survivors regrouped in the Sierra Maestra, where they were supported by the poor campesinos. The *guerrilla* forces attacked Rural Guard posts, prompting Batista to send army units to the mountains. At the same time, and perhaps more importantly, an urban underground carried out a sustained campaign of sabotage and subversion; they set bombs, cut power lines, and derailed trains. The urban and rural groups united as the 26th of July Movement, named for the ill-fated attack on Moncada. Finally, the United States withdrew its support and Batista fell. Triumphant revolutionary forces marched into Havana in January 1959.

Castro visualized a reformed Cuba, with agricultural cooperatives, industry, education, and health care for all, and broad segments of the Cuban population shared those goals. He looked to the rarely enforced 1940 constitution, which banned latifundia, discouraged foreign ownership of the land, and permitted the expropriation of property "for reason of public utility or social interest." The constitution authorized the state to provide full employment, claimed the subsoil for the nation, empowered the government to "direct the course of the national economy for the benefit of the people," and conferred on the state control of the sugar industry.

The nationalism that Castro promoted rested on a strong historical precedent, dating back to Cuba's struggle against Spain. The hated Platt Amendment, repeated U.S. intervention, support for Batista, and dominance of Cuba's economy left a bitter legacy. Cubans resented Washington's hegemony, and succeeding generations of patriots challenged it. And in the writings of Cuban independence hero José Martí, in the rhetoric of the forces that overthrew Machado, and in the Constitution of 1940, Fidel found much of the inspiration for the changes he proposed.

Despite Castro's intense nationalism, he tried to initiate good relations with the United States. In April 1959, he was invited to Washington by the National Press Club. In meetings with Vice President Richard Nixon and Secretary of State Christian Herter, Fidel did not seek an increase in Cuba's sugar quota, and U.S. officials did not offer to do so. Nonetheless, Fidel assured them that Cuba would remain a member of the Organization of American States, honor U.S. rights to the Guantánamo base, and protect U.S. investment and strategic interests. Nonetheless, officials offered no U.S. or World Bank aid, in contrast to generous support for Batista.

Fidel's attempts to reform the economy, however, brought him into conflict with Cuban elites and the United States. One of the revolution's first acts was to cut the exorbitant utility rates charged by U.S. companies, providing compensation via state bonds. Next was agrarian reform. Sugar companies owned 70–75 percent of the arable land, and 3 percent of the sugar producers controlled more than 50 percent of production. At the same time, although most arable land remained uncultivated, the country imported 50 percent of the food its people ate.

In June 1959, the Castro government issued the relatively modest Agrarian Reform Law. It limited farms to 1,000 acres, with exceptions, including highly productive sugar and rice plantations, which were limited to 3,333 acres. Even larger farms owned by foreigners would be permitted if the government decided they served the national interest. In fact, only 10 percent of the farms were affected by the 1,000-acre limit; they represented, however, 40 percent of farmland. The confiscated estates were worked as cooperatives under the management of the National Institute of Agrarian Reform. Eisenhower responded in September 1959 by recalling the U.S. ambassador, even though no property was nationalized without compensation.

In February 1960, Cuba negotiated an agreement with the Soviet Union for $200 million in trade over four years, less than Soviet trade had been with Batista. The United States, however, interpreted the agreement as a sign of the Cuban government's allegiance. In June, tension between the United States and Cuba intensified when Texaco, Royal Dutch Shell, and Standard Oil demanded payment for oil imports to their refineries rather than extending credit. Fidel canceled their exclusive contract and established a national petroleum institute to find other supplies. When no one in the West would sell to him, Fidel turned to the Soviets, who traded crude oil for sugar. The next day, Eisenhower cut Cuba's sugar quota. In retaliation, on October 12, 1960, Cuba expropriated the oil companies. On October 14, the United States announced a trade embargo on all goods except medicine. On January 4, 1961, Washington broke diplomatic relations.

By then the United States had labeled the Cuban Revolution communist and a tool of the Soviet Union. This judgment showed a profound misunderstanding of the domestic roots of the Cuban Revolution. Further, U.S. isolation of the revolutionary government gave Fidel only two options: turn to the Soviet Union or capitulate to U.S. demands. But the legacy of U.S. domination made the latter an impossible choice. And the genuine desire of the Cuban masses for egalitarian programs forced confrontations with international capital. Cuban social welfare and U.S. investment as constituted were not compatible.

Ironically, Fidel at that point did not consider himself or the revolution to be socialist or communist. He distrusted the Cuban Socialist Party, a traditional Communist party, which had allied with Batista in exchange for privileges. The Communists, in turn, had believed that a socialist revolution could not be made until the country had gone through the stages predicted by Karl Marx; therefore, no revolution could happen in Cuba until it became fully capitalist and industrialized. They dismissed Fidel as a bourgeois adventurer.

Fidel turned to the Cuban Communists because he needed administrative help. First the Batista property fell into state hands, followed by the expropriated property. The new state enterprises had to be managed, and much of the bourgeoisie with administrative skills had fled the island. Furthermore, the well-organized masses called for more social programs to meet their needs. As the masses pushed for more sweeping programs, the Communists provided organizing know-how. When the United States refused aid and pressured its allies to isolate Cuba, the Soviets became the only alternative. In part to facilitate relations with the Soviets, in 1965 the Cubans formed the Communist Party of Cuba, though no party congress met until 1975.

Castro did not declare Cuba's revolution to be socialist until the U.S. invasion in 1961. Washington was not content to break ties with the revolution—in Cold War mentality, the revolution had to be defeated and rolled back. The United States launched an invasion at the Bay of Pigs, using anti-Castro exiles trained by the CIA. The planners expected another Guatemala, a quick and easy government overthrow.

To prepare for the landing, CIA operatives strafed several airfields, killing seven. At the funeral the next day, Castro declared: "This is the Socialist and Democratic Revolution of the humble, with the humble, for the humble"—a cry for help to the Soviet Union, which had promised to aid socialist revolution anywhere in the world. Although Soviet help would be crucial in the future, it was Cubans who defeated the Bay of Pigs forces. Fidel himself was at Playa Girón, the beach at the Bay of Pigs, to lead the forces. His popularity soared.

Fidel became a heroic symbol of anti-imperialism throughout the Third World. But his domestic popularity also came from the new government programs. He regarded education as one key to the new future. Teacher-training institutes sprang up; in a decade, the number of teachers tripled; the number of schools quintupled; and young, eager volunteers traveled to the remotest corners of the island to teach reading and writing. Within a few years illiteracy virtually disappeared. By 1971, nearly one-quarter of the country's 8 million inhabitants were in school. Education was free from nursery school through the university. Reading became a national pastime. In 1958, Cuba published 100 different titles and a total of 900,000 books; in 1973, it published 800 titles and a total of 28 million books. Cuba has the highest per capita book production in Latin America and the highest literacy rate in the hemisphere.

Health care also became a revolutionary priority. Hospitals were built in remote cities, and doctors became available in the countryside for the first time. All medical services were free. By 1965, Cuba spent $19.15 per person per year for medical care, a figure that contrasted sharply with the $1.98 Mexico spent or the $.63 Ecuador spent. The health of the nation improved dramatically, and life expectancy lengthened. Cuban medical care became the envy of the hemisphere, and many Latin Americans traveled to the island to seek medical help.

The government dealt with Havana's housing problem by guaranteeing everyone housing at no more than 10 percent of his or her income. No one was allowed to own more than one dwelling, which could be passed on to family but could not be sold. Buildings abandoned by Batista supporters who fled were converted into housing units. Volunteer work crews were enlisted to build more housing, and a great deal of the focus was put on the Cuban countryside. By building housing and economic ventures in rural areas, the government kept more people from flooding into the capital. Unlike the rest of Latin America, Cuba does not suffer from homelessness; however, housing has not kept pace with needs, and many Cubans, particularly in Havana, live in substandard housing.

The government also encouraged the arts, and painting, literature, and music flourished, within revolutionary limits. Fidel famously said in 1961, "Within the revolution, everything; outside the revolution, nothing." The *nueva trova* movement, led by such singer/songwriters as Silvio Rodriguez and Pablo Milanés, brought revolutionary themes to 1960s music, combining rock and Latin sounds, leading a musical movement that swept Latin America. The National Ballet of Cuba emerged as one of the principal dance companies of the world, touring to critical and popular acclaim.

It was probably in film, however, that the Cuban Revolution reached its maximum cultural achievement. The first law of the revolutionary government in the field of culture created the Cuban Film Institute in March 1959. It has produced documentaries, newsreels, and feature-length films in addition to publishing the most serious Latin American journal on film, *Cine Cubano*. Cuban films have become regular features of film festivals, where they consistently win international awards.

Much of the Cuban filmmakers' attention focused on reinterpreting their country's past. In his alternatingly lyric and realistic *Lucia*, Humberto Solas studied the woman's role in three key Cuban struggles: the 1895 war for independence, the 1933 revolt against Machado, and the 1960s literacy campaign. The final segment's focus on continuing *machismo* shows a fascinating element of Cuban cinema—its capacity for criticism of the revolution. Double standards penalizing women were again criticized in *Portrait of Teresa* (*Retrato de Teresa*, 1979), the story of a woman trying to balance a job, union activism, cultural work, and a family. *Strawberry and Chocolate* (*Fresa y chocolate*, 1994) tackled the controversial issue of gay life and homophobia in Cuba. The treatment of the gay community—ranging from arrest and assignment to work camps in the 1960s to excellent but quarantined treatment for HIV/AIDS in the 1990s—has been one of the revolution's most problematic issues.

Women also found that their interests were often not taken seriously. Che talked about the selfless virtues of the revolutionary "new man." But women found that all too often the new man was much like the old man, despite some efforts made by the new government. Fidel targeted women as a group, encouraging formation of the Federation of Cuban Women (*Federación de Mujeres Cubanas*, FMC) just three months after the revolutionary triumph. Their first goal was to mobilize women to support the revolution through work and participation in the literacy campaign and in neighborhood projects. The government responded to women's increased activity outside the home by providing day-care centers. Their health needs were particularly targeted through perinatal care and the legalization and availability of abortions. Female participation in education soared, and women entered professions in record numbers. To balance work and home, legislation guaranteed women eighteen weeks of paid maternity leave.

But in the home, the double standard prevailed. The revolution responded with the 1975 Family Code, which required a revolutionary couple to be equal partners. Article 26 required both parents to participate in child rearing and housework. Of course, there were no revolutionary police knocking on doors to make sure that men did their part, but at least the moral weight of the revolution supported the idea of equality.

The change in women's roles was one of many social changes that discomfited the old oligarchy, most of whom fled within the first two years of the revolution. As changes were instituted, the traditional power of the old oligarchy, military, and Church vanished; the state and new mass organizations filled the vacuum. Castro frequently convoked people to mass meetings in which he spoke extemporaneously for hours to enthusiastic crowds. The old electoral system, which Cubans disdained for its corruption, was abandoned. New systems of power were formalized starting in 1976 with the formation of *Poder Popular* (People's Power). Neighborhoods elected representatives to municipal assemblies, which in turn elected provincial assemblies, deputies to the National Assembly, and judges for municipal courts. The National Assembly elected the Council of State. Fidel served as president of both the Council of State and Council of Ministers, his cabinet, until he stepped down temporarily in 2006 and permanently in 2008 because of ill health, turning power over to his brother, Raúl, the longtime defense minister.

By the early 1970s, Cuba's successes were irrefutable and recognized even by the revolution's critics. Pat M. Holt, chief of staff of the Senate Foreign Relations Committee, made a brief visit to Cuba and then wrote *A Staff Report Prepared for the Use of the Committee on Foreign Relations, United States Senate* in which he concluded that all Cubans enjoyed the necessities of life and indeed as a group had an impressive standard of living.

Cuba was able to make these changes in part because the Soviet bloc paid above-market prices for sugar, on the theory that the market price did not reflect adequate compensation for Cuban labor and profit. At first, Cuba vilified sugar and strove to diversify the economy and to industrialize via food processing and manufacture of fertilizers, sugarcane derivatives, agricultural supplies, pharmaceuticals, textiles, clothing, machinery, steel, and construction materials. By 1961, there was a renewed emphasis on sugar production to pay for the industrialization; nonetheless, sugar dropped from 90 percent of exports in 1975 to 65 percent in 1985. Domestically, the government allowed private peasant markets to function until 1986, when fears of exploitation by middlemen led to their closure. Other private businesses were discouraged, however, and in 1968 they were confiscated, sending a second wave of Cuban exiles to Miami.

Cuba's growth rate has generally been better than Latin America as a whole, averaging 6 percent in the 1970s and 5 percent in the 1980s. But the world economic crises eventually were felt on the island as well, leading to shortages. As growth rates fell, the revolution alienated the most marginal elements of the labor force, people who had sacrificed for the revolution and who could accept no more discipline in the name of socialism. They, too, wanted out. Castro opened the doors in mid-1980, and over 100,000 Cubans migrated to the United States from the port of Mariel. When the Soviet Union collapsed in 1991, Cuba lost 85 percent of its economy, and critics expected to see the island follow the path of Eastern Europe. After severe hardships in the early 1990s, culminating with a wave of people fleeing on rafts in 1994, the economy stabilized. Cuba opened to foreign investment, under government controls, using elements of capitalism to safeguard the revolution's socialist successes in education and health care.

9.3: Cuba's Impact

The Cuban revolution forced the United States to acknowledge that there were, indeed, many deep-rooted causes for revolution in Latin America. When the attempt to return to the Guatemala-style coup failed, the United States realized that new tactics were needed. If the United States could not overthrow the Cuban Revolution, perhaps it could prevent similar revolutions in Latin America. As President John F. Kennedy declared, "Those who make peaceful revolution impossible make violent revolution inevitable."

In 1961, Kennedy launched the Alliance for Progress, a small-scale Marshall Plan for Latin America designed to encourage economic development, which presumably would promote the growth of democracy. The Alliance charter, drawn up at the 1961 Inter-American Conference at Punta del Este, Uruguay, called for an annual increase of 2.5 percent in per capita income, establishment of democratic governments, more equitable income distribution, land reform, and economic and social planning. Latin American countries (with the exception of Cuba) were to invest $80 billion during the next ten years. The United States pledged $20 billion.

The Alliance struggled for ten years. The desired outcomes were often contradictory; for example, simultaneous goals were to raise agricultural productivity and to redistribute land in the countryside. Large landowners were loath to reduce their landholdings so that property could be redistributed. In fact, the aim of increased productivity was often achieved through further concentration of landholdings, with greater mechanization, leading to more rural unemployment and unrest. Alliance funds too often ended up paying for infrastructure that benefited large landowners and did nothing for the rural poor. The Alliance eventually failed because the traditional oligarchy had no intention of freely volunteering to give away or sell a portion of its lands, to tax itself more heavily, or to share power with a broader base of the population.

Throughout the 1960s, intellectuals and political activists debated the merits of reform versus revolution. Some political leaders genuinely believed that meaningful change in social and economic structures could be enacted within existing political systems. The solution was to be found in developmentalism, a focus on economic growth that would bring prosperity and the growth of a stabilizing middle class. But such changes entailed costs that many elites were not willing to pay. The Brazilian case was instructive.

President Juscelino Kubitschek (1956–1960) promised to give Brazil "fifty years of progress in five." He focused on creating a modern infrastructure, building highways, dams, and hydroelectric power plants. He even tried to open Brazil's interior by moving the capital to a new city, Brasilia, in the state of Goiás. Kubitschek encouraged development of domestic industry, but he also encouraged foreign investment to enlarge the industrial sector. The result was,

Development or Dependency?

Some Latin American intellectuals have wondered whether the core of the problem of economic development is not just Latin America's position in the international economy but the idea of development itself and the way development programs have been carried out since the end of World War II. At that time, U.S. officials were concerned about conditions in what became known as the Third World—everyone outside of the industrialized First World and the Soviet bloc's Second World. U.S. officials and businessmen had multiple concerns: Poor people make poor markets, and U.S. postwar strategy for rebuilding the world economy depended on trade.

Furthermore, poor people are likely to rise up to protest their conditions, and that instability is bad for business. Unrest might also result in a successful revolution. At its worst, the revolution could result in the elimination of a market and source of cheap labor and primary goods—the communist threat. At best, it might result in a government trying to mitigate the ills of the marketplace by insisting on higher prices for products, higher wages for labor, and the freedom to choose from multiple trade partners instead of being locked into exclusive relationships.

There were also many sincere people in the industrialized world who wanted to improve the standard of living for the masses in Latin America. With the best of intentions, they went to work for the Agency for International Development and the Alliance for Progress in the United States, or for the many new development-oriented agencies of the United Nations, such as the World Health Organization, the International Labor Organization, and the Food and Agriculture Organization.

Latin Americans debated how to help their countries achieve economic development. Some elites, in an echo of nineteenth-century patterns, wanted to "modernize" their countries to be like the United States in the classic combination of wanting to make money and to appear "modern" (no one said "civilized" anymore). At the same time, many Latin Americans wanted to improve the lives of the majority. The question was how to do it.

A blueprint was provided by U.S. economic historian W. W. Rostow in *The Stages of Growth: A Noncommunist Manifesto* (1960). Aid and investment programs were underpinned by Rostow's contention that modernization followed stages from traditional to advanced societies, and that all of the world could develop by simply following the lead of their predecessors. The problem, in his argument, was that traditional societies needed to become more capitalist. His book launched entire schools of thought about modernization and served as the basis for many government programs.

The reaction in Latin America, however, was a bit different. Starting with the analyses by Raúl Prebisch and ECLA, Latin Americans concluded that the problem was not a lack of capitalism but the way that capitalism unfolded in Latin America. Because the U.S. and Europe already had highly industrialized economies, Latin America could not compete; limited to production of primary products, the region was trapped into low growth, low wages, and declining terms of trade.

Out of this analysis came the dependency school, which contended that the problem was not too little capitalism but too much and in too unequal a form. *Dependentistas* described Latin America as the periphery, trapped by desires of the metropolis, usually the United States. Dependency challenged traditional paradigms that the so-called Third World could, through proper programs and economic policies, become like the First World. Dependency theory took the focus off the metropolis and turned the spotlight on the periphery. The theory was articulated by Latin Americans themselves: Fernando Henrique Cardoso, Enzo Faletto, and Theotonio Dos Santos (joined by U.S. and European scholars, most notably Andre Gunder Frank). Instead of prescribing more capitalism, dependentistas advocated either disconnecting from the international market or turning toward some form of socialism. Their arguments disputed the modernization theory that all countries could become prosperous through capitalism and Marxist theory that predicted socialism could come only after countries became fully capitalist and proletarianized.

Since the 1950s, people concerned about Latin America have focused on the issue of development, but its definition has not always been clear. For many, it was just another word for progress or modernization. As we have argued in this text, development often described mere growth in gross national product. Others defined it as structural change in the economy, especially via industrialization. We have focused on development as providing the most good for the most people.

for example, a Brazilian automobile industry—the seventh-largest producer in the world—but 69 percent of it was owned by foreigners. They created jobs, but they sent their profits out of the country.

Meanwhile, Brazil remained dependent on coffee as its main generator of income, and agriculture remained unchanged. Thousands of desperate Brazilians left the countryside for the cities, but they found limited opportunities there. Capital-intensive industrialization was highly mechanized and created few new jobs. The unskilled and unemployed rural folk constructed *favelas*, shantytowns with cardboard houses that crawled up the hillsides ringing Rio de Janeiro and São Paulo, a testament to the limits of developmentalist plans. Slum dweller Carolina Maria de Jesus summed up the contradictions well: "What our President Senhor Juscelino has in his favor is his voice. He sings like a bird and his voice is pleasant to the ears. And now the bird is living in a golden cage called Catete Palace. Be careful, little bird, that you don't lose this cage, because cats when they are hungry think of birds in cages. The *favelados* are the cats, and they are hungry."

By the time that Kubitschek left office in 1961, inflation and corruption were rampant. His successor, Janio Quadros, the quirky governor of São Paulo, campaigned with a broom, promising to sweep away corruption. Quadros tried to establish a foreign policy that was more independent of the United States, but, on domestic issues, he had no qualms about accepting help from the International Monetary Fund and the Alliance for Progress to keep the economy afloat. Impatient, he resigned after a scant seven months in office.

His vice president was João Goulart, secretary of labor in populist leader Getúlio Vargas's second administration. Goulart launched reforms to rectify Brazil's vast economic inequality: He urged land reform, a limit on foreign profit remittances, tax reform, wage raises for labor, and the granting of the vote to illiterates. The rural and urban masses supported Goulart's proposals and responded with demonstrations and strikes. But his policies, and the mass mobilizations they encouraged, drew the wrath of elites and frightened the middle class. Front-page editorials in middle-class newspapers called for his ouster. Rio de Janeiro's *Correio da Manhã* trumpeted the headlines "Enough" and "Out." The military hierarchy, which had been criticized by Goulart, was happy to oblige.

Military coups certainly were not new phenomena in Latin America. There were ninety-nine successful coups from 1920 to 1960, occurring in every country

Table 9.1 Latin American Military Coups, 1961–1964

Country	Date
El Salvador	January 24, 1961
Ecuador	November 8, 1961
Argentina	March 29, 1962
Peru	July 18, 1962
Guatemala	March 31, 1963
Ecuador	July 11, 1963
Dominican Republic	September 25, 1963
Honduras	October 8, 1963
Brazil	March 31, 1964
Bolivia	November 4, 1964

NOTE: The proclaimed motivation for these coups varied greatly, but the threat of "more Cubas" was a backdrop to each of them.

SOURCE: Brian Loveman, *For la Patria: Politics and the Armed Forces in Latin America*. Copyright 1999. Reprinted by permission of SR Books, now an imprint of Rowman & Littlefield Publishers, Inc.

except Mexico and Uruguay. But it is striking that there were ten coups in the sliver of time from 1960 to 1964. The rationale was usually the threat of communism, but even the reform efforts sanctioned by the United States and the Alliance for Progress smacked of communism for many in the Latin American military.

Brazil, torn between the demands of the impoverished masses and the intransigence of the elites, was but an extreme example of Latin America in the 1960s. The hopes that the middle class would be a moderate voice for change disintegrated amid the fear of communism, which strengthened the military. As the Alliance for Progress sputtered along, the fears of U.S. officials were realized—revolutionary movements proliferated throughout Latin America.

U.S. officials blamed revolutionary movements on Soviet and Cuban provocateurs. But Latin Americans did not need foreign agents to foment revolution—they were well aware of their own problems of poverty and oppression. What Cuba provided was an example and an inspiration. Latin Americans were inspired by Cuba's success in throwing off a brutal dictator and flouting U.S. domination. But they were also inspired by the ultimate revolutionary, Che Guevara.

Here was a middle-class Argentine medical doctor who had turned his back on personal economic opportunity to adopt the cause of another country as his own. Furthermore, Che turned the July 26 Movement's accidental tactics of warfare in the Sierra Maestra into a formal strategy in his manual, *Guerrilla Warfare* (1960). He argued that revolutionaries did not have to wait for ideal conditions to emerge, as traditional communist parties argued. According to

In 1960, Cuban photographer Alberto Korda snapped the most famous image ever taken of Che Guevara. Titled "Guerrillero Heróico," the photograph seemed to capture Guevara's idealistic quest for revolutionary change. (Alberto Korda, " Guerrillero Herioco," photo of Che Guevara, 1960. (ITAR-TASS Photo Agency / Alamy)

his theory, revolutionaries by their very actions could draw others to the cause and topple an entrenched dictatorship.

Che also lived up to his word, leaving Cuba to try to bring revolution to Bolivia, where he was killed in 1967. Two famous photographic images appeared that year: one was of Che's dead body, in an almost Christ-like pose. The other was the famous Alberto Korda photograph of a hopeful-looking Che in his dashing beret, a photograph snapped in 1960 but unpublished until Giangiacomo Feltrinelli, an Italian publisher, acquired it. The romantic image of the young, handsome revolutionary, eyes gazing upward, turned up on walls of young would-be rebels around the world, along with Che's idealistic pronouncement: "Let me say, at the risk of seeming ridiculous, that the true revolutionary is guided by great feelings of love."

His message did not fall on deaf ears.

Starting in the 1960s, guerrilla movements developed throughout Latin America. Although there had been only three armed uprisings in the 1950s, including the Cuban revolution, there were twenty-five armed groups founded in the 1960s. By the 1980s, armed groups appeared in seventeen of the nineteen countries in Latin America. Some groups were ephemeral, disappearing in a year or less. Others struggled for a decade or more, sometimes going underground only to appear again when conditions seemed more propitious. They were generally met by the arm of the Alliance for Progress that the Kennedy administration did not publicize—government counterinsurgency forces armed and trained by the U.S. and designed to combat armed revolutionary groups.

Revolutionary groups appeared even in Mexico, where the government still claimed to be revolutionary. The official party changed its name in 1946 from the Mexican Revolutionary Party to the Institutional Revolutionary Party (*Partido Revolucionario Institucional*, PRI). By the 1950s, as railroad workers were repressed and arrested, the emphasis seemed to be far more on the institution than the revolution. In 1968, students organized a series of peaceful protests demanding democratic reforms, including autonomy for the country's universities, the freeing of political prisoners, and social justice. The protests culminated in a rally on October 2 at Mexico City's *Plaza de las Tres Culturas* that included students and workers. They were attacked by government troops, killing hundreds and imprisoning dozens.

But in 1970, one country offered an alternative—a peaceful revolution.

Table 9.2 Latin American Guerrilla Groups

Country	Group	Years
Argentina	Tiger-Men [*Uturuncos*]	1959–1960
	Guerrilla Army of the People [*Ejército Guerrillero del Pueblo* (EGP)]	1963–1964
	Peronist Armed Forces [*Fuerzas Armadas Peronistas* (FAP)]	1967–1974
	Revolutionary Armed Forces [*Fuerzas Armadas Revolucionarias* (FAR)]	1967–1973
	Montonero Peronist Movement [*Movimiento Peronista Montonero* (*Montoneros*)]	1969–1977
	Armed Forces of Liberation [*Fuerzas Armadas de Liberación* (FAL)]	1969–1974
	People's Revolutionary Army [*Ejército Revolucionario del Pueblo* (ERP)]	1970–1977
Bolivia	Army of National Liberation [*Ejército de Liberación National* (ELN)]	1966–1970
	ELN/Commando Nestor Paz Zamora [*ELN/Nestor Paz Zamora Commando*]	1989–1990
Brazil	October 8th Revolutionary Movement [*Movimiento Revolucionario de Outubre 8* (MR-8)]	1960s–1970s
	National Liberating Action [*Ação Libertadora Nacional* (ALN)]	1968–1971
	Popular Revolutionary Vanguard [*Vanguardia Popular Revolucionaria* (VPR)]	1968–1970s
Chile	Movement of the Revolutionary Left [*Movimiento de Izquierda Revolucionaria* (MIR)]	1965–1990*
	Manuel Rodríguez Patriotic Front/Communist Party of Chile [*Frente Patriótico Manuel Rodriguez/Partido Comunista de Chile* (FPMR/PCC)]	1980–1990*
Colombia	Movement of Workers, Students and Peasants [*Movimiento de Obreros, Estudiantes y Campesinos* (MOEC)]	1959–1961
	Army of National Liberation [*Ejército de Liberación National* (ELN)]	1964–present
	Revolutionary Armed Forces of Colombia [*Fuerzas Armadas Revolucionarias de Colombia* (FARC)]	1966–present
	Popular Army of Liberation [*Ejército Popular de Liberación* (EPL)]	1967–1984

(continued)

Country	Group	Years
	April 19th Movement [*Movimiento 19 de Abril* (M-19)]	1974–1990
Costa Rica	The Family [*La Familia*]	1981–1983
	Santamaria Patriotic Organization/Army of Democracy and Sovereignty [*Organización Patriótica Santamaria/Ejercito de la Democracía y la Soberanía* (OPS)]	1985–1988
Cuba	July 26th Movement [*Movimiento 26 de Julio* (M-26)]	1953–1959
Dominican Republic	June 14th Movement [*Movimiento 14 de Junio* (M-14)]	1963, 1970
Ecuador	Alfaro Lives, Damn It! [*¡Alfaro Vive, Carajo!* (AVC)]	1981–1992
El Salvador	People's Revolutionary Army [*Ejército Revolucionario del Pueblo* (ERP)]	1970–1992
	Popular Forces of Liberation–Farabundo Martí (FPL) [*Fuerzas Populares de Liberación–Farabundo Martí*]	1970–1992
	Armed Forces of National Resistance [*Fuerzas Armadas de Resistencia National* (FARN)]	1975–1992
	Revolutionary Armed Party of the Workers of Central America/Armed Revolutionary Forces of Popular Liberation [*Partido Revolucionario de Trabajadores de Centroamerica/Fuerzas Armadas Revolucionarias de Liberación Popular* (PRTC/FALP)]	1976–1992
	Armed Forces of Liberation [*Fuerzas Armadas de Liberación* (FAL)]	1977–1992
	Farabundo Martí National Liberation Front [*Frente Farabundo Martí de Liberación National* (FMLN) (Union of ERP, FPL, FARN, PRTC/FALP, FAL)]	1980–present*
Guatemala	Rebel Armed Forces [*Fuerzas Armadas Rebeldes* (FAR)]	1960–1996
	Revolutionary Organization of the People in Arms [*Organización Revolucionaria del Pueblo en Armas* (ORPA)]	1971–1996
	Guerrilla Army of the Poor [*Ejército Guerrillero de los Pobres* (EGP)]	1972–1996
	Guatemalan Labor Party/Revolutionary Armed Forces [*Partido Guatemalteco del Trabajo/Fuerzas Armadas Revolucionarias* (PGT-FAR)]	1968–1996
	Guatemalan National Revolutionary Unity [*Unidad Revolucionaria Nacional Guatemalteca* (URNG) (Union of FAR, ORPA, EGP, PGT/FAR)]	1982–present*

Country	Group	Years
Honduras	Morazanist Front of Honduran National Liberation [*Frente Morazanista de Liberación Nacional Hondureña* (FMLNH)]	1967–1991
	Revolutionary Party of Central American Workers-Honduras [*Partido Revolucionario de Trabajadores Centroamericanos-Honduras* (PRTC-H)]	1977–1983
	Popular Revolutionary Forces Lorenzo Zelaya [*Fuerzas Populares Revolucionarias Lorenzo Zelaya* (FPR-LZ)]	1981–1991
	Popular Movement of Liberation "Chinchoneros" [*Movimiento Popular de Liberación "Chinchoneros"* (MPL-Chinchoneros)]	1980–1990
Mexico	Revolutionary Party of Workers and Peasants/Party of the Poor [*Partido Revolucionario de Obreros y Campesinos/Partido de los Pobres*]	1969–1974
	Communist League September 23rd [*Liga Comunista 23 de Septiembre* (L-23)]	1973–1976
	Zapatista Army of National Liberation [*Ejército Zapatista de Liberación National* (EZLN)]	1994–present†
Nicaragua	Sandinista National Liberation Front [*Frente Sandinista de Liberación National* (FSLN)]	1961–present*
Paraguay	United Front of National Liberation [*Frente Unido de Liberación National*]	1960
Peru	Revolutionary Leftist Front [*Frente Izquierdista Revolucionario* (FIR)]	1961–1963
	Army of National Liberation [*Ejército de Liberación National* (ELN)]	1962–1965
	Movement of the Revolutionary Left [*Movimiento de Izquierda Revolucionaria* (MIR)]	1962–1965
	Tupac Amaru Revolutionary Movement [*Movimiento Revolucionario Tupac Amaru* (MRTA)]	1975–1993
	Communist Party of Peru Through the Shining Path of the Thought of José Carlos Mariátegui [*Partido Comunista del Peru por el Sendero Luminoso del Pensamiento de José Carlos Mariátegui (Sendero Luminoso)*]	1980–present‡
Uruguay	Movement of National Liberation–Tupamaros [*Movimiento de Liberación National Tupamaros*]	1962–1972

(continued)

Country	Group	Years
Venezuela	Movement of the Revolutionary Left/Armed Forces of National Liberation (MIR-FALN) [*Movimiento de Izquierda Revolucionaria/Fuerzas Armadas de Liberación National*]	1960s
	Communist Party of Venezuela/Armed Forces of National Liberation [*Partido Comunista de Venezuela/Fuerzas Armadas de Liberación Nacional* (PCV-FALN)]	1961–1968

*Once guerrilla groups, these organizations have become political parties.

†The Zapatistas ceased military actions after the peace negotiations of 1996, but the organization has been forced to defend their communities from paramilitary attack. The organization continues as the main social, political, and cultural organization in many areas of Chiapas, but it has not chosen to form a political party and participate in Mexican elections.

‡Sendero Luminoso, commonly known as Shining Path, was primarily active from 1980 to 1992. Supposedly its last leader was arrested in 1999. However, there have been isolated incidents—a car bombing in 2002 and a 2008 attack on a military convoy. Its most recent leader was captured in 2012, however, it is unclear whether the organization continues in some form.

SOURCE: Compiled from information in Liza Gross, *Handbook of Leftist Guerrilla Groups in Latin America and the Caribbean* (Boulder: Westview Press, 1995).

9.4: Chile

Chile has always been a little different from its neighbors. From its earliest independence days, the country enjoyed greater stability than the rest of the region. It also enjoyed a long democratic tradition: From 1833 to 1973, Chile had only two constitutions. They provided for a strong presidency but also a strong voice from Congress. Chile's constitution gave political parties proportional representation, fostering an active political process.

Chile first became a regional power after the War of the Pacific (1879–1883), in which it fought Bolivia and Peru over the nitrate-rich region of the Atacama Desert. At the end of the nineteenth century, Chile was positioned as a serious aspirant to world power. It had a stronger navy than the United States and tremendous wealth from the export of nitrates, which were used for fertilizer and for making explosives. During World War I, Germany developed synthetic nitrates, and Chile lost its position in the market. During World War II, Chile accelerated development of its copper industry to take the place of nitrates. But by the twentieth century, Chilean entrepreneurs had exhausted the easily accessible copper deposits and needed foreign capital and technology to access the rest. Two U.S. companies came to dominate the industry: Anaconda and Kennecott.

By 1960, some 30–35 percent of the Chilean population had reached the upper or middle class. This sector was even larger in Santiago and Valparaíso, where almost half of Chile's population lived. Seventy percent of the country's population was urban, and 90 percent of them were literate. Nationwide, literacy was 84 percent. Yet the image of Chile as urbane, educated, and well to do hid gross inequality: While the per capita income of the top 5 percent was $2,300, the lower 50 percent earned only $140. Copper earnings paid for limited industrialization, whereas the countryside was little changed from the nineteenth century. Change was clearly needed.

Chile in the 1960s became a showcase for the Alliance for Progress, which invested heavily in the government of President Eduardo Frei (1964–1970). Frei represented a new political trend of the era, the Christian Democrats, who stressed the moral responsibility of government to mitigate societal ills, but stopped short of advocating revolutionary change. The Frei government promised "a revolution in liberty," and it did make some substantial changes. By 1970, the state had acquired ownership of most of Chile's copper mines. An agrarian reform gave land to 30,000 families. Farm laborers were unionized and given a minimum wage. Much of the change was funded by U.S. assistance and high prices for copper fueled by the Vietnam War.

Frei's reforms had two political results: Some felt that he did not go far enough and ended up joining the left; others felt he had gone too far and joined the right. The center lost its strength, and in 1970, the presidency was won by Salvador Allende, a medical doctor and committed Marxist who had been one of the founders of the Socialist Party in 1931. He was the candidate of six parties that had joined in *Unidad Popular* (UP, Popular Unity). Allende was a dedicated democrat: He had served as president of the Chilean Senate, served in Congress for thirty years, and ran for president four times. He had won a plurality of the vote, but not a majority, typical in Chilean elections because of the large number of parties participating.

Allende's election redefined revolution. Chileans in 1970 believed they had shown the world that a socialist revolution could be made without resorting to force of arms. Chile had acted within its 140-year democratic tradition. But the experiment was doomed to failure. Three years later, the presidential palace was in flames, Allende was dead, the military controlled the streets, and General Augusto Pinochet established a brutal dictatorship that ruled Chile until 1989. Allende's defeat came at the hands of a familiar alliance: Chilean right-wing forces and the United States, who objected to the new government's policies.

Allende's program was a radical one, calling for abolition "of the power of foreign and national monopoly capital and of large units of agricultural property, in order to initiate the construction of socialism." His economic advisors were influenced by Marxist theory but also by the structuralist theories popular in Latin America after World War II and propounded by Raúl Prebisch and ECLA, which was based in Santiago. The structuralist analysis contended that the economy could not grow because the market was limited by lack of buying power. The answer was to redistribute income, thereby increasing consumer demand. Income was to be redistributed by increasing wages while holding prices down and by expropriating large businesses, especially those that were foreign owned, so that the government could rechannel profits into social spending. Finally, a more widespread agrarian reform would make agriculture more efficient, producing enough low-cost food to feed the workforce without expensive imports. The government increased wages, gave low-interest housing loans to the poor, and created new social programs, including daycare centers, health and welfare programs, and school lunches.

For the first year, Allende's policies were effective, fueled by a foreign-reserve surplus from the high prices of copper. The government also simply printed more money and borrowed money to pay for imports of food and consumer goods. But in 1971, the economy took a sharp decline. First, copper prices dropped precipitously, then world food prices rose, and Chile had hit the limits of industrial expansion possible without increased investment. Then the United States stepped in.

The United States was wary of Allende from the start. Secretary of State Henry Kissinger commented tartly, "I don't see why we must sit with our arms folded when a country is slipping toward Communism because of the irresponsibility of its own people." President Richard Nixon gave orders to "make the economy scream." Some $8 million of CIA funds were used to fund Allende's opponents, including support for a truckers' strike that paralyzed the country and infiltration of the newspaper, *El Mercurio*, which kept up a drumbeat of denunciations, blaming the Allende government for limits imposed by the world market and the United States. The United States cut off loans and influenced other Western sources to do so as well.

By 1973, middle-class housewives were marching in the streets, banging pots and pans, complaining that they could not buy enough food for their families. At the other extreme, radical workers were taking over factories and urging Allende to take even more drastic action. The right wing, the elites, and the middle class clamored for change. The military became restive, convinced that civilian rule was ruining Chile and that only military control could restore order. General Carlos Prats, chief of the army, was loyal to Chilean constitutional structures, but he resigned when it became clear that he had lost military support. The next in command was Pinochet, who coordinated with the navy and ordered the occupation of the cities of Valparaíso, Concepción, and Santiago. On 11 September 1973, the presidential palace, *La Moneda*, was bombed. Allende, who was inside, committed suicide rather than be taken by the Pinochet forces.

Chileans were shocked. Theirs had been a peaceful, democratic country. Suddenly, it was a country of brutal military rule. Santiago's two stadiums were filled with prisoners, many of whom were brutally murdered. Between 3,000 and 10,000 people were killed during and after the coup. Left-wing political parties were outlawed, and Pinochet systematically eliminated rivals in the military and removed both the left wing and the moderates in unions, universities, and other institutions. Chile descended into a state of siege.

For the rest of Latin America, the lesson was clear: Revolutions were not won at the ballot box. As Augusto César Sandino said, "Liberty is not won with flowers but with gunshots."

9.5: Nicaragua

The dominant reality of twentieth-century Nicaraguan history was the lack of opportunity the people had to govern themselves. The United States intervened in 1909 to overthrow President José Santos Zelaya and occupied the country, with brief interruptions, until early 1933, a period characterized by disintegration of the state and the economy. During its final years of occupation, the United States created the Guardia Nacional and appointed as its commander Anastasio Somoza García. The maintenance of order and stability was the primary duty of the Guardia, and it received U.S. training, financing, and equipment to fulfill it. The Somoza dynasty—Somoza was succeeded in turn by his sons, Luis and Anastacio—ruled from the mid-1930s until 1979. Even the Kissinger Commission declared that the last Somoza's regime "gave new meaning to the term *kleptocracy*, that is government as theft." By 1979, the Somozas owned 20 percent of the arable land (the best lands with good soil and ready access to roads, railroads, and ports), the national airline, the national maritime fleet, and a lion's share of the nation's businesses and industries. One of the great symbols of Somoza rapacity was Plasmaferesis, a blood plasma factory where the poor sold their blood for as little as $1 per liter; Somoza then sold the blood to the United States for ten times that amount.

The Somozas were loyal to only the Guardia, a small coterie of relatives and members of the old elite who supported the regime, and the United States. Nicaragua's voting record in both the Organization of the American States and the United Nations, for example, conformed 100 percent with the United States. The senior Somoza provided airstrips to support the overthrow of Arbenz in Guatemala. The Bay of Pigs invasion was launched from Nicaragua's Atlantic coast; indeed, Luis Somoza was on hand to send them off and asked that they bring him back a hair from Fidel's beard.

During the long Somoza decades, the Nicaraguan people exercised neither political power nor influence, and the Guardia dealt swiftly and brutally with anyone courageous or foolish enough to try. The voice of the Nicaraguans remained muted from 1909 to 1979.

In 1961, an opposition group emerged to oppose the third member of the dynasty, Anastasio Somoza Debayle. It was inspired, like so many others, by the Cuban Revolution. But they also looked to their own history and particularly to the struggle of Augusto César Sandino, who fought against U.S. Marine occupation from 1927 to 1933, when he was killed on Somoza García's orders. The *Frente Sandinista de Liberación Nacional* (FSLN; Sandinista Front for National Liberation) advocated democracy, agrarian reform, national unity, emancipation of women, the establishment of social justice, and an independent foreign policy.

The struggle of the Sandinistas against the well-armed and trained Guardia Nacional proved to be long and bloody, but the greed and brutality of the last Somoza were their most effective allies. One example is Somoza's handling of the massive 1972 earthquake, which destroyed much of Managua. Some $250 million in aid was rushed to Nicaragua, only to be stolen by Somoza and his cronies. Opposition was met by brutal torture in Somoza's prisons. By the late 1970s, even the small but potent middle class and members of the tiny elite joined the opposition, and Somoza's military arsenal could no longer repel popular wrath. As in Cuba, the United States withdrew its support when it was clear Somoza could not win. Somoza fled, and the victorious rebels led by the FSLN entered Managua triumphantly on July 19, 1979.

The challenge of reconstruction surpassed even the task of overthrowing the regime. The Sandinistas inherited an economy in shambles. Somoza had bombed the cities and industries. War damages amounted to approximately $2 billion; 40,000 were dead (1.5 percent of the population), 100,000 were wounded, 40,000 children were orphaned, and 200,000 families were without homes. The national treasury was empty, and foreign debt exceeded $1.6 billion. Add to these woes the reality of an underdeveloped country: Fifty-two percent of the population was illiterate, life expectancy was slightly more than fifty-three years, infant mortality was 123 per 1,000 live births, and malnourishment plagued 75 percent of the children.

During the early months after the victory, a wide variety of political groups supported the revolution. However, Sandinista emphasis on redistribution of wealth via agrarian reform and support for labor was not the goal of many middle- and upper-class Nicaraguans. They withdrew their cooperation and became

Thousands of Sandinista supporters crowd the Plaza de la Revolucion to commemorate the anniversary of Augusto César Sandino, the anti-imperialist hero from whom the revolutionaries took their name. (Photograph by Julie A. Charlip)

critical, some even hostile. Such behavior followed the classic model of revolutions: United to overthrow a common enemy, a heterogeneous group brings about the final victory. Later, the temporary alliance disintegrates as some realize that their own interests will not be served by revolutionary change. As a last resort, the disaffected try to stop the revolution by force, usually allying with some sympathetic foreign power to enhance their strength. Events in Nicaragua followed this predictable pattern.

The Catholic Church was divided over the revolution. The hierarchy had always been allied with Somoza, but the brutality of the final years led the Church to split from the regime. Thus, the religious hierarchy gave its blessings to the overthrow of Somoza, an extremely unusual position for the Church in Latin America. But though the hierarchy opposed Somoza, they did not support many of the changes proposed by the Sandinistas. They, like the middle class and elite opposition, wanted what the Sandinistas called "Somocismo without Somoza."

Although the church hierarchy feared the revolution, many lower-ranking priests and nuns embraced it. The religious represented a new wave of Catholicism, liberation theology, which had its roots in the Second Vatican Council (1962–1965), convoked by Pope John XXIII and continued under Pope Paul VI. The result of Vatican II was the Pastoral Constitution on the Church in the Modern World, urging Catholics not to shirk their responsibility to the poor and oppressed. Priests and nuns around the world responded by becoming involved in social organizing.

The greatest impact of Vatican II was felt in Latin America, where economic and social inequalities were extreme. Latin American theologians developed a new doctrine, called liberation theology, drawing on Jesus's first sermon in Luke (4:18–4:21), in which the gospel says Jesus had come "to liberate those who are oppressed." Liberation theology was given impetus by the Second Latin American Bishops Conference in Medellín, Colombia, in 1968. Pope Paul spoke at the conference, the first pope to visit Latin America. He told the bishops, "We wish to personify the Christ of a poor and hungry people." The bishops accepted the "preferential option for the poor," that is, they prioritized social justice. They also chose to deal with the shortage of trained religious by giving their blessing to the organization of Christian base communities (*comunidades de base*), which were lay Bible groups. Many Sandinistas came to the revolution through these communities and other Christian youth groups, and several priests joined the government. The religious component of the revolution set the Nicaraguan revolution apart from the revolutions in Mexico and Cuba.

The Sandinista government embarked on an ambitious program to improve the quality of life of the majority. A 1980 literacy campaign reduced illiteracy from 52 to 12 percent. The government followed up by doubling the number of schools in four years and making education free from preschool through graduate studies. In 1988, more than 1 million Nicaraguans (40 percent of the population) were studying.

Attention was also focused on expanded health care. Nicaragua eliminated measles, diphtheria, and polio, diseases that once took a heavy toll among children. Clinics and hospitals sprang up in the countryside and in small towns. The number of health centers multiplied from twenty-six to ninety-nine, and Cuban physicians and nurses volunteered to staff them. Infant mortality fell 50 percent. Despite the shortage of medicine, medical equipment, and physicians, health-care delivery was so impressive that the World Health Organization cited Nicaragua in 1983 as a model nation. A proper diet further explained the improving health of Nicaraguans. Caloric intake rose because more basic foods were now available to larger numbers of people.

One of the three founders of the FSLN, Carlos Fonseca, had vowed, "In Nicaragua, no peasant will be without land, nor land without people to work it." A far-reaching agrarian reform law based on Sandino's hopes, Fonseca's promise, and global experiences with rural restructuring was promulgated in 1981. It provided land for anyone who wanted it and would work it. As the largest nation in Central America, Nicaragua (54,864 square miles) had more than enough land for its relatively small population (3.2 million). The principal goal was to put the land into use so that Nicaragua could both feed itself and export crops for needed foreign exchange. For six years, the government distributed only land that once belonged to the Somozas or their closest allies, abandoned lands, and unused lands. In 1986, after successfully distributing nearly 5 million acres of land to 83,000 families, the government made significant changes in the Agrarian Reform Law, permitting the expropriation of any unused land without compensation, and began to distribute lands already in production in areas where rural workers demanded land. However, the government continued to guarantee the right to private property, and the majority of the land remained in private hands.

There was also an important cultural component of the revolution. Popular culture centers offered classes in folk dance. The Ministry of Culture supported the primitivist art movement, which began at Father Ernesto Cardenal's community on the island of Solentiname. Popular Poetry Workshops sprang up in barracks, factories, cooperatives, and neighborhoods, encouraging Nicaraguans to emulate national hero Rubén Darío, the father of Spanish modernism. Musicians such as the brothers Luis Enrique and Carlos Mejía Godoy wrote music dedicated to the revolution, joining similar *nuevo canción* movements in Cuba and Chile. Murals adorned the buildings, painted by Nicaraguans and the many foreigners who came to support the revolution.

Women played an important role in the Sandinista revolution, starting during the armed conflict, when they constituted 30 percent of the people in arms. To help mobilize women, the revolution opened day-care centers and broadened employment opportunities. Although no women served on the Sandinista directorate until the 1990s, there were several women in high positions, including Dora Maria Telles, the secretary of health, who was second in command in the daring takeover of the National Palace in 1978 and headed the column that took León in 1979. However, many women chafed against the continuing double standard and tendency to prioritize defense over domestic needs. For example, to avoid further conflict with the Church the Sandinistas refused to legalize abortion; but the leading cause of death among women of child-bearing age was self-induced abortions. Despite these conflicts, many women felt that the Sandinista revolution provided more opportunities for them and benefits for society than any previous system.

Nationalism, socialism, and Christianity converged in Nicaragua to offer a different model for change than other revolutions had presented. The revolution drew international attention, including the wary gaze of the U.S. government. President Jimmy Carter had tried to prevent the Sandinistas from taking power in the waning days of the war. But under President Ronald Reagan, Nicaragua was labeled a communist threat.

Washington allied itself with the remnants of the discredited National Guard, the dispossessed elite, and members of the frightened middle class to drive the Sandinistas from power. The CIA financed, trained, and armed counterrevolutionaries, known as the *contras*, short for the Spanish term *contrarevolucionario*, who invaded Nicaragua as early as November 1979 from Honduras and later from Costa Rica. The contras became known for their brutality; they particularly took aim at schools and clinics, the symbols of Sandinista change. As the war intensified, the Sandinistas lacked the money to rebuild these targets, and teachers and doctors were afraid to enter the war zone. In 1984, the CIA mined the harbors of Nicaragua in violation of international law, drawing the condemnation of the international community. In 1985, President Reagan imposed a trade embargo on Nicaragua, which plunged the economy into crisis.

Daniel Ortega, one of nine comandantes in the leadership of the Sandinista National Liberation Front, became Nicaragua's first democratically elected president. (Photograph by Julie A. Charlip)

In 1984, Nicaragua held its first free elections, and eight political parties participated. The Sandinistas won 67 percent of the vote. In early 1990, the revolution celebrated its second presidential election, and like the earlier election, it was open and democratic. However, much had occurred during the intervening years. In 1984, Nicaraguans were buoyant, hopeful, and proud of the achievements of the revolution; by 1990, they were exhausted from the prolonged war pressure from Washington. The FSLN lost the 1990 presidential elections to a heterogeneous political coalition of fourteen parties, the National Opposition Union (*Unión Nacional Opositora*, UNO), financed and advised by Washington. The Sandinistas were still the strongest single party, garnering 41 percent of the vote. The new president, Violeta Chamorro, had promised that her close ties to the United States would benefit

Nicaragua by bringing in aid and investment. But once the United States had achieved the goal of unseating the Sandinistas, Nicaragua receded from the U.S. agenda. Subsequent Nicaraguan governments steadily rolled back Sandinista achievements.

The revolutions in Cuba, Chile, and Nicaragua were landmark events in Latin American history. Cuba showed what was possible, but it also showed the limits of being a small country torn between superpowers. Chile tried to show that a peaceful path was possible, but as the revolution died in flames, Latin American revolutionaries contended that it could have survived had Allende dismantled the armed forces and armed the workers and farmers. Nicaragua tried to show a new way via political pluralism and a mixed economy, but it met the usual intransigence from Washington, which was unwilling to let Latin Americans decide their own fate. By 1990, many wondered whether the era of revolutions had indeed come to an end. And if indeed there was no revolutionary option, what chance was there for political, economic, and social change?

Questions for Discussion

1. What did revolutionaries want to achieve and why? How did they go about it?
2. Why has the United States so vigorously opposed these revolutions?
3. Why did the Cuban Revolution have so great an impact on Latin America?
4. Why did reform programs fail to appease those who supported revolutionary movements?
5. Can we consider the Allende government to have represented a revolution in Chile?
6. How did the Sandinista revolution exemplify the lessons learned from earlier revolutions and the particular strains of the 1980s?

10

Debt and Dictatorship

The central dynamic of Latin American history in the late twentieth century was the tension between equity and efficiency, democracy and dictatorship. Leaders searched for ways to make their economies grow and sometimes for ways to redistribute some of the wealth. Reformers tried to appease the masses to forestall revolution, but all too often their failures led to unrest, reaction, and repression. Throughout the 1970s and 1980s, Latin American economies careened from boom to bust, democratic openings alternated with brutal military regimes, and the specter of revolution continued to haunt and inspire.

In many ways, the 1970s and 1980s reenacted the problems of the 1920s and 1930s. Extensive loans fueled economic growth, but the debt became a trap when an international recession made repayment impossible. Economic distress fueled popular mobilization, which was crushed by military dictatorship. But the late twentieth century brought a new kind of military dictatorship that focused on economic development and carried repression to new levels. The military eventually were defeated, not so much by a new political movement as by their ultimate economic failures.

10.1: Changing Economic Patterns

As early as 1965, some countries in Latin America were experiencing the limits of import-substitution industrialization (ISI) that had dominated the region by default or conscious policy for much of the twentieth century. Import substitution was blamed for chronic inflation because tariffs and overvalued exchange rates artificially raised prices of foreign goods, while local producers had no incentive to produce more efficiently. Although many countries produced more consumer goods, they still needed to import capital goods, the sophisticated machinery needed in manufacturing. Capital-intensive industry, which relied more on machines than on people, did little to increase urban employment. But increased mechanization and production of export crops drove more people to the cities. The shantytowns that had started to circle Latin America's capitals in the 1950s became massive centers of poverty in the 1970s. Even the middle class was in straitened circumstances as policies meant to foster industrialization concentrated income among the upper sectors. The result was a limited market for the new consumer goods. But the goods, inefficiently and at times shoddily produced behind tariff barriers, could not be exported.

One theory held that Latin America needed more money to invest in modernizing infrastructure and to improve and expand industrialization for export. The multilateral and official lending organizations, such as the World Bank, did not offer the large amounts needed or carried too many restrictions on the use of the funds. Latin American leaders looked elsewhere and found that their interest in borrowing large sums of money fortuitously coincided with an increase in international banking and investment.

In the 1970s, the money poured in. Lending was fueled in part by the oil crisis of 1973, when the Organization of Petroleum Exporting Countries (OPEC) raised prices and restricted exports after the Arab–Israeli War in 1973. The OPEC countries deposited the enormous sums of money they earned (petrodollars) in U.S. and European banks (Eurobanks), which eagerly sought investment opportunities. Latin America seemed an ideal place for such ventures, leading to a new dance of the millions, echoing the boom of the 1920s. According to Mexico's minister of finance David Ibarra, "I had many bankers chasing me trying to lend me more money." Interest rates were lower than inflation, meaning the loans would be paid back, in real terms, at less than they had cost.

In the late 1960s, Latin America was borrowing around $300 million a year from private lenders; by the early 1970s, the region's debt was $34 billion. As long as the money yielded productive investments, debts could be paid. World trade was growing, commodity prices were at an all-time high, and Latin America was prospering.

The prosperity depended on several key factors: a supply of foreign funds to invest; healthy international trade to generate repayment income; and, perhaps most importantly, a docile, low-cost workforce. Wage limits would enable Latin American governments and capitalists to increase corporate profits, making the region more attractive to foreign companies. The concentration of wealth would, in theory, increase money for investment; however, in Latin America, elites pocketed their increased income—often sending it to safer foreign banks—and depended instead on borrowing to fuel investment.

The constraint of the labor movement was most easily accomplished by authoritarian governments, who contended that labor organizers were part of an international communist movement threatening western democracy. Given the dominant Cold War fears, these authoritarian contentions easily won the support of Latin American elites and middle classes and the backing of the United States.

The success of these policies was measured in the spectacular growth rates of Latin American economies in the 1970s. The "Brazilian miracle" produced average annual growth rates of 11 percent in gross domestic product, and Mexico boasted growth of 6–8 percent a year. Such rates are staggering, considering that 3 percent growth in gross domestic product had been considered healthy.

Most of the debt was taken on by governments, which passed the money on to private ventures, funded state-run business, or created a combination of the two. Much of the money went to industrial investment, particularly in Brazil, where there were significant increases in the production of steel, automobiles, and machinery. Even more went to the creation of new infrastructure: Brazil's Itaipu hydroelectric power plant, Venezuela's Guri Dam, and the Yacyretá Hydroelectric Dam on the Argentina–Paraguay border. Even Guatemala built the Chixoy Dam. The huge development projects required enormous investment but would take years to complete before producing income.

The crunch came with the second oil crisis in 1979, following the fall of the Shah of Iran. In an attempt to reign in galloping inflation, the U.S. Federal Reserve raised interest rates above 20 percent. The increase in U.S. interest rates echoed in European financial markets, where interest rates were set as a percentage above the London Inter-Bank Offered Rate (LIBOR). LIBOR rose to 17 percent in 1981 and 14 percent in 1982. The loans made to Latin America had floating, not fixed, interest rates, generally set as a percentage above LIBOR. Suddenly, Latin American governments found their interest rates soaring from 9 to 19 percent.

Then came the final blow: Because of worldwide recession, buyers abroad stopped or drastically reduced their purchases of Latin American goods. Without sales, there were no earnings to repay the spiraling debt. In 1981 and 1982, Latin American governments borrowed frantically, often at high short-term rates, to simply meet the debt service on existing loans. According to the Bank for International Settlements, the region borrowed $4 billion a month during the latter half of 1981.

Even oil-rich Mexico, which discovered more oil reserves in the 1970s, had dipped into the international lending pool. Rather than rely on its own income from new oil discoveries, the Mexican government borrowed, using the money to fund petrochemical development, steel production, and upgrading of infrastructure. Much of the spending was nonproductive, including subsidies to private industry and lavish lifestyles among the elites.

On August 12, 1982, Mexican Finance Minister Jesús Silva Herzog reported to the United States and IMF that Mexico could not repay its loans. The next day, debt negotiations began. In November, Brazil was in the same boat, and the crisis spread throughout the Third World in the next year. The specter of nonpayment on loans threatened the entire international banking system. Banks depend on interest income from loans to pay interest on depositors' capital and to repay other banks from which they borrowed. This was particularly true of U.S. banks that had borrowed petrodollars from Eurobanks. They needed to keep interest income flowing into the banking system.

Negotiations led to debt refinancing rather than to forgiveness. A key player in the negotiations between Latin American governments and international banks was the IMF, which promised to provide funds to help Latin American countries meet their obligations. But there was a catch: Borrowing countries needed to cut government spending. The final effect was to increase debt even further, with new loans to repay interest on the old loans. The cost of repaying those loans was then transferred to the Latin American masses because the IMF imposed austerity programs, which cut social spending.

Mexico's foreign debt jumped from $14.5 billion in 1975 to $85 billion in early 1984, surpassing $110 billion by 1989. The price was paid by the masses. In 1983, President Miguel de la Madrid, following IMF recommendations, cut government spending in half, eliminating subsidies for some basic foods and for transportation. While inflation continued in the 75–100 percent range, the president held wage increases between 15 and 25 percent. From 1981 to 1987, consumer prices increased fourteen times and unemployment and underemployment rose to approximately 45 percent. The IMF-imposed austerity program created no jobs, even though Mexico needed 700,000 new jobs a year to absorb new entrants into the workforce.

Throughout the 1980s, per capita income fell an average of 1.1 percent a year, leading the Economic Commission on Latin America and the Caribbean (ECLAC) to dub those years "the lost decade." By 1990, the region's foreign debt exceeded a staggering $420 billion, several times Latin America's annual income. Payment of interest alone came to absorb more than 40 percent of all export earnings. A significant percentage also went to fund the military.

The leadership that adopted the disastrous economic strategies of the 1970s and 1980s were mostly military regimes. Their initial success, along with their willingness to use brutality, kept them in power. The failure of their economic policies led to their downfall.

10.2: Military Models for Change

The shadow of the military has loomed large over Latin America. First, it was caudillos of the early independence era who ruled by force. Professionalization of the armed forces in the late nineteenth century, generally with training from Great Britain or Germany, seemed to successfully subordinate the institution to its proper role as defender of national interests, led by civilian governments. Then the unrest of the Depression era brought a new military figure, a caudillo-like dictator who rose from the professional armed forces.

Yet another military model emerged in the 1960s–1980s—a bureaucratized military that believed it could rule better than civilians. Military leaders believed they could restore order and bring development. These new military regimes tended to rule as an institution, rather than under the leadership of a caudillo-like personalistic dictator, with the notable exception of Chile's Augusto Pinochet. It was a military that distrusted civilians and was willing to use violence to enforce its agenda.

The traditional concern of armed forces is to defend the country against foreign attack. Though some Latin American countries resorted to arms to settle border disputes, the region did not face foreign attacks in the late twentieth

Table 10.1 Antipolitical Military Regimes, 1964–1990*

Country	Years
Ecuador	1963–1966; 1972–1978
Guatemala	1963–1985
Brazil	1964–1985
Bolivia	1964–1970; 1971–1982
Argentina	1966–1973; 1976–1983
Peru	1968–1980
Panama	1968–1981
Honduras	1972–1982
Chile	1973–1984
Uruguay	1973–1984
El Salvador	1948–1984†

*In some cases, dating the beginning of these regimes is difficult inasmuch as military governments succeeded one another with changes in policies and personalities but maintained, overall, an antipolitical outlook and national security rationale for direct military rule. Sometimes brief civilian interludes seemed to interrupt military domination (e.g., in Guatemala from 1966 to 1969, and Panama after General Omar Torrijos's death in 1981).

†El Salvador was dominated by military regimes from 1948 onward. It may be more appropriate to date the last episode from 1979, although the developmental focus and antipolitical themes of the junta headed by Major Oscar Osorio in 1948 anticipated a "Peruvianist" (or populist) version of antipolitics.

SOURCE: Brian Loveman, *For la Patria: Politics and the Armed Forces in Latin America*. Copyright 1999. Reprinted by permission of SR Books, now an imprint of Rowman & Littlefield Publishers, Inc.

century. But the military believed that it faced an internal threat from the forces of the political left. They were encouraged in this view by the United States, which poured money into the coffers of Latin American armed forces. The Cold War between the Soviet Union and the United States after World War II came to be fought not in their own countries but in developing nations seen as vulnerable to communist propaganda. The Cuban Revolution became the proof of a clear and present danger.

These Latin American generals, however, went even further in developing a doctrine of national security and development. According to this doctrine, a strong economy was essential for a country's security, making economic development as important an issue as military protection. Furthermore, it was impossible for a country to create a strong economy, especially by attracting foreign investment, if there was domestic unrest and subversion. By linking these two ideas, the suppression of supposed subversion gave the military a role that linked national defense to the economy.

Brazil was the first country in the region to fall under the rule of the new military model, after the overthrow of the government of João Goulart in 1964. The new regime was different from any that was in the region before. It was not led by one charismatic or all-powerful military caudillo; there was no Perón, no Somoza, not even a Vargas. The officers agreed that there would not be one figure as dictator; instead, Brazil had a succession of military presidents, all generals (Humberto de Alencar Castello Branco, 1964–1967; Artur da Costa e Silva, 1967–1969; Emílio Garrastazu Médici, 1969–1974; Ernesto Geisel, 1974–1979; and João Baptista Figueiredo, 1979–1985).

Using extensive foreign loans, the generals oversaw the transformation of Brazil's economy. In 1960, industrial goods constituted a mere 3 percent of the country's exports, but they jumped to 30 percent by 1974. Brazil's factories turned out steel, automobiles, military equipment—including tanks and submarines—and computers. The results: GNP grew 10 percent a year. *The Economist* called it the "Brazilian miracle."

Because a strong economy was essential to Brazil's national security, anyone who disagreed with the agenda was viewed as a threat to the nation's security. And disagreement was not long in coming. The generals did indeed produce a Brazilian miracle, achieved by shifting income from lower to upper sectors, concentrating wealth at the top to create investors and a market, and contracting large foreign loans to make up for the lack of investment by elites, who largely chose to pocket their newfound wealth. The loans were also necessary because Brazil's low rate of taxation limited government funds available for investment.

As the *Los Angeles Times* pointed out in an editorial on July 21, 1974, "Despite Brazil's impressive growth rate, the gap between the rich and the poor is wider than ever." The statistics were grim. Of the total gain in Brazilian income during 1964–1974, the richest 10 percent of the population absorbed 75 percent, whereas the poorest 50 percent got less than 10 percent. Compounding that inequity was a regressive tax system putting the heaviest burden on the working class.

The military government, determined not to let union organizers stand in the way of the miracle, formed secret police and death squads to torture and eliminate people they saw as threats. The government solicited the help of major U.S. and European corporations with subsidiaries in Brazil, which routinely provided lists of suspected union activists. As a result, there were no strikes against major companies in Brazil from 1969 to 1978. The military governments continued in power until 1985.

Sadly, Brazil was not alone in its turn to brutal military dictatorship. In South America, military dictatorships also came to power in Argentina, Bolivia, Chile, Ecuador, and Uruguay. Paraguay had already been under the iron hand of General Alfredo Stroessner since 1954.

As in Brazil, military leaders in Argentina believed that civilian control was irresponsible and left the country in chaos. At first, the military would step in to restore order and then turn government back to civilians. In 1955, the armed forces ousted Juan Domingo Perón, who fled into exile. The Peronist party was outlawed as the military ruled from 1955 to 1958. The military then allowed elections, won by Radical Civic Union candidate Arturo Frondizi (1958–1962), who concentrated on building the steel and oil industries. Following the common pattern, economic restructuring led to economic difficulties, which in turn led to unrest and military intervention. A new election in 1962 brought Radical Arturo Illia to power, but in 1966, he was ousted. This time the military kept power until 1973, when Perón returned from exile to reclaim his mantle as leader of the workers. His vice president was his new wife, Isabel, a former nightclub dancer who had been his secretary while in exile.

When Perón died in July 1974, Isabel found herself trying to solve problems that would have challenged an experienced politician. Economic chaos reigned, armed guerrillas challenged the government, and finally the 1973 oil crisis crippled the economy. This time the military stepped in with a vengeance. From 1976 to 1983, a series of generals ruled the country (Jorge Rafael Videla, 1976–1981; Roberto Eduardo Viola, March–December, 1981; Carlos Alberto Lacoste, December 1981; Leopoldo Galtieri, December 1981–June 1982; Alfredo Oscar Saint-Jean, June–July 1982; and Reynaldo Bignone, July 1982–December 1983).

Like the Brazilian generals, they worried about national security. And like the middle classes of Brazil who had urged the 1964 coup against Goulart, the Argentine business sector turned to the military. The business community was concerned about guerrilla kidnapping of foreign businessmen (170 kidnappings in 1973), violence at factories, and inflation rising to 30 percent despite price controls. The result was a meeting in September 1975 between the leading business organization, the Argentine Industrial Union, and Army Chief of Staff General Jorge Rafael Videla, in which the two groups agreed on a coup that took place six months later. "The 1976 coup was either silently accepted or overtly supported, since virtually no political party tried to mobilize society in defense of the democratic system," according to Luis Moreno Ocampo, who later prosecuted the generals. "The coup was not simply a case of the military imposing its will upon a reluctant civil society but the result of a civic–military alliance, which found support in the international community."

The generals disbanded Congress and the Supreme Court and launched a campaign of murder and torture that killed perhaps 30,000 people. However, the generals were concerned about their international image, especially as the brutal regime of General Augusto Pinochet across the Andes in Chile was drawing international outrage. The Argentine generals wanted to keep their targeting of dissidents clandestine. In Argentina, supposed subversives were kidnapped and tortured in some 340 secret detention centers. The bodies of more than 10,000 who were killed were dropped into the ocean from planes or buried in mass graves. Because civilian government had been destroyed, and all means of communication were censored, the government was simply able to deny that anything had happened. A new noun entered the Argentine vocabulary— *desaparacido* (disappeared).

The military–business alliance supported economic liberalization, which consisted largely of dropping tariff barriers established under ISI and forcing local industry to compete with lower-priced imports. Government reduced the public sector, bloated by years of populism and ISI, and sold public enterprises, cut wages, and increased taxes. Some domestic manufacturers collapsed in the face of competition, and their laid-off workers swelled the ranks of the unemployed, allowing surviving firms to pay lower wages. New investments were funded by massive foreign borrowing.

Finally, the generals were brought down in 1983 by their failure in the one legitimate role they could claim—defenders of the nation. In an attempt to divert domestic attention from the crumbling economy and increasing repression, the generals launched a war to reclaim the Malvinas Islands from the British, who had de facto control of the island since the 1830s. The Falklands, as the British called it, were populated by fewer than 2,000 English-speaking residents and 600,000 sheep. The generals assumed that Great Britain would not fight over the insignificant and distant possession and successfully rallied the population around the flag. But British Prime Minister Margaret Thatcher chose to meet the Argentine challenge to rouse British patriotism and deflect attention from Britain's economic crisis. The much stronger British forces easily defeated the Argentines, killing thousands of young, ill-trained troops. It was the last straw for the Argentine military government.

A 1977 poster from a New York protest against the Chilean dictatorship is an example of the solidarity movements that arose in the United States in the 1970s. (Library of Congress)

Though Argentina had suffered more than its share of military coups throughout the twentieth century, Chile had a long history of rule by democratic institutions. Even Chile's path to revolution had taken a democratic, electoral route. Because of that history, the military coup that toppled Allende in 1973 was particularly shocking. Like the other military dictatorships, Chile under Pinochet was characterized by repression and brutality. Unlike Argentina and Brazil, however, the Pinochet government sought to limit governmental involvement in the economy.

The Pinochet government subscribed to monetarist economic policies. Popularized by the conservative economists of the University of Chicago, nicknamed "the Chicago boys," monetarism advocated minimizing governmental control over the economy, extensive budget cuts, reduction of tariffs, and incentives for foreign investments. The dictatorship divested the state of more than 550 businesses. The budget was cut by virtually dismantling social security and pension plans. Between 1977 and 1980, Pinochet balanced the budget, reduced inflation, and witnessed substantial economic growth. In 1982, Milton Friedman, Chicago's theoretician of monetarism, wrote, "Chile is an economic miracle." Friedman conveniently confused growth with development: Chilean economic growth certainly did not improve the living standards of the majority.

Chile's indisputable economic growth depended largely on a substantial inflow of foreign credit, which financed the import of luxury goods for the elites. Pinochet did not expand Chile's limited manufacturing sector, focusing instead on export of primary products, expanding the export of grapes and wine. Much of Chile's growth depended on high prices for copper, which accounted for 95 percent of export earnings. But the world price of copper plummeted from $1.70 in 1975 to $0.52 per pound in 1982. Chile's export model was no more successful in the long run than was Brazil's industrialization model. The recession of the 1980s sunk all boats.

For many, Chile's miracle was a myth. Rising foreign debt accompanied by falling export earnings revealed the fragility of Chile's economic growth. In 1982, the GNP fell 14 percent; in 1983, it dropped another 3 percent. Chile holds a record for the steepest drop in GNP in Latin America. By the mid-1980s, the economy lay in ruins. Foreign debt, by 1988 in excess of $20 billion, gave Chile the dubious distinction of having one of the highest per capita debts in the world. Servicing that debt consumed 80 percent of the nation's export earnings. Bankruptcies multiplied; unemployment soared. The *Los Angeles Times* (February 22, 1983) reported, "Businesses are going bankrupt at a record pace, the banking industry has all but collapsed, and the country is dangerously near default on its foreign debt."

The poor and the middle class felt the brunt of the economic deterioration. Chile's Cardinal Raúl Silva Enriquez observed, "I could be wrong, but never in my life have I seen such a disastrous economic situation." One Chilean wryly observed, "The free-market policies of the Chicago Boys destroyed more private enterprises in the past year than the most radical sectors of Allende's coalition dreamed of nationalizing in three years; they have turned more middle-class people into proletarians or unemployed than any Marxist textbook ever described." British economist Philip O'Brien labeled the Chilean economy in 1984 as a "spectacular example of private greed masquerading as a model of economic development."

The military dictatorship's economic experiments left 7 million out of 12 million Chileans impoverished. Observers estimated a sevenfold increase in the number of people living in substandard conditions. They concluded that 10 percent of the population benefited from the "miracle" at the expense of the other 90 percent.

The military rulers of these countries did not restrict their authoritarian policies to their own national territories. In 1976, Argentina, Bolivia, Brazil, Chile, Paraguay, and Uruguay joined in what became known as Operation Condor, organized by the Pinochet government. The purpose was to help like-minded governments track down dissident refugees and assassinate them in other countries. Chile provided the leadership of Secret Police Director Manuel Contreras as well as computers and centralized services. The United States facilitated communications among South American intelligence chiefs. Victims of Operation Condor included Chilean General Carlos Prats and his wife, Sofia Cuthbert, killed in Buenos Aires in 1974, and the car-bombing assassination of former Allende government diplomat Orlando Letelier and Ronni Moffitt, U.S. research associate, in Washington, DC, in 1976.

10.3: War in Central America

In fundamental ways, Central America serves as both an extreme and instructive example of the crises of underdevelopment besetting late twentieth-century Latin America. The region can be described by the classic dependent growth model, an emphasis on exports rather than internal production and on profits rather than on wages. The economies grew but could not develop. Growth depended on highly cyclical external demand for its primary products: coffee, bananas, cotton, sugar, and beef. Any benefits from periodic growth accrued to a small portion of the population; the majority remained marginalized and voiceless in a society that was run neither by nor for them. Conservative estimates place 60 percent of the population in the impoverished category. No amount of economic growth could help—or had helped—the masses achieve a satisfactory standard of living. In late 1986, in a homily during Mass at the cathedral in San Salvador, Archbishop Arturo Rivera y Damas stated, "The real cause of underdevelopment is the ideological, economic, and political dependence of our countries."

Most of Central America afforded its citizens few democratic rights. Occasionally the elite, and much later the middle class, exercised the formalities of democracy for their own enhancement, but the military always kept a sharp eye on such rare experiences and readily intervened if the ballot box suggested even a hint of social, economic, or political change. Electoral farces in Guatemala, El Salvador, Honduras, and Nicaragua in the 1960s and 1970s confirmed traditional forces in power or served as excuses for further military intervention. To those who wanted to challenge underdevelopment, poverty, dependency, authoritarian rule, and social inequities, elections offered no opportunity, prompting guerrilla movements in Guatemala and Nicaragua in the 1960s.

Like the rest of the region, Central America attempted economic change in the 1960s. The five nations agreed to create the Central American Common Market (CACM) in 1960–1963 with hopes that a regional market could sustain some import-substitution industries that no single Central American nation could support. But the CACM had weaknesses from the start. Industrialization was capital intensive, creating few jobs. Heavy expenditures for machinery and technology drained hard currencies. Many plants assembled rather than manufactured goods—importing component parts, putting them together, and exporting the finished product. Such assembly plants did little for the local economy except employ a few workers at modest salaries. Profits from those few industries left the region because 62 percent of all industries were in foreign hands.

The CACM was also characterized by unequal industrial growth, which ignited regional rivalries. Guatemala and El Salvador boasted the lion's share of industrialization, to the annoyance and economic disadvantage of Honduras, Nicaragua, and Costa Rica. Nonetheless, El Salvador's assembly plants provided little employment. Tiny El Salvador had insufficient land for its population, with more than 6.5 million people inhabiting 8,260 square miles—twice the population of Nicaragua in less than one-sixth the area. Most of that land was controlled by a group of elites so small that they were called "the 14 families," though their numbers were certainly far greater. Roughly 2 percent of the population owned 60 percent of the arable land. The Salvadoran escape valve was the Honduran border; by 1968, some 300,000 Salvadorans lived illegally in Honduras.

Honduras, in contrast, has a relatively low population density. For many years, there was little competition for land in the Pacific zone of the country because the economy centered on the Atlantic region's banana enclave. Hondurans tended to have access to land, though not all was of good quality, and landholdings tended to be small. Subsistence farming came under pressure in the 1950s, however, as the government encouraged agricultural modernization and export diversification. For the first time, beef, cotton, and coffee became significant export products, pushing Honduran subsistence farmers off their land. By 1965, landlessness had become a problem. As Honduran campesinos began to organize, politicians sought to deflect their anger by using Salvadoran immigrants as a scapegoat. Salvadorans were targeted as squatters in the Pacific zone and low-wage workers in the Atlantic. Salvadorans began to leave Honduras in droves, some expelled by the government, others escaping the new antipathy of Hondurans. El Salvador, which could ill afford the immigrants' return, claimed that Honduras was brutalizing the immigrants.

Conflict between the two countries came to a head in June 1969 when riots broke out during soccer matches leading up to the World Cup games. In July, El Salvador invaded Honduras; the "soccer war" lasted only four days before the Organization of American States succeeded in ending the violence, but it took ten years to heal the diplomatic wounds, and the CACM never recovered. Limited industrialization clearly was not going to be the answer to Central America's problems.

Like the rest of the region, Central American governments availed themselves of foreign loans during the 1970s, though they funded little in the way of productive investment. Much of the borrowing paid for increased fuel costs because none of the countries produced its own oil. In 1975, prices for Central America's exports plummeted, while the price of imports steadily rose. Economic reverses sparked political unrest, particularly among rural populations hungry for agrarian reform. An alliance of landowners, commercial bourgeoisie, and the military forcefully repressed protest. To do so, they turned increasingly to the United States for military support.

The United States historically viewed Central America as the U.S. sphere of influence. Interest increased when Central American elites claimed unrest was caused by Communists. It was in this context that the Sandinistas came to power in Nicaragua in 1979, and bloody wars raged through El Salvador and Guatemala, spilling over into Honduras and Costa Rica.

El Salvador had a bleak history of violent repression dating back to the 1932 *matanza* carried out by General Maximiliano Hernández Martinez. The military ruled thereafter, in close alliance with Salvadoran elites, but cracks in the system began to appear in the 1960s. The Christian Democratic Party (*Partido Demócrata Cristiano*, PDC) was founded in 1960, and by 1964, it had won significant legislative and regional representation. Further, the war with Honduras opened the debate on agrarian reform, culminating in a national conference in January 1970. This was followed by urban strikes, particularly on the part of teachers, and a new militancy among Salvadoran university students.

The oligarchy responded by using the ferocious National Guard to form violent paramilitary forces, focused on terrifying peasant leaders, and by stealing the presidential election from PDC candidate José Napoleon Duarte in 1972. As even moderate Christian Democrats were targeted for repression, more radical Salvadorans turned to armed struggle. By the end of 1975, there were three guerrilla organizations in the countryside. When years of frustration came to a head over the military's rigging of the 1977 presidential elections, the nation disintegrated into civil war. A military coup staged by junior officers on October 15, 1979, turned the government over to a civilian–military junta.

In early 1980, the junta, even though it had become more conservative in its composition, made efforts to institute reforms by nationalizing the banking system and sale of coffee. Most significantly, the junta promulgated Decree 153, a land reform. At first the government nationalized estates exceeding 1,250 acres and turned them into cooperatives run by the workers. The government promised payment in bonds to former owners. A second phase was to have nationalized estates exceeding 375 acres, later raised to 612 acres. A third phase, called "Land to the Tiller," was intended to benefit the campesinos directly by transforming renters of small plots into owners. Over 80 percent of such plots measured fewer than five acres each. The land reform looked good on paper, but efforts to implement it failed. Jorge Villacorta, former undersecretary of the Ministry of Agriculture, observed, "In reality, from the first moment that the implementation of the agrarian reform began, what we saw was a sharp increase in official violence against the very peasants who were the supposed 'beneficiaries' of the process."

Later in 1980, after the junta of reformers had been pushed aside by more traditional figures, advocates of change formed a political alliance, the FDR (*Frente Democratico Revolucionario*; the Revolutionary Democratic Front), and a united military front, the FMLN (*Frente Farabundo Martí para la Liberación Nacional*, the Farabundo Martí Front for National Liberation). Together, they created a viable infrastructure for revolution. The FDR issued a broad program calling for national independence, profound economic and social reforms to guarantee human welfare, nonalignment in international affairs, democratic government, a new national army, support for private enterprise, and religious freedom.

Until 1979, the military and the elites were able to deal effectively with any challenge to their authority. The events of 1979–1980 demonstrated they could no longer frighten their opponents, regardless of how violent their death squads became. For the first time, they had to reach out to the United States for direct support. Their tactic was simple and already proven effective: Latin American elites identified any longing for change, no matter how modest, with communism. Once the alert to a "Communist threat" in El Salvador had been sounded, military aid from Washington flowed.

U.S. intervention exacerbated the civil war. By the mid-1980s, FMLN forces, estimated at 10,000, occupied one-third of the country as liberated territory. They collected their own taxes along national highways, and their attacks nearly destroyed the national economy. Successful in penetrating the largest cities, they were attacking military and economic targets in San Salvador by early 1989. Pentagon officials concluded that the demoralized army could not fight without U.S. support. The United States provided funding, arms, and training, which caused state-sponsored violence to escalate, exemplified by the actions of the Atlacatl Battalion, a counterinsurgency squad responsible for the massacre of the village of El Mozote in 1981 and the murder of six Jesuit priests who were professors at the Central American University, their housekeeper, and her daughter, in San Salvador in 1989. With more than 60,000 people killed in eight years—many of them civilian victims of right-wing death squads—El Salvador suffered one of the bloodiest civil wars in Latin American history.

Although the revolutionaries proved themselves capable of dispatching the Salvadoran army, they could not defeat the United States. In response to the lengthening, bloody stalemate, the FMLN signed peace accords with the U.S.-sponsored government in 1992. They were influenced in no small part by the defeat of the Sandinistas at the polls in 1990. The Sandinistas had won their revolution in 1979, but the counterrevolution launched by the Nicaraguan right wing and its U.S. allies cost 30,000 lives and destroyed the economy. Under threat of continued war, the exhausted Nicaraguans voted the Sandinistas out. By 1990, the days of Soviet or Cuban support for Latin American revolution—to the limited extent that it had ever existed—was long gone.

Guatemala was the other scene of brutal warfare in Central America in the 1980s. The roots of the struggle were in the overthrow of Jacobo Arbenz in 1954 and the string of military governments that ruled thereafter. The wars began with an unsuccessful 1960 revolt by nationalist Guatemalan military officers, angered that the government allowed the CIA to use national territory to train Cuban exiles for the unsuccessful invasion at the Bay of Pigs. Two officers, Marco Antonio Yon Sosa and Luis Turcios Lima, became radicalized by their experiences in the countryside and ended up heading revolutionary guerrilla groups. In 1966, the military allowed the election of the civilian government of Julio César Méndez Montenegro (1966–1970), prompting the guerrillas to suspend action. The military took advantage of the unofficial truce to launch a brutal counterinsurgency campaign.

A U.S. advisor instructs a Salvadoran air force cadet in the use of an M-60 mortar. The United States funded the Salvadoran government's war against the Farabundo Martí National Liberation Front and the repression carried out against the civilian population. (UPI/CORBIS-NY)

Although the military had a free hand during the Méndez government—indeed, he was forced to sign an agreement with the military before taking office—the military decided that it needed direct control over the government, so it retained military officers in the presidency from 1970 to 1986. At the same time, peasant leagues formed to organize for social change, and guerrilla forces regrouped and worked to recruit the mobilized campesinos, especially the indigenous majority.

In 1982, several guerrilla groups united under the umbrella organization *Unidad Revolucionaria Nacional Guatemalteca* (URNG; Guatemalan National Revolutionary Unity). In response, the military pursued a scorched earth policy, burning villages of suspected guerrilla sympathizers, and torturing and killing thousands. In 1982 alone, the government's counterinsurgency campaign claimed 75,000 lives and razed 440 villages. But that same year, the military rethought its strategy, moving from blind brute force to more selective targeting, a strategy of killing 30 percent of the population and pacifying 70 percent with government programs. The political–military project was designed to bring government, through the military, to formerly isolated villages.

Some 200,000 Guatemalans were killed in the struggles after the fall of Arbenz, most of them by death squads, while another 250,000, mainly indigenous, fled to Mexico to escape the holocaust. By the time the military allowed a civilian president to take power, Vinicio Cerezo in 1986, the state bureaucracy had been reconfigured to create a seamless alliance between military and government. The military remains as the architect of national stability and arbiter of democratic limits.

Finally, Guatemalan peace talks began in 1990, and accords were signed in 1996. The limits to peace, however, were clearly shown two years later when Auxiliary Bishop Juan Gerardi Conedera was beaten to death two days after presenting a scathing report on human rights violations that blamed the Guatemalan army and civilian paramilitary groups for nearly 80 percent of rights abuses in the civil war.

The neighboring Central American countries were unable to avoid the struggles in El Salvador, Guatemala, and Nicaragua. Costa Rica became involved against its will when one group of counterrevolutionaries set up base along the Costa Rican side of the border with Nicaragua. The United States also exerted considerable pressure on Costa Rica to support the counterrevolutionaries and oppose the Sandinistas. Confronting a faltering economy exacerbated by regional war, President Oscar Arias seized the diplomatic initiative after his inauguration in 1986, calling for a broad, Central American peace settlement. He won the Nobel Peace Prize in 1987 for his efforts. But the peace plan was of limited effectiveness, partially because the United States only gave it lip service.

More than Costa Rica, Honduras became completely involved in the crises. It has the uncomfortable geographic distinction of being bordered by the three nations in crisis: simmering guerrilla warfare in Guatemala, civil war in El Salvador, and revolutionary change in Nicaragua. At the same time, monumental economic problems challenged Honduras, among the poorest states in Latin America. Its population of 5.1 million grew at an annual rate of 3.5 percent, one of the highest in the world. All of its economic statistics indicated widespread social injustice. Approximately 53 percent of the population was illiterate; infant mortality was 118 per 1,000 live births. Nearly 90 percent of the rural population and 66 percent of the urban population lived below the poverty level. The economy rode a roller coaster of rising fiscal deficits and foreign debts and falling export income and foreign reserves.

Honduras exemplified the classic enclave economy. In the first half of the twentieth century, three companies—United Fruit, Standard Fruit, and Rosario Mining—dominated the economy. The first two grew and exported bananas, and the third extracted gold and silver. Foreigners owned those companies; they shipped their products abroad, contributing only low wages to Honduras. These enclaves did not contribute to the creation of support industries or related businesses. Those companies owned and operated the railroads as well as several of the principal ports. The banana companies controlled the oil, beer, and tobacco industries. In 1950, the three companies earned sums equal to the entire Honduran budget. Occasionally Hondurans made efforts to regulate the companies. A major strike against the banana companies in 1954 strengthened the unions and increased salaries and benefits for the workers. The victory instilled a stronger sense of nationalism among Hondurans and made the unions a new social and economic force. President Ramón Villeda Morales (1958–1964) tried to further curb the fruit companies and institute a modest land reform and social security program.

The victorious strike and Villeda Morales reforms unnerved landowners and the military. The generals overthrew him during his final days in office and, with a brief exception in 1971–1972, ruled Honduras directly until 1982. Corruption and increasing conservatism characterized the military governments. When the military turned the government over to an elected civilian in 1982, the economy was in a shambles.

At that moment, the administration of President Ronald Reagan realized the strategic location of Honduras and resolved to use it in order to destabilize Nicaragua and contain the Salvadoran rebels. Consequently, U.S. military aid to Honduras increased dramatically. The United States held joint military exercises with Honduras throughout the 1980s. Honduras provided ample opportunity for the United States to construct air bases and strips, a seaport, radar sites, military encampments, and tank traps. After 1983, Honduras was for all intents and purposes an occupied country. When the war finally ended in 1990, Honduras was left with hundreds of landmines in the border area and a significant problem with HIV/AIDS as a result of prostitution linked to U.S. military installations and contra camps.

10.4: The Church Under Attack

The Catholic Church became a significant adversary of right-wing governments in Latin America in the 1970s and 1980s. Throughout Central America, Christian base communities formed, and priests adopted the option for the poor expressed at Medellín in 1968.

In 1972, Salvadoran Jesuit Rutilio Grande was one of several priests sent to the countryside to train lay leaders, delegates of the word, to organize for labor rights and agrarian reform. Local landowners were outraged, and priests were targeted by death squads, who distributed flyers exhorting, "Be a patriot! Kill a priest!" In a 1977 wave of repression, Grande was assassinated just three weeks after Oscar Arnulfo Romero was selected as a compromise choice for San Salvador's new archbishop. Romero was viewed as a scholarly, conservative figure likely to be apolitical. Grande's death and the continuing repression transformed Romero.

The archbishop traveled El Salvador, visiting parishes and observing poverty and repression first hand. He addressed these realities in his weekly sermons, broadcast on Church radio. He even resorted in 1980 to writing to U.S. President Jimmy Carter, begging him to stop sending aid to the Salvadoran government, by then $1.5 million a day, because "it is being used to repress my people."

Two months later, Romero spoke directly to the army in his Sunday homily. "Brothers, you are from the same people . . .," Romero preached. "No soldier is obliged to obey an order that is contrary to the will of God. . . . In the name of God then, in the name of this suffering people I ask you, I beg you, I command you in the name of God: stop the repression." The next day, Romero was assassinated while saying Mass. The crowds of mourners at his funeral were greeted by gunfire.

Salvadoran religious were not the only targets of the Salvadoran military. In December 1980, a shallow grave on an isolated road revealed the tortured bodies of Maryknoll Sisters Maura Clarke and Ita Ford, Ursuline Sister Dorothy Kazel, and lay missioner Jean Donovan. The women, all from the United States, had been working with the archdiocese of San Salvador to help refugees fleeing rural violence.

Brazil was also a site of significant participation in the liberation theology movement, and the religious paid a high price. From 1968 to 1978, the worst years of the dictatorship, more than 120 bishops, priests, and nuns were arrested, along with nearly 300 Catholic lay workers. Seven were killed, and most were tortured. Churches were raided, and Church-run media were censored or closed.

But liberation theology was never accepted by all of the Church hierarchy; indeed, many fought actively against it. Some supported the dictatorships, whereas many more simply believed that the Church should not be actively involved in the struggle. The latter were given support in 1978, when Cardinal Karol Wojtyla became Pope John Paul II. One of his first acts in 1979 was to attend the third Latin American Bishops Conference in Puebla, Mexico. He endorsed the option for the poor and critiqued the poverty created by unrestrained capitalism. But he also emphasized two concerns: maintenance of theological orthodoxy, interpreted as a defense against Marxism and liberation theology, and conservation of traditional institutional authority. His distrust of the socialist implications of liberation theology was rooted in his youth in Soviet-dominated Poland, and his traditionalism translated to lack of support for delegates of the word and Christian base communities.

Lest there be any doubts about John Paul's views, he made them clear on an official visit to Nicaragua in 1983. The pope refused to allow Father Ernesto Cardenal, the minister of culture, to kiss his ring and advised Nicaraguan priests to leave the Sandinista government. As he lectured about the importance of traditional Church hierarchy, he ignored the chants of the Mothers of Heroes and Martyrs (*Madres de Héroes y Mártires*), who begged him to say a prayer for their children killed in the contra war.

By 1985, the Vatican attack on liberation theology had reached new levels. The Vatican's Congregation for the Doctrine of the Faith, headed by Cardinal Joseph Ratzinger, silenced Brazilian priest Leonardo Boff, a founder of liberation theology. When he was silenced again in 1992, Boff left the Franciscan order and became a lay worker. In 2005, Ratzinger succeeded John Paul II as Pope Benedict XVI.

As liberation theology began to decline, the Catholic Church braced for another struggle—competition from evangelical Protestantism. In 1986, the Vatican published a study titled "Sects or New Religious Movements: Pastoral Challenges." The Vatican agreed that powerful U.S. economic and political institutions were backing Protestant proselytizing in Latin America, but the study also reflected on the genuine appeal of evangelicals, who were filling "needs and aspirations which are seemingly not being met in the mainline churches. The [Catholic] church is often seen simply as an institution, perhaps because it gives too much importance to structures and not enough to drawing people to God in Christ."

In 1988, Catholic Church surveys claimed that 400 Latin Americans converted to evangelical churches every hour. Among the reasons for conversion was the scarcity of Catholic priests and the view that the Church does not address people's real spiritual needs. Evangelical theology demands personal responsibility, which succeeded in transforming husbands who had been drinkers and womanizers, to the delight of many women in Latin America.

The Catholic Church has borrowed a page from the evangelicals through the Catholic Charismatic Renewal (CCR). Like Pentecostals, CCR uses baptism of the Holy Spirit, speaking in tongues, and faith healing. The CCR also invokes the Virgin Mary, which evangelicals eschew in favor of a focus on Jesus. The Virgin already has a strong appeal in Latin America, particularly for women, who have always made up the majority of church congregations.

By 2005, an estimated 20–30 percent of Latin Americans identified as evangelical Protestants, most frequently as Pentecostals. Their numbers are strongest in Guatemala, Brazil, and Colombia, but the phenomenon includes all countries in the region. Some social activists fear that the movement is conservative, focusing on personal salvation rather than community action. They also fear the evangelicals' links to right-wing U.S. movements through evangelical missions.

10.5: The New Social Movements

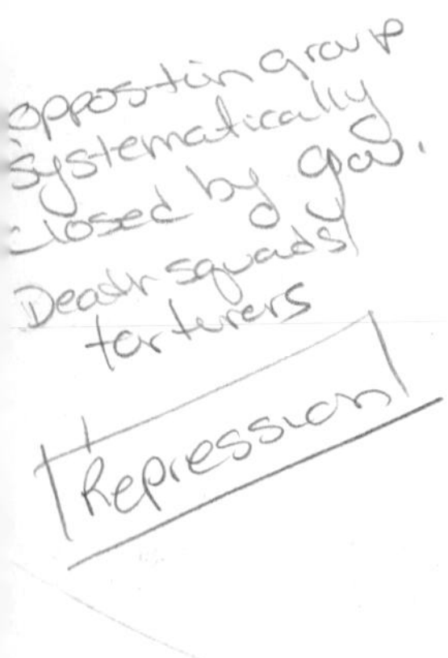

During the long years of repression, Latin Americans watched as traditional twentieth-century means of opposition were systematically closed off. Political parties were outlawed and their leaders assassinated. Labor unions and student groups were targeted by death squads and torturers. The news media were censored.

But repression did not stop people from organizing. Instead, it resulted in a change of actors and kinds of organizations. Frequently, the organizers were women, who were already outside the traditional forms of organization because they had been ignored or marginalized by unions and political parties. They tended to organize around specific issues, such as the need for water in a community, establishment of neighborhood kitchens to feed the hungry during the worst years of the recession, and the struggle to protect environments endangered by reckless economic policy. Most importantly, they organized around the issue of human rights.

In Santiago, Chile, according to one survey, 20 percent of the urban poor were involved in popular organizations in the 1980s. Among the groups were 201 soup kitchens, 20 community kitchens, 223 cooperative buying organizations, 67 family garden organizations, 25 community bakeries, and 137 health groups, claiming 12,956 active members.

Many of these organizations were formed under the protective arms of the Catholic Church, which also criticized the Pinochet regime's human rights violations and economic policies. One way that the Church helped was by providing a space where women could gather. At one parish, women were encouraged to document what happened to their families by making *arpilleras*, patchwork tapestries telling their stories. Eventually there were some 230 churches with secret workshops making thousands of arpilleras, which were smuggled abroad to expose the Pinochet dictatorship.

The most famous of the groups to emerge among the new social movements was Argentina's *Madres de la Plaza de Mayo* (Mothers of May Plaza), who marched in front of the Casa Rosada on the Plaza de Mayo once a week for many years to demand the return of their disappeared family. The women came together as they crossed paths while relentlessly questioning police and military officials about the whereabouts of missing relatives, especially their children. The mothers were mostly traditional housewives who had never considered entering politics. It was that very maternal role that led them into the streets to search for missing loved ones, to keep their families intact. The traditional maternal role initially protected the Madres from military repression. First, they were not seen as a threat. Later, the military hesitated in its use of repression because it would be hard to justify attacks on apparently harmless older women. Eventually, several of the Madres were disappeared, and one general said it had been a mistake to leave them alive.

The Madres have not weathered the transition to electoral government well. First the group split on the issue of whether to support Raúl Alfonsín, the first civilian president after the fall of the military. Some were satisfied with the return of electoral politics and the trials of Argentine generals and their subordinates. Others demanded a full accounting of the disappearances or that their children be returned alive. Some childless women resented the group's emphasis on motherhood, although the Madres argue that they represent a radicalized and socialized maternity, defining the entire population as children who must be fed, housed, educated, and protected. In this paradigm shift, they emphasize maternal values in place of traditional patriarchal values exalting economic competition and authoritarianism.

New social movements also emerged in countries with civilian governments, which were no more responsive to community needs than the military dictatorships. In Mexico City, people organized self-help groups after the September 19, 1985, earthquake that killed at least 8,000 people. The government response was so ineffectual that several organizations formed, including *Coordinadora Única de*

The mothers of the Plaza de Mayo formed when Argentine women discovered that there were many of them searching for their missing relatives, who were "disappeared" during Argentina's Dirty War. Here they display kerchiefs with the names of their missing relatives while demonstrating in Buenos Aires in 1982. (CORBIS-NY)

Damnificados (CUD; Victims' Coordinating Council). Leslie Serna described how the CUD began: "We didn't have much of an idea of what to do, and we didn't have any great plan; we did one thing at a time, and things had their own dynamic. Suddenly we were an organization and we had named commissions and a board of directors." They cleared rubble and looked for people trapped in the wreckage. They built squatter settlements for the homeless; collected clothing, furniture, and blankets; and created nurseries. They also organized demonstrations and lobbied for government aid.

The largest social movement to emerge in the 1980s was Brazil's *Movimento dos Trabalhadores Rurais Sem Terra* (MST; Landless Workers Movement), representing hundreds of thousands of landless workers. According to the MST, 60 percent of Brazil's agricultural land is idle, while 25 million people lack year-round agricultural work and access to land. Since the MST organized in 1985, with the help of the Catholic Church, more than 250,000 families have occupied 15 million acres of idle land under MST auspices and won land titles from the government. These victories were hard fought. Land occupations have been greeted with violence, from both landowners and the police. From 1989 to 1999, more than 1,000 people were killed in land conflicts, but as of August 1999, only 53 suspects had been brought to trial.

The MST adopted the kinds of strategies carried out by revolutionary governments—creation of food cooperatives, small agricultural industries, and a literacy program—and their efforts have been supported by such international organizations as UNESCO (United Nations Educational, Scientific, and Cultural Organization). Within Brazil, the group also enjoyed widespread support. A 1997 public opinion poll showed that 77 percent approved of the MST and 85 percent supported their nonviolent occupation of idle farmland.

The MST made a point of linking with international movements to fight globalization and its effects on local economies, as well as with Brazil's *Partido dos Trabalhadores* (PT; Workers' Party). In January 2001, the PT sponsored the first World Social Forum, which attracted an estimated 5,000 scholars, politicians, labor union leaders, and representatives of grassroots organizations from around the world to challenge the business elite and political leaders meeting at the same time at the World Economic Forum in Switzerland.

A highlight of the World Social Forum, which gathered progressive groups and politicians, came when 1,300 members of the MST, joined by forum participants, invaded the grounds of the U.S. biotechnology firm Monsanto. The protesters were led by João Pedro Stedile, head of the MST, and French trade unionist José Bové, a French sheep farmer who became famous when his Farmers Confederation stormed a McDonald's restaurant in rural France to draw attention to the risks of genetically modified crops to farmers, consumers, and the environment. The forum showed the possibilities of linking social movements with progressive political parties to work for change. Such links were clearly important, even with the return to electoral politics in Latin America.

10.6: Do Elections Make Democracies?

One by one, from the mid-1980s to 1990, the countries of Latin America returned to electing civilian governments. Washington officials applauded the new but fragile "democracy" in the region. But the mere existence of electoral politics is not all that is required for a country to be democratic. True democracy requires a change in the power balance among various groups in society and institutions responsive to the majority. For Latin America in the 1990s, true democracy was elusive.

In Brazil, the transition to civilian government was ushered in during the government of Ernesto Geisel (1974–1979), when opposition to the military in power mounted. The Catholic Church raised its voice against social injustice and demanded economic opportunities for the masses and freedom for all. Students became vocal again, organizing important demonstrations in 1977. Labor showed renewed independence: In May 1978, a strike of 50,000 workers in São Paulo was the first in a decade.

More significantly, the business community began to voice criticism. In November 1977, some 2,000 businesspeople gathered in Rio de Janeiro and called for democratic liberties. In July 1978, eight wealthy industrialists advocated a more just socioeconomic system. It can be argued that elite criticism mattered most: As a key part of the power structure that the military upheld, elites could not be repressed with impunity. Finally, reform sentiment was growing within the military, and it became increasingly difficult for the governing generals to disguise the cracks in the facade of unity.

Unlike his military predecessors, Geisel did not consult his colleagues in selecting a consensus candidate for the next president. Arbitrarily he picked General João Baptista Figueiredo, a relatively unknown figure who had directed the National Intelligence Service. Beginning a six-year presidency in 1979, Figueiredo declared his desire to preside over the political transition from dictatorship to democracy. In 1985, the government enfranchised illiterates (perhaps as many as 40 percent of the adult population), and an electoral college selected a civilian president, Tancredo Neves, who died shortly after his election and was succeeded by José Sarney. In 1989, the first direct presidential election was held since the election of Goulart. But the transition to electoral politics was not an easy one: The winner, Fernando Collor de Mello, was impeached after a scandal in 1992.

In Argentina, the generals had been discredited by their defeat in the Malvinas war and the downward spiral of the economy. They agreed to allow elections in 1983, and the winner was Radical Party candidate Raúl Alfonsín. In an historic decision, Alfonsín vowed to prosecute the generals who carried out the repression. In 1985, after detailed testimony about disappearances and torture, four men were convicted: General Jorge Rafael Videla and Admiral Emilio Eduardo Massera were sentenced to life imprisonment. General Roberto Viola was sentenced to seventeen years in prison, whereas Brigadier Ramón Agosti was sentenced to only four and half years. Brigadier Omar Graffigna, air force commander of the second junta, was acquitted, along with three members of the third junta, General Leopoldo Galtieri, Admiral Jorge Anaya, and Brigadier Basilio Lami Dozo. But Argentina remained a deeply divided country. During the first term of Peronist Carlos Saúl Meném (elected in 1989 and in 1995), he responded to a military uprising by pardoning Videla, Viola, Massera, and top commanders of the Buenos Aires police force who had been convicted.

The demise of the Pinochet regime was even more startling than the fall of the other regimes. Unrest began in 1982; by then Chile had accumulated the highest debt in Latin America, $16 billion, and one-third of the workforce was unemployed. In 1983, labor leaders organized mass protests, and even sectors of the right began to fear an uprising that might be avoided by removing Pinochet as a rallying point. Pinochet declared a state of siege in 1984, but in 1985, he was presented with the National Accord for Transition to Full Democracy, an agreement brokered by Cardinal Juan Francisco Fresno among eleven political groups, ranging from center right to socialist. The call for legalization of political activity was signed by enough groups to indicate that Pinochet could no longer claim that he enjoyed majority support. But the general refused to acknowledge the accord or meet with the groups it represented.

By 1987, the economy improved, and Pinochet had survived an assassination attempt. Feeling that his popularity was assured, he agreed to a plebiscite on his continued rule. The election was announced only five weeks in advance, but a coalition of fourteen parties and many social movements organized, finally even gaining access to television. In a stunning defeat, 57 percent voted against Pinochet. Perhaps more frightening for the future, however, was that 43 percent voted for him to continue.

The dictator agreed to step down, although he retained his military rank and a position as a senator for life. In 1989, Christian Democrat Patricio Aylwin was elected. In his campaign, Aylwin assured the business community that there would be no change in the basic structure of the economy. He also campaigned on a human rights platform, and the National Commission on Truth and Reconciliation was formed. But the emphasis was on reconciliation, not on justice. Although the report revealed many abuses, the conclusions called for apologies and forgiveness, not for legal action.

The civilian regimes simultaneously increased and narrowed the spaces for political participation and the development of an involved, politicized civilian society. Clearly, there was more freedom of speech and freedom to organize for social change. But there was little support from these governments for social movements that appeared in the 1970s and 1980s. The new governments saw such groups as the Madres as necessary only during the struggle against dictatorship. The civilian leaders presented themselves as the logical successors of the aspirations of the new social movements. But though the new leaders were opposed to repression and dedicated to electoral democracy, they were not interested in changing the economic and social programs of the dictatorships.

Demands for economic and social democratization fell on deaf ears, especially as the new governments represented the same economic interests as did the old. This was seen most clearly in Chile during the Aylwin government. Six months before the election, Aylwin and colleague Alejandro Foxley met with Eduardo Matte, president of Cape Horn Methanol, one of the largest transnational corporations in Chile. Matte later told an interviewer, "[T]hey gave us all sorts of guarantees that economic policy would continue to be the same as what we had known before: an open market, favorable investment terms, in sum, all the good things that we inherited from the military government. And a year later, there is no doubt in my mind that the warranty that this country has, in fact, is precisely Alejandro Foxley as Minister of Finance." Another businessman observed that Foxley, a Christian Democrat, and economics minister Carlos Ominami, a member of the Socialist Party, "could perfectly well have been members of Pinochet's cabinet."

Quite simply, the new governments were still run by members of the economic elites. Their interest in pushing social change from the top was minimal, and pressures for change from the bottom came from a weakened, disorganized mass whose political parties and representative groups were destroyed by the dictatorships. The social movements that arose in their place, usually specific to issues of the dictatorship, frequently were not in a position to take up the struggle.

Meanwhile, institutions of governance—and even many of the individuals within them—remained unchanged. Military officers, Supreme Court justices, and bureaucratic appointees were frequently holdovers from the dictatorships. In most of the transitions to elected government, guarantees were given to the outgoing governments about just such continuity. The strength of such institutions was made clear with the military uprising in Argentina that prompted Meném's pardons. Clearly, the military still had a key role to play. Furthermore, history has shown that human rights violations are not restricted to military regimes. Thousands of Latin Americans were killed by death squads during the rule of civilian regimes in the Dominican Republic, Guatemala, Peru, El Salvador, Brazil, and Mexico.

Much of the open violence to which people were subjected during the dictatorships had disappeared. But the violence of everyday life that comes from unequal economic structures, that causes malnutrition and high infant mortality, had not changed. In Brazil, under the Sarney and Collor governments, agrarian reform proposals were blocked, a program to provide low-income housing was first retargeted to the middle class and then dropped, and emergency distribution of food to the poor became an opportunity for corruption and political patronage.

As the 1990s began, Latin American elites continued to confuse their own well-being with that of the nation.

Questions for Discussion

1. How did the oil crises of the 1970s contribute to changes in Latin America's economic strategies?
2. Why were Latin American development programs so often linked to dictatorial regimes?
3. How did the dictatorships in Central America differ from the bureaucratic authoritarian governments of the Southern Cone?
4. How did the 1980s economic crises help bring down Latin American dictatorships?
5. What was the significance of liberation theology and changes within the Catholic Church?
6. What were the new social movements, and what impact did they have on new democracies?

11 The Limits of Liberalism

Historians frequently use centuries as markers for the dominant characteristics of an era. And, indeed, across long periods of time, there are significant patterns that distinguish one era from another. But these patterns do not neatly follow the calendar. Latin America's twentieth century, it can be argued, began in 1910 with the Mexican Revolution, which ended traditional nineteenth-century patterns and opened the era of revolutionary demands for change.

As the twentieth century came to an end, Mexico again took center stage. Like an echo of the past, a new movement arose there in 1994, resurrecting an historic name and claim—Zapata and the land. As in the 1910 Revolution, the new Zapatistas' demands were specific to Mexico but also illuminated issues that affected the entire region.

Poor Latin Americans were once again further impoverished by elites' policy choices as neoliberalism and reliance on the world market reprised the free-trade liberalism of the late nineteenth century. Of course, the world context had changed since 1910. Historic demands for economic equality still topped the agenda, but the demands were framed not just in terms of class but also of gender and ethnicity.

The Mexican Revolution, herald of the revolutionary century, was followed by the Russian Revolution in 1917, and the Soviet challenge to Western hegemony resulted in the Cold War that dominated the latter half of the twentieth century. At the end of the century, the Soviet Union collapsed, leaving a seemingly mono-polar world system. But just as the Mexican Revolution predated the Soviet, the Cuban Revolution outlasted the Soviets. Unlike the Eastern European satellites, and to the surprise of the United States, Cubans did not overthrow their government.

Although Cuba renewed its commitment to socialism, Communism was no longer a credible threat in Latin America. Nonetheless, the United States was determined to maintain its presence in the region, especially as the United States focused via neoliberalism on freer international trade. Hopes of a reduced U.S. military presence in the region were dashed, as continued military aid and involvement was justified by a renewed conflict—the war on drugs and "narcoterrorism."

11.1: The End of the Cold War

The Western world was not the only region beset by the economic problems of the 1980s. In 1985, when Soviet leader Mikhail Gorbachev came to power, the Soviet Union was wrestling with economic stagnation and political discontent. Gorbachev responded with two policies: economic reforms (*perestroika*) and a political opening (*glasnost*). But the economic changes—a loosening of central control with limited market incentives—did not bring prosperity, while glasnost, the encouragement of free speech, created space for criticism of the government. These pressures were complicated by ethnic and nationalist sentiments that never disappeared under the Soviet mantle. In 1987, Estonia sought autonomy; other Soviet states followed suit. In December 1991, Gorbachev resigned. In January 1992, the Soviet Union ceased to exist, and the European countries in its orbit turned away from Communism. After decades of tension, the Cold War was over.

As each Communist government in the Eastern Bloc fell, Washington leaders waited for Cuba to follow suit. Cuba had depended for decades on Soviet trade, which paid above market prices for Cuba's sugar and nickel in exchange for oil and discounted Soviet manufactures. With the collapse of the Soviet Union and its satellites, Cuba lost 85 percent of its foreign trade practically overnight. Oil imports dropped from 13 million tons in 1989 to less than 7 million tons in 1992. The impact on Cubans was staggering: Gross domestic product fell 35 percent, consumption 30 percent, and caloric intake 38 percent. For the first time in many years, Cubans were hungry.

Fidel Castro declared a "Special Period in Peacetime"—in other words, Cubans were asked to live with the kind of deprivation that most would only see during wartime. People were asked to mobilize and increase volunteerism—for example, urban gardens sprang up throughout Havana. The bywords were scarcity and austerity. Without oil, factories did not run; the fuel-less buses were replaced by bicycles. There were declines in productivity, and budget deficits grew. The hardships led to a new exodus from the island in 1994, when some 35,000 Cubans fled on flimsy rafts (*balsas*). The *balseros*, however, did not receive the warm welcome from the United States that previous waves of Cuban emigrants received. Most were turned back or interned at the U.S. military base at Guantanamo.

The *balseros* left just as the Cuban government's economic measures were showing promise. Cuba's first priority was to find new trade partners to replace the Soviets. In 1992, Cuba allowed up to 49 percent foreign ownership in new enterprises, and in 1995, the government issued guarantees against nationalization. New free-trade zones welcomed 240 foreign enterprises. Cuba found willing investment partners in fifty-seven countries, including Canada, France, Germany, Great Britain, and Russia. Investments included mining, textiles, and tourism.

Tourism is a particularly thorny issue for Cuba, which had been a destination for drinking, gambling, and prostitution in the 1950s. Fidel famously said that Havana had been a U.S. brothel. The opportunities provided by the revolution—and the limited tourism—largely eliminated prostitution. But as tourism expanded in the 1990s, prostitution reappeared. The tourist sector also offered higher wages, drawing Cubans away from working for the state.

In 1993, the Cuban government legalized the dollar, established dollar stores, and legalized remittances from abroad. Since the majority of Cuban exiles before 1994 were white, the remittances that they sent bolstered the economic status of white Cubans, destabilizing the measures of equality that had been achieved for Afro-Cubans during the previous decades. The government also authorized self-employment and reopened peasant markets in 1994.

Additional revenue was gained by increasing prices for cigarettes, alcohol, gasoline, electricity, public transportation, and mail service, and even charging, albeit nominally, for sporting events, school lunches, and some medicines. The result was a reduction of budget deficits to 7.4 percent of gross domestic product by 1994. That year, the GDP grew by a slight 0.7 percent, but it signaled the beginning of a recovery. From 1993 to 2000, GDP grew 29.7 percent, while exports grew 47.4 percent, dominated by nickel and citrus production.

In hopes that economic hardship would provoke Castro's overthrow, the United States tried to tighten the economic blockade of Cuba via the Helms-Burton law, signed by President Bill Clinton in 1996. The law allows the United States to sanction companies from foreign nations that did business in Cuba. The European Union, Canada, and Mexico declared they would ignore the law, arguing that the United States violated international law by attempting to have jurisdiction over other countries. Even the Organization of American States, which at U.S. urging ousted Cuba in 1962, condemned the Helms-Burton law. Carlos Lage, then vice president of Cuba's Council of State, shrugged off the new law: "Not only do [foreign companies] have an opportunity to do business with Cuba, but they will find in Cuba the only market in the world where they don't have to compete with U.S. companies."

The United States failed to stop foreign investment in Cuba and also failed to see an uprising overthrow the Cuban government, despite continued U.S. support for dissidents. Although much changed on the island, the government safeguarded what it saw as the essence of the revolution—education, health care, and the meeting of everyone's basic needs. Those programs were paid for by opening up to foreign trade.

Investments and trade also remained at the top of the U.S. agenda. As Rear Admiral James B. Perkins, acting commander in chief of the Southern Command, told the Senate Arms Services Committee in 1996: "In the 1990s U.S. exports to the region grew by $36 billion while over the same period, U.S. exports to the European Union, Japan, and China increased by just $12 billion. Latin America purchases over 40 percent of its imports from the United States. For instance, Costa Rica with a population of just 3 million buys more U.S. goods than does all of Eastern Europe with some 100 million people. In 1994, the United State sold more to Chile's 14 million people than to India's 920 million people."

Perkins argued that the increased trade with the region needed to be protected. "America's support of democracy and free market economies makes it a full partner with the democratic nations of the region. The political instability that formerly marked Latin America will be relegated to the past only if democratic and free market institutions continue to succeed."

11.2: Free Trade and the Zapatistas

The original liberals were Adam Smith and David Ricardo, who touted free markets and comparative advantage as the routes to prosperity. Neoliberalism, beginning in the 1980s, was a return to that unfettered capitalism, and in Latin America, its achievement meant undoing years of state involvement in the economy. Liberalism returned to Latin America in the form of structural adjustment policies demanded by the IMF and the World Bank so that the region could repay its debts and pull out of the tailspin of the "lost decade" of the 1980s. Structural adjustment policies were designed to cut government spending and encourage growth of the private sector. Adjustment required cuts in social spending, privatization of public activity, encouragement of market solutions to social problems, and deregulation of private activity. These neoliberal policies were backed by the United States in the "Washington consensus," a term coined by economist John Williamson in 1989 and reflected in the policies of the Clinton administration.

One of the hallmarks of the new liberalism was pressure from the United States to form free-trade areas. U.S. officials saw these agreements as an interim step towards the freeing of trade worldwide. Negotiations regarding trade and tariffs had been carried out since 1948 under the auspices of the General Agreement on Tariffs and Trade (GATT), one of the agencies created at Bretton Woods in 1944. The GATT was superseded in 1995 by the World Trade Organization (WTO), which provides the institutional framework for multilateral trade.

According to the Congressional Budget Office, "One reason for the recent U.S. pursuit of [free-trade areas] is that progress in multilateral trade negotiations has become more difficult. The increasingly large membership of the GATT/WTO over time means that more countries must reach agreement in each subsequent round of negotiations. The newer members are generally developing countries that see their interests as being different from those of the United States and other industrialized countries that were more dominant in the earlier rounds."

Just as Mexico kicked off the debt crisis in 1982, it was also the first country in the region to adopt structural adjustment policies and reconsider foreign trade. In 1986, the Mexican government decided to join the GATT, a move leaders had always said they would never consider. In 1990, Mexican officials acknowledged that they had begun negotiations with the United States for a free-trade agreement. The negotiations culminated in 1994 in the North American Free Trade

Agreement (NAFTA), encompassing the United States, Canada, and Mexico. Mexican elites saw it as their salvation, but others saw it as an attack.

On January 1, 1994, the day that NAFTA took effect, armed guerrillas occupied government buildings in San Cristobal in Chiapas. They stated, "This is our response to the implementation of the North American Free Trade Agreement because this represents a death sentence for all of the indigenous ethnicities in Mexico." They called themselves Zapatistas, harkening back to the Mexican Revolution. The name is significant—Emiliano Zapata's cry of Land and Liberty resonated again after 1992, when Mexico's congress amended Article 27 of the constitution, essentially ending land reform by allowing privatization of ejido plots.

The location was significant as well. Chiapas was both the poorest and the most indigenous state in Mexico. In 1994, only 11 percent of the population earned a moderate income, compared with 24 percent nationwide. Less than 50 percent had running water, compared with 67 percent nationwide. While Chiapas had only 3 percent of Mexico's population, it produced 54 percent of the nation's hydroelectric power, 13 percent of its gas, and 4 percent of its oil—yet nearly half of Chiapas was without electricity.

Many say that the Mexican Revolution never reached Chiapas. Large estates were left intact, while the indigenous majority competed for the remaining land. As the Mexican economy industrialized in the 1950s, state economists advocated production of cheap food for urban workers. Low food prices meant low prices for small farmers and low wages on large farms. Commercial farmers in western Chiapas expanded, encroaching on subsistence farming in the east. Desperate indigenous farmers colonized the Lacandona forest. Subsistence farmers then struggled to define their position as the government created bio-reserves in the forest. In 1979 the government designated the Montes Azules Biosphere Reserve, which comprises 331,200 hectares of the forest. In 1990, another 119,177 hectares was designated as the El Triunfo Biosphere Reserve in the Sierra Madre de Chiapas; it comprises 331,200 hectares in the Lacandona forest. El Triunfo divides two hydrological regions of the state: the Pacific coast and the Grijalva-Usumacinta River, where two hydroelectric plants were built in the 1970s. In the 1980s, large landowners converted the land they had once rented to small farmers into cattle-grazing areas.

Chiapas seemed a fertile area for revolution and was targeted by the Forces of National Liberation (*Fuerzas de Liberación Nacional*, FLN), one of the Marxist guerrilla organizations formed in the 1970s in the aftermath of the massacre of student demonstrators in 1968. In 1983, six organizers—three indigenous and three mestizo—came to Chiapas, all urbanites who looked to the local population to learn how to survive in the jungle. But the FLN cadres assumed they were the vanguard who would teach the indigenous about revolution. In 1984, the revolutionaries were joined by a man known as Marcos, who found the indigenous unmoved by "the absurdities that we had been taught; of imperialism, social crisis, the correlation of forces and their coming together, things that nobody understands, and of course neither did they."

The FLN began to listen to the indigenous, who explained their history of exploitation and racism and shared stories told by their elders, with a nonlinear conception of time and a sense of place unrelated to a concept of nation. The result, described Marcos, was "the idea of a more just world, everything that was socialism in broad brushstrokes, but redirected, enriched with humanitarian elements, ethics, morals, more than simply indigenous.... More than the redistribution of the wealth or the expropriation of the means of production, the revolution began to be the possibility for a human being to have ... dignity."

A new organization was created—the *Ejército Zapatista de Liberación Nacional* (EZLN, Zapatista Army for National Liberation). Subcomandante Marcos was chief military commander and spokesman, but the EZLN became largely an indigenous organization (primarily Tzeltal, Chol, Tzotzil, and Tojolabal). The 1994 armed battle lasted only twelve days, but the EZLN had not expected to win a military victory and take state power. They intended to capture world attention and build a movement.

Especially in the early years, the Zapatistas became international media darlings, largely because of the media savvy of the charismatic and witty Marcos, who eyed the press, pipe in mouth, from behind a balaclava. The Zapatistas said they wore ski masks to be seen, whereas their indigenous faces were ignored. Marcos distributed his eloquent Zapatista declarations online, an enormous shift from just a decade earlier, when internationalists in solidarity with the Nicaraguan revolution carefully carried Sandinista brochures and posters back to their home countries. The use of the new media prompted many to call the Zapatistas the world's first postmodern guerrillas.

The Zapatistas reject vanguardism, the idea that only a few revolutionary elite know the way to effect change, and although the media have focused on Subcomandante Marcos, the organization runs on communal decision making. While the majority is indigenous, the EZLN is not just an ethnically based movement. The Zapatistas have urged all Mexicans to join them, not by taking up arms but by calling on the Mexican government to create true democracy, to develop the country rather than just grow the economy, and to give indigenous regions autonomy.

Women have played a particularly strong role in the Zapatista movement, beginning with the women's revolutionary law calling for equal rights, among them protection from rape and domestic violence. Women have frequently represented the Zapatistas: Comandanta Ramona was sole Zapatista delegate to the First National Indian Congress in 1995; Comandanta Ana gave the main speech at the First Intercontinental Encounter for Humanity and Against Neoliberalism in 1995, and Comandanta Ester was the main speaker before the Mexican Federal Congress in 2001.

The Mexican government responded initially to the uprising by stationing some 40,000 troops, one-quarter of the nation's armed forces, in Chiapas. The army launched an assault against the guerrillas in February 1995, but then agreed to enter peace negotiations. Throughout the negotiations, however, guerrillas and campesinos were attacked by paramilitary groups linked to landowners and ignored or supported by the government. By 2000, there were at least ten such groups, and more than 20,000 people were displaced by paramilitary terror from 1995 to 2000. The worst episode was a massacre in the town of Acteal in December 1997, when forty-five people, mostly women and children, were driven from the town church and killed. Negotiations from 1995 to 1996 resulted in the San Andrés Accords, which were never implemented, although in a March 1999 referendum, 3 million Mexicans went to 15,000 Zapatista polling places and voted their support for the agreement.

The intriguing Zapatista military leader, Subcomandante Marcos, became a media darling, although he insisted that he was but one of many leaders in the struggle in Chiapas, Mexico. (AP Wide World Photos)

In 2001, the Zapatistas marched from Chiapas to Mexico City, drawing crowds along their trek and culminating with Comandanta Ester's historic speech before the Mexican congress. In response, congress finally approved an autonomy plan, but it was so weak that the states with the largest indigenous populations voted against it. "The March for Indigenous Dignity" showed that although the Zapatistas enjoyed widespread support, they still could not achieve change on a national level.

The Zapatistas instead focused their efforts on the local level, setting up autonomous communities. They have worked to create self-sufficient production and social service projects, including community gardens, rabbit raising, beekeeping, and candle making. In 2003, they reorganized their territorial structure, grouping municipalities into five regional units known as *caracoles* (snails or conch shells, a Mayan symbol), each run by a Council of Good Government, with the motto: "Here the people rule and the government obeys." Zapatista projects have inspired groups throughout Latin America and become a magnet for nongovernmental organizations (NGOs). Perhaps most important, the Zapatistas have created their own schools.

Meanwhile, NAFTA's performance has been as dire as the Zapatistas predicted. By NAFTA's fifteenth anniversary, trade among the three countries had increased 213 percent, but benefits in Mexico were seen by only 10 percent of the population. Simultaneously, migration to the United States, which NAFTA was supposed to curtail, increased from an average of 235,000 a year to a new high of 500,000. Migration was fueled by the loss of Mexican jobs as businesses collapsed under the onslaught of less expensive U.S. goods. More than 1 million jobs were lost in Mexico during the first year alone.

The situation was even more precarious in the Mexican countryside because NAFTA ended farm subsidies in Mexico but not in the United States. In 2002, the United States paid $10 billion in subsidies to agricultural giants Cargill and Archer Daniels Midland, which tripled their corn exports to Mexico. By 2009, at least 2 million farming jobs and 8 million small farmers had been driven out of business by the influx of cheap grain. By 2014, Mexico's real gross domestic product per person ranked eighteenth of twenty Latin American countries.

When jobs did appear, they were in the *maquiladoras*, assembly plants that proliferated along the border. By 2001, there were 2,000 maquiladoras employing more than 1.3 million workers. But the health of the plants was tied to the U.S. market, and sales plummeted in the 2001 recession. More than 400,000 workers lost their jobs. Most of those who managed to keep their jobs earned the minimum wage for border work, $4.20 a day, which had not changed since 1994. Income disparity between the two countries grew by nearly 11 percent, and Mexico's real wages actually were halved.

Critics of NAFTA have not just come from the left. Center-right political scientist Jorge Castañeda commented that NAFTA was "an accord among magnates and potentates: an agreement for the rich and powerful … effectively excluding ordinary people in all three societies."

11.3: Neoliberalism and Its Discontents

Globalization was initially greeted in Latin America with great euphoria. Latin American governments opened their economies wide, dropping tariff barriers and inviting foreign investment, which flowed into the region for the first time since the wave of defaults that began with Mexico in 1982. But by the end of the 1990s, many Latin Americans were disappointed and pessimistic. Free markets were supposed to bring economic growth, but growth levels were far lower than expected. Whereas the world's developing economies grew on average 4.7 percent a year, Latin America mustered only 3.3 percent, with less than 2 percent growth in Ecuador and Venezuela. Growth in per capita income lagged even further, growing only 1.3 percent.

From 1990 to 1997, Latin American economies grew, and poverty fell, although it was still higher than it was in 1980. But starting in 1997, growth rates plummeted and poverty increased. By 2002, economists were talking about a "lost half decade," much like the lost decade of the 1980s. ECLAC warned that to meet United Nations' goals of halving poverty by 2015, Latin America would have to grow by unrealistically high percentages. It was time, the agency recommended, for the region not just to rely on growth but to also carry out government-led redistribution of wealth. But government intervention in the economy was anathema in the 1990s as governments adhered to neoliberal precepts with an almost religious fervor.

Globalization also had a profound impact on the structure of Latin American economies. For much of the twentieth century, government leaders focused on industrialization as the way to modernize their economies. But during the 1990s, productivity in manufacturing either declined or remained stagnant. In fact, most of Latin America deindustrialized, according to the United Nations Commission on Trade and Development (UNCTAD). Although Mexico and several Central American nations increased manufacturing, it was only through the addition of assembly plants using imported components. South America, with the exception of Brazil, returned to the export of natural resources. The prices of those products declined during the decade, just as the terms of trade on commodities had declined for much of Latin America's history.

The new policies were shocking not just because of the economic changes but because they seemed to reject economic nationalism, one of Latin America's most important philosophies of the twentieth century. While Mexicans in 1938 had rallied behind their government in support of a nationalized PEMEX as a symbol of Mexican sovereignty, in 1998 they rallied to make sure that the state would not sell PEMEX off to the highest bidder. Their fears were justified, as the government went from owning 1,050 enterprises in 1983 to 210 in 2003, with many companies purchased by foreigners.

Economists insisted that privatization would eliminate problems of inefficiency and corruption in state-run corporations. Officials were so enthusiastic that Latin America led the world in privatization in the 1990s, representing nearly half of the value of sales worldwide.

The effects of privatization, however, were problematic. The sales concentrated wealth in the hands of even fewer people and corporations because few had the resources to buy these companies. The sales also increased foreign ownership of the economy because foreign corporations had more resources to buy the state assets. Between 1990 and 2002, multinational corporations acquired 4,000 banks, telecommunications, transport, petroleum, and mining interests in Latin America. Furthermore, privatization directly affected the services provided to Latin Americans, who saw utility rates rise and quality of services deteriorate.

By the end of the decade, nearly two-thirds of Latin Americans polled were opposed to privatization. The reasons were exemplified by the case of Aguas Argentinas. In 1993, the Argentine government sold the Buenos Aires water company to a consortium comprising the Suez group from France, the world's largest private water company; Spain's Aguas de Barcelona; the World Bank, which bought a small stake in the consortium; and several small Argentine banks that included relatives and business partners of government officials. The consortium slashed its workforce in half and raised water rates by 88 percent from 1993 to 2002, earning profits of 20 percent—compared with typical water company profits of 6 percent in Europe and 6–12 percent in the United States. By 1997, the company had failed to honor 45 percent of its commitments to improve and expand services, and it was accused of dumping sewage into the Río de la Plata.

The reduction of the size of government was also managed through cutbacks in public spending, which meant that rights won through years of struggle—labor laws, public maintenance of water and sewage, and education—were eroded or abandoned. One example was Nicaragua's health budget, slashed in half from 1990 to 1991 to meet demands from multilateral lenders.

The structural adjustment programs were greeted by protests from Latin Americans throughout the region. In April 1996, thousands of Venezuelan teachers were on strike for more than a month; in September, austerity measures led to a general strike in Argentina. The following year, hundreds of thousands of Ecuadorans participated in a two-day general strike. In 1998, thousands protested the devaluation of the *sucre*. There were eighty protests against austerity programs in Latin America from 1976 to 1989. From 1995 to 2001, there were 281 "popular movement campaigns," which sponsored 961 separate events.

Nearly 56 percent of the demonstrators came from working-class groups, and another 24 percent were public employees. They were followed by students (17 percent), campesinos (16 percent), and teachers (14 percent). All of these groups might be considered the usual protesters throughout the twentieth century. But the protests were also characterized by new groups with new techniques.

The unemployed and workers in the informal economy, who have traditionally been considered impossible to organize, led protests in Argentina. From 1997 to 2003, there were on average 137 protests every year, nearly 1,000 for the entire period. The roots of their organization came via a national program, *Plan Trabajar*, which directed NGOs and municipal governments to develop programs to build needed infrastructure or provide community services, and to staff these program with the unemployed. The problem was that the program only served 8 percent of the unemployed. Brought together by the programs, the unemployed were able to organize. Their demonstrations—including roadblocks and building occupations—forced the government to increase resources.

The indigenous emerged as protesters in two significant ways—as the organizers of movements focused specifically on indigenous rights, as well as participants in broader coalitions to protest neoliberal policies and to defend the environment. Indigenous rights issues have focused on cultural recognition, such as Bolivia's *Movimiento Indio Tupac Katari* (MITKA, Tupac Katari Indian Movement); land rights, as demonstrated by the Wauja of Brazil; and forestry management demands by the Quechua of Ecuador.

Revolutionary groups in the 1960s and 1970s at times attempted to organize indigenous people, but usually as campesinos and workers—a class focus—rather than on a cultural basis. In fact, Marxist groups were sometimes dismissive of indigenous culture. At the same time, indigenous groups tended to remain isolated from each other. But as the quincentennial of the conquest approached in 1992, many indigenous groups sought to organize against triumphalist official celebrations and emphasize the continued oppression of native peoples. The anniversary of the conquest no doubt played a role in the selection of Guatemala's Rigoberta Menchú, a K'iche woman, as the 1992 Nobel Peace Prize winner. Menchú became the international spokeswoman for the struggle against the genocide unleashed by the Guatemalan government against its indigenous majority.

The spotlight on Menchú helped lead to the designation of 1993 as the UN's International Year of the Indigenous Peoples of the World and in designating the International Decade of the Indigenous (1995–2004). These events built on the 1985 creation of the Working Group on Indigenous Populations of the UN's Sub-Commission for the Prevention of Discrimination and the Protection of Minorities.

Such international backing was crucial in supporting indigenous groups that previously had little standing with their national governments. With "democratization" in the late 1980s and early 1990s, new constituent assemblies created a space for the indigenous to press their demands. Constitutions in Bolivia, Colombia, Ecuador, Peru, and Nicaragua all included recognition of the indigenous as distinct groups; made indigenous languages official; committed to bilingual education; recognized indigenous common law; and acknowledged property rights, including communal land and the right to autonomous territories. However, as is often the case, legal standing did not necessarily translate to material reality.

Neoliberal reforms, in fact, attacked communal land much as the Liberal governments had in the nineteenth century. At the same time, the neoliberals' desire to diminish the role of the national government led to decentralization and strengthening of the role of municipal government. Local government was more vulnerable to pressure from indigenous majorities and the activist groups who mobilized them, and those groups then reached out to each other to form national networks. The most dramatic results were seen in Ecuador and Bolivia.

Ecuador was the scene of the National Indigenous Uprising in June 1990. It was organized by the *Confederación de Nacionalidades Indígenas del Ecuador* (CONAIE, Confederation of Indigenous Nationalities of Ecuador), a nationwide organization formed in 1986 to link the regional groups from the highlands and the coast. In the 1980s, CONAIE gained support by linking with a broad coalition rather than focusing just on indigenous concerns. Their slogan was "*Nada sólo para los indios*" ("Nothing only for Indians"). By 1990, their focus had shifted.

In six highland provinces, indigenous activists decided to protest government inaction on land claims by occupying Santo Domingo Cathedral in Quito, a historic colonial site. Their actions sparked massive nationwide protests as thousands of indigenous blocked highways, picketed along roads, shut markets, occupied government buildings. Their slogans this time had changed: "This is not a workers' strike. This is not a teachers' strike. This is not a students' strike. This is an Indian uprising." They succeeded in forcing the government to negotiate, though only two of their sixteen demands were granted: the allocation of land in the Amazon region and indigenous control over choosing the head of bilingual education.

Two years later, when the government tried to impose more neoliberal reforms, CONAIE joined the United Workers Front (*Frente Unitario de Trabajadores*, FUT) call for a general strike. By 1995, activists formed the *Coordinadora de Movimientos Sociales* (CMS, Coordinator of Social Movements), which coordinated indigenous and nonindigenous actions. In 1997, the indigenous movement led the nationwide uprising that culminated in the overthrow of President Abdalá Bucaram.

In Bolivia, unexpected consequences of 1980s neoliberal reforms coupled with the U.S.-sponsored war on drugs led to joint mestizo and indigenous organization. Because of neoliberal policies, many miners—who were well organized—lost their jobs. Many of them moved to the distant, primarily indigenous regions of Chapare and Achacachi, where coca was grown legally for domestic use. These workers brought their union militarism to a region where the coca growers had already organized and found themselves under attack by U.S. coca eradication efforts.

Coca production has historically been an important part of Andean indigenous life. Long before the Spanish conquest, they cultivated and used coca leaves for both religious and medicinal purposes. These uses continued during the colonial period, especially to combat altitude sickness, provide an energy boost, and stave off hunger for those working in the silver mines. After independence, those traditional uses continued. Modern cocaine production began in the late nineteenth and early twentieth centuries as a legal industry with medicinal or anesthetic purposes.

U.S. authorities, however, became concerned about the increasing popularity of illicit drugs. U.S. pressure led to the 1948 criminalization of Peru's legal cocaine production, which in turn led to the birth of an illegal industry. Repression in Peru—and lack of U.S. influence on the Bolivian revolutionary government in the 1950s—led to the spread of illegal cocaine manufacturing to Bolivia. U.S.-encouraged antidrug campaigns in the early 1960s pushed cocaine production into more remote areas of Peru and Bolivia.

In 1969, President Richard M. Nixon declared a "war on drugs," aimed primarily at marijuana, which was popular in the counterculture, and against heroin, feared as a problem among returning Vietnam veterans. In 1973 Nixon authorized formation of the Drug Enforcement Agency. Drug trade was curtailed in the 1970s by the attack on the "French connection" heroin trade, and by Operation Intercept, which targeted the Mexican border. The resulting decline in marijuana and heroin trade created an opening for cocaine.

In the mid-1960s, cocaine trafficking was so limited that seizures of cocaine globally amounted to no more than tens of kilograms a year. By 1975, trade had surged to an estimated four tons, and by 1980 hundreds of tons of cocaine were brought into the United States from Colombian traffickers. President Ronald Reagan pursued the drug war with renewed vigor. Between 1980 and 1987, U.S. spending on attempts to eliminate the drug supply tripled. In 1989, President George H. W. Bush raised the stakes by launching the Andean Initiative, authorizing the expansion of military aid to Latin America to fight the drug war. Bolivia was awarded $15.7 million for law enforcement and $39.1 million for the armed forces, as well as $7.8 million in Department of Defense equipment, services, and training. Further, the United States threatened to cut off economic aid unless the Bolivians cooperated with U.S. troops in eradication efforts, which included aerial spraying of coca areas. These eradication efforts came as well-organized former miners and coca growers joined forces.

In keeping with the neoliberal challenge to the national government, in 1994 Bolivia adopted the Law of Popular Participation, which provided for the first direct municipal elections. As a result, 464 indigenous and campesino leaders were elected as mayors and council representatives in 1995, and more than 500 were elected in 1999. The indigenous pushed their own issues but also joined with others who felt disenfranchised by neoliberal programs: workers, teachers, intellectuals, artists, the middle class, and religious groups.

Indigenous organizing and political participation laid the groundwork for what became known as the water war of Cochabamba. In 1999, the Bolivian government sold the water system in Cochabamba to a multinational consortium owned by a subsidiary of Bechtel Corporation, which drastically raised prices, refused to provide service to poor neighborhoods, and tried to even prohibit collection and use of rainwater. Massive protests led to the canceling of the Bechtel contract in April 2000.

Demonstrations were also sparked by another element of the 1990s neoliberal turn: government corruption. Of course, corruption in Latin America was nothing new, but the context had changed. The privatization process led to more opportunities for corruption, while democratization gave more independence to newspapers to report on accusations of corruption. In Brazil in 1992, President Fernando Collor de Mello was accused of receiving $6.5 million rerouted from campaign contributions. Venezuela's President Carlos Andrés Pérez in 1992 was accused of receiving $17 million earmarked as "secret security funds." In 1997, Ecuadoran President Abdalá Bucaram was accused of corruption, embezzlement, nepotism, and influence peddling.

The corruption was so widespread that in 1995 Venezuelan writer Moises Naim coined the phrase "corruption eruption." Latin Americans, struggling with social service cuts, rising prices, and unemployment, were incensed by reports of corruption, and they took to the streets in massive demonstrations.

Congressional leaders responded by appointing committees to investigate, and newly independent courts were the setting for corruption trials. Nine Latin American presidents or former presidents faced charges of corruption. Three were impeached—Brazil's Collor, Venezuela's Pérez, and Peru's Alberto Fujimori. Paraguay's Raúl Cubas chose to resign under threat of impeachment, and Ecuador's congress voted Bucaram incapable and removed him from office. The wave of impeachments was unprecedented and indicated the new importance of a mobilized civil society along with democratic congressional and judicial practices.

11.4: Colombia

The Longest War

Although Latin Americans were taking to the streets to protest in the 1990s, revolutionaries throughout the region were laying down their arms. Yet in Colombia, an intractable guerrilla war had been waged since 1964, and it showed no signs of ending. While the Zapatistas waged a war of words, in Colombia the war of bullets escalated, pitting the Colombian military, with U.S. aid, against the *Fuerzas Armadas Revolucionarias de Colombia* (FARC; Revolutionary Armed Forces of Colombia), and the much smaller *Ejército de Liberación Nacional* (ELN; National Liberation Army).

Colombia's history has been characterized by lack of control over its territory—the loss of Panama in 1903 being the most extreme example—and the use of violence to resolve political disputes. Colombia's nineteenth-century conflict between Liberals and Conservatives was unusually long and bitter, culminating in the War of a Thousand Days (1899–1902), in which 100,000 people died before the Conservatives won control. Liberals regained power in 1930, but elites of both parties agreed on an economy based on exports of coffee and bananas grown primarily on large landholdings. From 1931 to 1945, elites privatized about 150,000 acres of public land each year, leading to the increasing impoverishment of most Colombians.

The one leader who opposed such policies was a dissident Liberal, Jorge Eliécer Gaitán, who had led the investigation of a military massacre that ended a 1928 banana workers' strike against United Fruit Company—an incident later immortalized by Gabriel García Márquez in *One Hundred Years of Solitude*. Gaitán was expected to win the upcoming presidential election in 1948 when he was assassinated, sparking riots known as the Bogotazo, beginning a ten-year period of armed struggle between Liberals and Conservatives so bloody that it came to be known simply as *La Violencia*. Some 200,000 people were killed from 1948 to 1958, when the parties finally agreed to share power. The FARC was an outgrowth of Liberal guerrillas that emerged during this period.

The guerrillas were supported by peasants who were losing access to more land with each passing year. Land concentration worsened with privatization of public land in the 1950s, and workers and tenants were evicted from haciendas that mechanized in the 1960s. In the 1970s, production of cocaine began to edge out traditional farmers; coca growth expanded in the 1980s, as drug cartels invested 45 percent of their earnings in land, acquiring an estimated 7.5–11 million acres. In 1960, farms of more than 500 hectares constituted 29 percent of cultivated land; by 1996, they accounted for 61 percent. At the same time, the share of the smallest proprietors went from 6 to 3 percent of cultivated land.

From 1964 to 1980, FARC fought in the countryside, allied with campesinos and demanding land reform. The failure of the government to carry out a serious agrarian reform produced more guerrillas: From 1970 to 1982, FARC's numbers rose from 500 to 3,000. The government responded with a state of siege, beginning in 1978, and human rights abuses escalated. By the 1980s, the guerrillas agreed to peace talks. Although they did not give up their weapons, they put them down and opted to form a political party, *Union Patriótica* (UP, Patriotic Union). In the 1986 congressional election, UP won a surprising fourteen seats. The response was a series of assassinations by paramilitary groups, and FARC took up its weapons again.

In the 1980s, FARC took advantage of increasing drug production as a way to finance its guerrilla struggle. FARC provided law and order in drug areas, and in exchange for protection of coca cultivation and production, they charged a 10 percent tax. With these profits, FARC was able to train and equip a professionalized army, with some 17,000 troops by the 1990s. By the end of the decade, FARC had a presence in 622 of Colombia's 1,071 municipalities. In 1998, they were powerful enough to convince the government to cede control over an area the size of Switzerland. In areas under FARC control, the guerrillas essentially were the state, providing education, courts, health, road construction, and even loans to small farmers.

FARC's exploitation of the narcotics trade gave the United States a new rationale to aid the Colombian government's efforts to defeat the guerrillas. The United States originally claimed to be fighting communism in Colombia, a view that dated from the Bogotazo, which exploded while the OAS happened to be meeting in Bogotá. Many U.S. diplomats believed that communist agitators caused the riots to undermine the OAS conference. During La Violencia, when peasant groups were encouraged by the Communist Party to struggle for land, the United States rushed military aid to the government. In 1962 the United States sent a team from the Special Warfare Center at Fort Bragg to evaluate Colombia's counterinsurgency capability. More advice and aid were to follow: 10 percent of the Latin American soldiers trained by the United States from 1950 to 1979 were Colombian. From 1950 to 1966, Colombia was the fifth-largest recipient of U.S. military aid in Latin America. In the late 1960s, it moved up to fourth place and in the 1970s and 1980s to third.

But with the fall of the Soviet Union and the weakening of Cuba, the United States was hard-pressed to justify continued aid to fight communists in Colombia. The U.S. military had been reluctant to play a significant role in the drug war, despite the desires of the Reagan administration, which in 1982 linked drug policy to national security. However, in 1988, facing the possibility of shrinking military budgets, suggestions were made that the drug war could substitute for the Cold War. William J. Taylor, vice president for political and military affairs at the Center for Strategic and International Studies, put it bluntly: "There will be new initiatives to reduce U.S. military troop commitments worldwide. If the DOD [Department of Defense] leadership were smart about the coming environment, they would approach the Congress with a military 'social utility' argument which says that military manpower should not be further reduced because the Congress is mandating increased military involvement in the 'war on drugs'."

The drug industry came to pervade all aspects of the Colombian struggle. Drug production began in the late 1970s; by 2000, more than 300,000 acres of land was cultivated in coca. By 2004, Colombia produced 80 percent of the world's coca as well as significant amounts of opium poppies. Many of the producers were small farmers who turned to coca because they could make a 49 percent profit, whereas with traditional crops, they actually lost money. Coca is easy to produce, grows in poor soil, and gives a high return. The coca grower, however, makes only a tiny percentage of the profit in the illegal industry; most of the profit is made from distribution in the United States, which is the primary market.

The United States now labeled the FARC as narcoterrorists, accusing them of trafficking drugs. However, there is no evidence that FARC controlled distribution nor that FARC leaders have personally profited from the drug business. The guerrillas have used their control over drug regions to fund their continued struggle, based on the same demands since the 1960s: economic equality and democracy.

The guerrillas were countered not just by the Colombian army but by nineteen right-wing paramilitary groups, united starting in 1997 in the *Autodefensas Unidas de Colombia* (AUC; United Self-Defense Forces of Colombia). According to Amnesty International, the paramilitaries, numbering 8,000–11,000, were responsible for 75 percent of the rampant human rights violations in Colombia. Although the AUC supposedly demobilized in 2006, death threats continued to be issued by paramilitary groups.

Poverty, drug dealing, and war combined to make Colombia the most violent country in the hemisphere. More than 50,000 people were killed in political violence between 1985 and 2000, and 2 million fled their homes.

11.5: The End of Revolution?

Colombia was the exception to the general pattern in Latin America in the 1990s. Although Latin Americans filled their streets in protests, the era of revolution seemed to be over. In 1990, the Sandinistas lost elections in Nicaragua, and peace agreements ended the revolutionary wars in El Salvador in 1992 and Guatemala in 1999. Peru's *Sendero Luminoso* (Shining Path), a self-described Maoist group that was most active in Peru from 1980 to 1992, seemed to have ceased activity when its leader was captured in 1999. (There were isolated attacks in 2002 and 2008, followed by the capture of its last known leader in 2012.) Even the Zapatistas had used armed uprising more as an attention-getting technique than an attempt to overthrow the government and take state power.

But there was one last attempt at an armed uprising in Venezuela in 1992. Its origins would not be in the guerrilla groups of the 1960s, which had been defeated, but from within the armed forces. Its leader was a lieutenant-colonel in the army named Hugo Chávez, the commander of a parachute regiment. Chávez, who came from a poor background, joined the army in 1971 at seventeen years old. As a cadet, he travelled to Peru in 1974, where a reformist military government had been in power since 1968. Chávez was impressed with the possibility of the military as a force for social change.

Chávez spent two years serving in a counterinsurgency battalion. In 1976 the battalion was sent to put down a guerrilla outbreak, and he began to feel some sympathy for the guerrillas' point of view. From 1980 to 1985, he was an instructor at the military academy, originally teaching sports but moving on to history and politics. In this setting, starting in 1982, he began to organize the *Movimiento Bolivariano Revolucionario-200* (MBR-200, Bolivarian Revolutionary Movement) which initially amounted to little more than a study group. The would-be revolutionaries planned to move up in the military ranks so that they would have troops to command at the right moment. But when the first opportune moment came, the military conspirators were not prepared.

That moment was the 1989 *Caracazo*, a popular uprising over the increase in bus fares that was part of a neoliberal package introduced by President Carlos Andrés Pérez. The rebellion was spontaneous, beginning when passengers boarded buses only to find that fares had increased 30 percent, the first step in a planned 100 percent increase. Rioting and looting resulted in fierce repression, with some 2,000 killed and many more wounded.

Many of the soldiers who were ordered to shoot poor Venezuelans were ready to join Chávez when in February 1992 he led five army units in a coup attempt. The conspiracy, however, had been betrayed, and Chávez was thwarted from capturing the president. Pérez's forces were crushing the rebellion when Chávez decided to turn himself in to prevent more bloodshed. He was granted a request to speak on television to call on his followers to put down their arms. Chávez took responsibility, apologized, and urged his compatriots to avoid a bloodbath by abandoning the cause "*por ahora*" (for the moment). Venezuelans were captivated by Chávez, and "por ahora" became a rallying cry, showing up as graffiti during the two years he spent in prison.

While he was in prison, Chávez had determined that change would need to be made at the polls rather than with arms. Just four years after his release, he won a landslide electoral victory, ending dominance by the two parties that shared power for forty years and ushering in a new political era in Latin America.

Questions for Discussion

1. What is neoliberalism, and what impact has it had on Latin America?
2. What were the causes of the Zapatista uprising, and how does it differ from earlier revolutionary movements?
3. How has the drug war affected Latin American countries and their relationship with the United States?
4. How has the end of the Cold War affected Latin America? How has it changed the U.S. role in the region?
5. Compare the protest movements of the 2000s with previous movements for change in twentieth-century Latin America.

12

Forward into the Past?

In 1999, Hugo Chávez was sworn in as president of Venezuela after winning a landslide victory a mere four years after his prison term for his failed coup attempt. In his inaugural speech, he implored Venezuelans to join together to make "the necessary revolution, because it is necessary in the social, the economic, the political, and the ethical."

The 1998 election of Chávez arguably marks the beginning of Latin America's twenty-first century. Far from an anomaly, Chávez proved to be the first of a new wave of leaders who tapped into the zeitgeist of the millions of Latin Americans fed up with neoliberalism. Latin Americans demanded radical change, but they sought it with ballots, not bullets. The new trend showed surprising resilience, marking an historic break from the past.

The new national leaders, brought to power in what has been called the "pink tide," are a diverse group who share a common commitment to reduction of economic inequality and greater use of the state to solve social problems. Chávez spoke of socialism for the twentieth century, and all of the new leaders have sought to mediate rather than replace the role of the market.

The rightwing would call the leaders populist demagogues, but the presidents have all encouraged greater popular participation in a variety of democratic structures in addition to periodic elections. The new political movements have also supported greater social inclusivity for ethnic groups and gender orientations. The new governments have sought greater Latin American solidarity, displayed independence from U.S. tutelage, and welcomed new investors, particularly the Chinese.

Although there appears to be a fresh approach in the new century, which is still young, the problems that Latin America faces are old. Despite the growth of the middle class and the inclusion of more sectors in the polity, the region is still the most unequal in the world. And so the historic questions remain: How do societies achieve equity and efficiency, development and democracy? Why do so many poor people inhabit rich lands?

12.1: Latin America Swings Left

Throughout Latin America at the end of the twentieth century, people formed organizations based on their very marginalization—women, indigenous, landless, unemployed. They protested in the streets and at the polls, where they swept out the parties and leaders who had overseen the neoliberal agenda. In their place, they elected candidates who promised a more radical platform, from Hugo Chávez in Venezuela in 1998, running as a candidate of his new party, *Movimiento V [Quinta] República* (MVR, Fifth Republic Movement), to Salvador Sánchez Cerén in El Salvador in 2014, a former guerrilla leader for the FMLN, now a political party.

The first of the new left presidents to be elected—Venezuela's Chávez—proved to be a key leader of the new approach to politics and economics. His importance came from the vast oil resources that made it possible for him to bankroll change in the region, as well as his willingness to openly challenge neoliberalism and U.S. policy. He embraced Fidel Castro and proclaimed a new twenty-first-century socialism—his MVR party was replaced in 2007 with the *Partido Socialista Unido de Venezuela* (PSUV, United Socialist Party of Venezuela).

Chávez tapped into popular discontent that had been brewing for years in a country where 3 percent of the population owned 77 percent of the land, while nearly 60 percent of the population lived in poverty, despite Venezuela's substantial petroleum reserves and annual oil executive salaries of $100,000 to $4 million.

One of Chávez's first actions was to convene a National Constituent Assembly to rewrite Venezuela's constitution. Various sectors of society, including Venezuela's twenty-eight indigenous groups, elected representatives to the assembly, which produced a constitution that protects minority rights, permits people to claim title to their farms and homes, and expands political participation at the grassroots level.

Chávez then set out to reassert government control over the semiautonomous oil industry, putting a halt to privatization, reinvigorating OPEC to improve oil prices, and charging foreign producers output royalties. By raising and taking control of oil industry revenues, Chávez was able to fund an ambitious social agenda: a literacy campaign, job training, land reform, food subsidies, and small business loans. He also expanded health care, partly by providing Cuba with low-cost oil in exchange for the services of 13,000 Cuban doctors.

The new policies, part of what Chávez called his Bolivarian agenda, drew the wrath of local elites, particularly oil executives, and their U.S. allies. In April 2002, Chávez was ousted for forty-eight hours in a coup attempt that was quickly endorsed by the U.S. government and condemned by Latin American leaders. He was returned to power in a surge of support by a majority of Venezuelans. His opponents failed again in a 2004 recall election that gave Chávez 59 percent of the vote.

Table 12.1 Latin America Elects Leftists

Date	Country	President
1998	Venezuela	Hugo Chávez
2000	Chile	Ricardo Lagos
2001	Venezuela	Hugo Chávez*
2002	Brazil	Luis Inácio "Lula" da Silva
2003	Argentina	Néstor Kirchner
2004	Panama	Martín Torrijos
2004	Uruguay	Tabaré Vázquez
2004	Venezuela	Hugo Chávez†
2005	Bolivia	Evo Morales
2005	Chile	Michelle Bachelet
2006	Brazil	Luis Inácio "Lula" da Silva
2006	Ecuador	Rafael Correa
2006	Nicaragua	Daniel Ortega
2006	Venezuela	Hugo Chávez
2007	Argentina	Cristina Fernández de Kirchner
2008	Bolivia	Evo Morales‡
2008	Paraguay	Fernando Lugo
2009	El Salvador	Mauricio Funes
2009	Uruguay	José Mujica
2009	Bolivia	Evo Morales
2010	Brazil	Dilma Rousseff
2011	Argentina	Cristina Fernández de Kirchner
2011	Nicaragua	Daniel Ortega
2012	Venezuela	Hugo Chávez
2013	Venezuela	Nicolás Maduro**
2013	Ecuador	Rafael Correa
2013	Chile	Michelle Bachelet
2014	El Salvador	Salvador Sánchez Cerén
2014	Bolivia	Evo Morales
2014	Brazil	Dilma Rousseff

*Chávez was reelected under the new 2000 constitution.

†Chávez won a recall referendum with 59 percent of the vote.

‡Morales won a recall referendum with 67 percent of the vote.

** Chávez won reelection in October 2012, while battling cancer. Before he died in March 2013, he endorsed his vice president, Nicolás Maduro, who had been serving as acting president. Maduro won election by a narrow margin.

A new path was also taken by Argentina's Néstor Kirchner (2003–2007), whose government renationalized many corporations that had been privatized, including the controversial Aguas Argentinas, and overturned amnesty laws that shielded hundreds of former officers from charges of human rights abuses during the dictatorship. Kirchner's most important action was to restructure Argentina's foreign debt. In December 2001, Argentina had defaulted on more than $100 billion in foreign loans. In the ensuing crisis, the country went through five presidents in two weeks as desperate Argentines rioted. When Kirchner took office in 2003, he addressed the economic problems with a landmark achievement: He convinced bankers to restructure some 70 percent of Argentina's debt at 35 cents on the dollar, lowering the nation's debt payments and creating breathing room to rebuild the economy. The IMF objected to the debt arrangement and tried to pressure Kirchner to reopen negotiations with the banks, while also trying to force another austerity program on Argentina. Kirchner refused to follow the IMF plan and prioritized domestic spending to stimulate the economy rather than payment to creditors. "We want to pay off our debt; we don't want to owe anything anymore to the IMF," Kirchner explained. "Because of the debt, the Fund wants to impose domestic and foreign policies on us, but Argentina is a sovereign country." In another interview, Kirchner quipped, "There is life after the IMF, and it's a very good life.... [B]eing in the embrace of the IMF isn't exactly like being in heaven."

However, even such solutions to economic problems may be vulnerable. Kirchner was succeeded in office by his wife, Cristina Fernández de Kirchner, who was elected in 2007 and reelected in 2011. When Kirchner struck the original deal in 2005, 75 percent of Argentina's creditors accepted the new agreement. In 2010, a similar offer was agreed to by the holders of another 17 percent of the debt. But 8 percent held out—and a U.S. federal judge upheld their demands to be paid in full. Most of the holdouts were not the original creditors but "vulture capitalists" who buy discounted bonds at low prices and then sue for full payment. While the judge could not control the Argentine government's behavior, he could control the New York bank that held Argentina's funds that were ready for disbursement to the majority of its creditors. In 2014, the judge ruled that the bank could not pay any creditors unless all were paid. Fernández faced a choice of defaulting on all the loans—even though the money was there to pay—or capitulating to the holdouts. Fernández, with the support of most Argentines, chose default.

Fernández is also part of a small but significant new wave of women heading Latin American countries. Although her husband preceded her in office, Fernández had a well-established political career in her own right. She was elected to two terms as a deputy in the Santa Cruz provincial legislature, elected to the Senate in 1995, to the Chamber of Deputies in 1997, and again to the Senate in 2001, where she served till becoming first lady in 2003. Fernández's political career distinguishes her from Argentina's only other woman president, Isabel Perón, who became president when her elderly husband, the populist Juan Perón, died in 1974. Isabel was Latin America's first woman president, though

The leftward swing of Latin American politics was represented by Venezuela's Hugo Chávez, who had oil wealth to help fund his social programs, and Evo Morales, the first indigenous president of Bolivia. In January 2006, Chávez joined Morales in Bolivia to sign accords promising aid with energy, education, agriculture, and social development. (*The Economist*)

she had no political experience, and it was her overthrow in 1976 that ushered in Argentina's "Dirty War." The success of women politicians in Argentina, as in many other Latin American countries, is in part due to legislation requiring political parties to field at least a percentage of women candidates.

Perhaps the most remarkable leader of the new left was Evo Morales, who in 2005 became the first indigenous president of a country where the indigenous constitute nearly 60 percent of the population. Morales spent ten years as head of Bolivia's indigenous coca growers' union before becoming a leader of the *Movimiento a Socialismo* (MAS, Movement to Socialism), the multiparty coalition that brought him to the presidency. The day before his inauguration in January 2006, more than 50,000 indigenous people—not just from Bolivia but from around the Americas—gathered at the ruins of the ancient indigenous city of Tiwanaku to confer the "Power of the Original Mandate" on Morales.

Morales came to power in a country where the two previous presidents were toppled amid hundreds of demonstrations, most significantly protests against foreign control over resources. The first dramatic protests were the "water wars" of 2000, when a popular uprising forced out Bechtel Corporation. In 2003, "gas wars" were sparked by a plan to export natural gas to the United States from a Chilean port that Bolivia lost in the War of the Pacific. First, Aymara campesinos in the highlands set up roadblocks, and the government responded with

Table 12.2 Women Presidents in Latin America

Years	Name, Country
1974–1976	Isabel Perón, Argentina*
1979–1980	Lidia Gueiler Tejada†
1990–1997	Violeta Chamorro, Nicaragua‡
1997	Rosalia Arteaga, Ecuador**
1999–2004	Mireya Moscoso, Panama
2006–2010	Michelle Bachelet, Chile
2007–2015	Cristina Fernández de Kirchner, Argentina
2010–2014	Laura Chinchilla, Costa Rica
2011–2014	Dilma Rousseff, Brazil
2014–2018	Michelle Bachelet, Chile
2014–2018	Dilma Rousseff, Brazil

*President Juan Perón had insisted that his wife, Isabel, be his vice president, and she succeeded him on his death.

†Gueiler, leader of the lower house of Congress, was named interim president when the previous president was overthrown. She oversaw new elections, but before the winning candidates could take office, she was overthrown. She served from November 16, 1979, to July 17, 1980.

‡Chamorro was the first elected woman head of state in the Americas. However, she had no political experience; her main fame and credential was as widow of national hero and martyr Pedro Joaquin Chamorro.

**Arteaga, who was vice president when President Abdalá Bucaram was removed from office, served only two days, February 9–11, 1997, and was succeeded by congressional leader Fabián Alarcón.

massacres that triggered uprisings in the mostly indigenous urban shantytown of El Alto. Miners marched to El Alto to join the struggle, which next descended to La Paz, where popular neighborhoods protested and the middle class joined in hunger strikes. A key demand was that natural gas be renationalized, which became a priority for Morales.

With revenues from gas, Bolivia's economy enjoyed phenomenal growth beginning in 2005, higher than any time during the previous thirty years, according to the Washington-based Center for Economic and Policy Research. The country's national reserves were 48.4 percent of gross domestic product in 2013, which was even higher than China's. With these resources, Morales was able to fund development projects, including subsidies to families who send their children to school, higher pensions for the elderly, and maternal care for the poor to reduce infant and child mortality.

Mass movements were less pleased with the performance of Brazil's Luis Inácio "Lula" da Silva. When he was elected in 2002, Lula was extremely popular as the country's first working-class president, a former metalworker, union leader, and a founder of the *Partido dos Trabalhadores* (PT, Workers' Party). But for the first two years, Lula's government followed strict IMF policies, including budget cuts and alliances with the center right. Transnational capital was welcomed, particularly for investment in government bonds at high interest. However, neoliberal policies were offset by government programs, including *Programa Bolsa Família* (Family Allowance), which paid poor families half a minimum wage per month for each child in school, and the *Fome Zero* (Zero Hunger) program, which included construction of water tanks in the arid northeast, agricultural loans, and direct food aid. The combination of broad alliances and popular poverty programs led to Lula's reelection in 2006, but the landless movement continued to challenge the government for doing little to help the country's 4 million landless families.

Nonetheless, Lula was succeeded by his party's candidate, Dilma Rousseff, who served as Lula's chief of staff from 2005 to 2010. Rousseff's background is even further to the left than Lula's. She was a member of the guerrilla group *Comando de Libertação Nacional* (COLINA, National Liberation Command). In 1970 she was arrested, imprisoned for three years on charges of subversion, and suffered torture at the hands of the military government.

Rousseff is not the only former member of the armed left to be elected to the presidency. In 2009, Uruguayans elected José Mujica, a leader of the Tupamaros, who had been imprisoned for fourteen years, including extensive periods in solitary confinement and two years at the bottom of a well. Mujica donated most of his presidential salary to the poor and chose to live on a simple farm rather than in the presidential palace. In El Salvador, the FMLN finally succeeded in winning the presidency in 2009 with Mauricio Funes, who was not a historic member of the party and did not fight in the war. But in the subsequent election in 2014, Salvadorans elected Salvador Sánchez Cerén, a former FMLN guerrilla commander.

The most controversial former guerrilla to be elected president is Nicaragua's Daniel Ortega, who was first elected president in 1984 during the years of the Sandinista Revolution. Ortega lost his bid for reelection in 1990, under threat of a continued contra war and with the United States heavily funding opponent Violeta Chamorro. He lost again in 1996 (with 41 percent of the vote), in an election that most observers considered to be riddled with fraud. In 2001, Ortega had the lead in political polls, but after 9/11, the opposition characterized him as a terrorist and brought back images of the contra war as the United States bombed Afghanistan. Ortega lost with 38 percent of the votes. In 1999, he signed a controversial political pact with former President Arnoldo Alemán, who had been convicted of graft. The pact gave Alemán's Liberal Alliance (*Alianza Liberal Nicaragüense*, ALN) and the Sandinistas more control over the electoral process and lowered the percentage of votes needed to win from 45 to 35 percent of the vote. In 2006, Ortega finally won a new term with 38 percent of the vote, amid criticism about the low percentage. But in 2011 he was reelected by 62 percent. Ortega has brought back free education and health care and instituted antipoverty programs, while also earning the support of the business community. Many of Ortega's former comrades have left the FSLN and adamantly oppose Ortega, but their party, the *Movimiento de Renovación Sandinista* (MRS, Sandinista Renovation Movement), has never polled above single digits in elections.

There has been some concern expressed, especially by U.S. analysts, about the pattern of reelection in Latin America. Latin Americans have struggled since independence with *continuismo*, the tendency of leaders to continue in office. But Latin American commentators argue that in a democracy voters should be able to return leaders to office and to vote them out when they no longer have popular support.

Some analysts, particularly Jorge Castañeda, have tried to divide the new leaders into two groups: populists, seen as intransigent radicals and exemplified by Chávez, contrasted with social democrats, described as more reasonable, market-oriented leaders, represented by Lula. Critics argue that this analysis is a strategy to continue promotion of the neoliberal agenda, disparage those making the most sweeping changes, and weaken the region's newfound unity.

All the leaders have favored democratic mechanisms. For example, voters rejected Chávez's proposal to eliminate term limits in 2007, but supported the measure in a 2009 referendum. In 2014, former U.S. President Jimmy Carter, whose Carter Center is the leading international observer of elections, declared the Venezuelan process the best in the world.

None of the leaders has completely rejected market mechanisms or advocated centrally controlled economies. As political scientist Maxwell A. Cameron comments, "The backlash against neoliberalism does not signal a rejection of markets, but a repudiation of the ideology that places markets at the centre of the development model to the detriment of public institutions and their social context."

12.2: A Mobilized Population

As might be expected, the new left governments faced opposition from the righ-twing and from business interests. In 2006 in Bolivia's wealthiest state, Santa Cruz, residents led mass demonstrations opposing Morales's plans for an assembly to write a new constitution, took over the Viru Airport to try to collect the landing fees that go to the national government, and attempted to achieve autonomy. Their efforts were curbed by a recall election that Morales won with an overwhelming 63 percent of the vote. Morales won reelection with a sizable majority in 2009 and in 2014.

But the new left governments have not been immune to pressures from popular organizations—after all, it was mass mobilization that helped bring these leaders to power. Morales's decision to build a 190-mile highway through the Isiboro-Secure Indigenous Territory and National Park was greeted in 2011 by a protest march that was set upon by riot police who wielded teargas and batons. While the original march was attended by about 1,000 protesters, hundreds of thousands protested against police brutality. Morales condemned the police reaction, withdrew plans for the highway project, and offered support for legislation to bar highways in the contested region. "This is leadership obeying the people," said Morales. The protest revealed some of the complexities of satisfying different constituencies. Protesters' varied demands included opposition to the highway, calls for land reform, and even defense of logging in the region. The highway, in turn, was supported by many coca growers and campesinos, who saw the project as a way to open access to more land.

In Ecuador, Rafael Correa's government faced popular protests over plans for oil drilling in Yasuní National Park in the eastern Amazon region, a protest that began on social media and ended up with thousands of people in the streets. Ecuador relies on oil revenues, and with production falling, Correa said new drilling was needed to raise money for antipoverty programs. In 2007, Correa launched an initiative calling on international donors to save the Yasuní environment by contributing the $3.6 billion that Ecuador would lose by not drilling. But after six years, Ecuador had received only $13.3 million. "The world has failed us," Correa contended as the government approved drilling. In 2010, there were protests against a new water resources law that the government said would decentralize and better regulate resources, and in 2012, indigenous groups also protested a contract with Ecuacorriente, a Chinese company, to mine copper in the Shuar territory of southern Ecuador. Amnesty International has accused the government of suppressing dissent and harassing protest leaders, but Correa continued to win sizable electoral victories for three terms in office.

In Brazil, Lula faced ongoing protests by the landless movement, as well as unsuccessful demonstrations against a proposed hydroelectric dam on the Xingu River (construction began in 2012), and conflicts between ranchers and the indigenous in the Amazon region. Under both Lula and Rousseff, police and residents fought in the vast favelas, the slums that surround São Paulo and Rio de Janeiro. In 2013, protests rippled across the country over increases in bus and subway fares; police responded to the escalating protests by firing rubber bullets and beating protesters. The police response prompted more protests, and Rousseff promised to meet demands for reform of the political system, curbs on corruption, and improved public services. There were also massive protests against preparations for the 2014 World Cup and 2016 Summer Olympics, which included construction of expensive sports venues and pacification of favelas.

In Chile in 2006, only a few months after President Michele Bachelet took office, thousands of secondary school students protested to demand free transportation passes and elimination of fees to take university admission exams. They also critiqued the quality of Chilean education, much of which had been privatized under previous neoliberal reforms. Some of the proposed reforms were enacted, but the overall system remained unchanged. In 2011, many of those same students—now at university—took to the streets in seven months of protests aimed at Bachelet's successor, Sebastian Piñera, a wealthy businessman whose appointee as Minister of Education was Joaquín Lavín, owner of a private university and supporter of neoliberal policies. Under the slogan "*Educar, no Lucrar*" (Educate, Don't Profiteer), the protesters occupied more than 200 schools and universities and, according to polls, won the support of 80 percent of the population in their demands for free and improved public education. The star of the protests was Camila Vallejo, a photogenic and articulate student leader who in 2014 won election to Congress as a Communist Party candidate. At twenty-five, Vallejo was the youngest member of Congress. In the same elections, President Bachelet was returned to office—with a strong agenda for educational reform.

The protests of the early twenty-first century were characterized overall by a mobilized populace, democratic political structures, and a strong preference for electoral change. Leftist presidents found that the same movements that helped bring them to power would demand that they keep their promises. But the new leaders also would find that dealing with competing national interests and international pressures, the need to grow their economies and provide for majority needs, was a difficult balancing act. As a result, the popular movements found that political rhetoric was not always matched by action. The key difference from earlier unsuccessful protests against governments seemed to be the new democratic context that brought greater government responsiveness to majority concerns.

One of the most successful movements in the region organized to demand gay rights and same-sex marriage. The first success was legalization of civil unions: in Buenos Aires (2002), Brazil (2004), Uruguay (2008), and Colombia (2009). In 2010, Argentina was the first Latin American country (and one of only ten countries in the world at that time) to adopt a law at the national level legalizing same-sex marriage. Argentina has since been followed by Uruguay in 2013. In Mexico, same-sex marriage has been legalized in Mexico City—where the mayor officiated at mass public wedding ceremonies—and in the states of Quintana Roo and Coahuila.

In Brazil in 2005, almost 2 million people—many in colorful Carnival costumes—paraded in São Paulo to celebrate gay pride and call for the legalization of civil unions. Two years later, millions again packed the streets of São Paulo in the world's largest gay pride parade, calling for an end to homophobia, racism, and sexism. In May 2011, Brazil's high court ruled that same-sex civil unions must be recognized, a decision followed a month later by a São Paulo state judge's ruling that two men could convert their civil union into a full marriage, the nation's first gay marriage.

These successes have been the result of tireless organizing by groups promoting LGBTQ rights, in an atmosphere of greater democracy and strengthening of civil society, coupled with increased secularization and the diminishing political power of the Catholic Church.

Demands for equality have also arisen in Afro-Latin American communities throughout the region. The dominant discourse had contended that race was not an issue in Latin America, especially in comparison with the United States. There was no official segregation and there were no laws that specifically applied to black communities. The issue is also complicated by the varying definitions of race and of blackness. Traditionally in Latin America, the term black was used only to refer to the darkest members of the population. The various shades of mulattos, indicative of a history of miscegenation, were never all grouped as black, as they are in the United States. Some Latin Americans complained that ideas about blackness were imports from the United States, much as feminism often has been dismissed as an imported ideology.

Brazil, the country with the largest black population in Latin America, has begun to address racism through the use of affirmative action programs. The *Movimento Negro Unificado* (Unified Black Movement) has emphasized the obvious racial discrimination in Brazil's social, economic, and political arenas. They also used the United Nation's World Conference against Racism, Racial Discrimination, Xenophobia and Related Intolerance in Durban in 2001 to pressure Brazil, which signed the conference statement on the recognition and eradication of racism. As a result, in 2002, then President Fernando Henrique Cardoso acknowledged that Brazil suffered from racism. Lula adopted the idea of affirmative action and appointed Afro-Brazilians to posts at all levels of government, including Joaquim Barbosa, the first Afro-Brazilian to serve on the Supreme Court. In 2012, the Supreme Court ruled that affirmative action programs enacted by several public universities did not violate equal rights. The ruling paved the way for a law mandating quotas at all federal universities and technical schools.

Central American black communities have been successful in organizing through both local and regional groups, partly because they tend to be primarily located on the Atlantic Coast, isolated from the rest of the country. The *Organización Negra Centroamericana* (ONECA, Central American Black Organization) coordinates groups throughout the region, stressing issues of transnationality. ONECA prioritized environmental justice and access to health care, particularly for HIV/AIDS, as well as preserving traditional medicinal practices. The Central American community includes the Garifuna, mixed African and indigenous peoples with a more distinct identity. Among the success stories is the founding in 1992 of *Universidad de las Regiones Autónomas de la Costa Caribe Nicaragüense* (URACCAN, The University of the Autonomous Regions of the Caribbean Coast of Nicaragua). Located in Bluefields, the university has played an important role in developing leadership in the black, Garifuna, and indigenous communities.

The Cuban revolution brought extensive benefits to the island's black community, providing equal opportunity for education and employment. But when the "Special Period" began, the government allowed the remission of dollars. Since most Cuban refugees have been white, the money they sent their families contributed to renewed racial inequality. Racism has been addressed by Cuban hip hop artists—leading the government to form the *Agencia Cubana de Rap* (Cuban Rap Agency) to promote, and perhaps try to coopt, the movement.

Even countries that have not typically been identified with race issues are beginning to recognize black populations. Mexico conducted a small sample survey in 2014 and was considering asking questions about race in its 2020 national survey. The survey asked about African descent and what term was preferred. Some chose *negro* (black), others *Afromexicana* (African-Mexican), others *prieto* (dark). The discussion of race is new for Mexico, which issued a postage stamp in 2005 with the image of Memín Pinguín, a black boy with wide eyes and exaggerated lips from a popular comic book. The stamp drew criticism from the United States, but defensiveness from Mexico. The stamp sold out—but it has not been reprinted.

12.3: The Conservative Exceptions

Not all Latin American countries made a left turn. Mexico seemed to democratize when the *Partido Revolucionario Institucional* (PRI, Institutional Revolutionary Party) relinquished control, after seventy-one years, in the 2000 election. But the subsequent twelve years revealed the limits of the democratic opening. Honduras suffered a coup in 2009, and the democratically elected president of Paraguay was forced from office in 2012.

Vicente Fox of the *Partido Acción Nacional* (PAN, National Action Party) became president of Mexico in 2000 when PRI conceded the election. Fox's greatest success was defeating the PRI. During his term, he made no inroads on the violence of Mexico's drug traffickers; he did not resolve the Zapatista conflict; and, largely because of 9/11, his plans for an immigration agreement with the United States collapsed. His successor, PAN's Felipe Calderón, used the military to try to curb drug trafficking, a strategy that escalated the violence, resulting in 50,000 deaths in six years, leading to massive protests.

In 2012, PRI made a comeback with Enrique Peña Nieto, but many believed that electoral fraud had returned as well. He defeated Andrés Manuel López Obrador, the popular mayor of Mexico City, who had instituted social welfare programs including financial support for single mothers and the unemployed, investments in urban redevelopment and transportation infrastructure, and educational outreach programs. A recount was halted, and Mexicans reacted with massive demonstrations.

In Honduras in June 2009, soldiers stormed the presidential palace and rousted democratically elected President Manuel Zelaya, and escorted him, still in his pajamas, to the airport and put him on a plane to Costa Rica. He was replaced that afternoon by congressional president Roberto Micheletti.

When Zelaya was elected president in 2006, no one predicted that he would become a symbol for progressive sectors. A rancher, Zelaya was a center-right politician and member of the establishment Liberal Party. But he struggled to make a difference in Honduras, which is among the hemisphere's poorest countries. In 2008, Zelaya turned to the economic initiatives led by Venezuela's Chávez. First, Honduras joined Petrocaribe, which was to supply the country with oil on favorable terms. Then, Honduras joined the *Alianza Bolivariana para las Américas* (ALBA; Bolivarian Alliance for the Americas), the Chávez-led project designed to counter such U.S. economic integration efforts as the Central American Free Trade Area (CAFTA). Zelaya said of joining ALBA: "This is a heroic act of independence and we need no one's permission to sign this commitment. Today we are taking a step towards becoming a government of the center-left, and if anyone dislikes this, we'll just remove the word 'center' and keep the second one."

Zelaya went on to raise the minimum wage, further infuriating the business community. He won the support of Honduras's poor majority by abolishing fees for primary education, providing free meals for poor schoolchildren, and increasing spending on education and health care, particularly vaccinations. His programs expanded electricity and increased production of basic food grains.

The immediate excuse for the coup was that Zelaya was trying to begin a process of constitutional reform. The day of the coup there was to be a nonbinding vote to determine whether people wanted to consider a binding vote in November calling for a constituent assembly to write a new constitution. The current constitution, written under the military government in 1982, provided no means to change the document and reinforced the power of elite-dominated institutions.

Both the Congress and Supreme Court ruled that Zelaya had no right to call for even a nonbinding poll and tried to stop distribution of ballots. Zelaya responded by ordering the military to carry out the balloting; when they refused, he fired the defense minister and senior commander. Zelaya opponents claimed the reform was designed to keep the president in office. But the constitutional assembly would not have met until he was already out of office.

Since Zelaya's ouster, which was endorsed by the United States, Honduras has suffered repression, with arbitrary arrests, threats, disappearances, and murder, particularly aimed at unionists, peasant activists, environmentalists, indigenous people, and the journalists, lawyers, and others who supported them.

A nonviolent coup took place in Paraguay in 2012, when the Senate voted to remove President Fernando Lugo, a left-leaning former liberation theologian who when elected in 2008 ended sixty-one years of rule by the rightwing Colorado party. Wikileak documents would later show that members of Congress were looking for an excuse to oust the president. They accused him of malfeasance in office because of an incident in which several people were killed when the military attempted to evict landless farmers who had occupied land. Lugo's removal led to the return of rightwing governance.

12.4: A New Regional Independence

Latin American leaders have long catered to U.S. demands, especially during the Cold War. In the twenty-first century, however, Latin American presidents are showing a new independence. For example, the leaders of Venezuela, Brazil, Argentina, and Uruguay united in their opposition to the U.S.-sponsored Free Trade Area of the Americas agreement, derailing the proposal in 2005. In its place, Chávez offered ALBA, fostering economic cooperation and integration of Latin American nations as a bloc without the United States.

In 2005, Chávez was also provided the impetus—and the funding—for Petrosur, a Latin American petroleum company with participation from Argentina and Brazil, and Telesur, a TV channel to provide news from a Latin American perspective to counter CNN in Spanish, the only continent-wide channel. Telesur was launched with funding from Venezuela, Argentina, Cuba, and Uruguay.

Latin American leaders created two new regional organizations—the Union of South American Nations, formed in 2008, and the Community of Latin American and Caribbean States in 2010. When Bolivia's Morales grappled with violent rightwing protests after winning the recall referendum, he did not turn to the Organization of American States but to the Union of South American Nations. Furthermore, Morales received support from Bolivia's historic rival, Chile, and from Brazil, even though Morales's nationalization of oil and gas removed concessions that had been given to Brazil's Petrobras.

The most significant challenge to the United States is on drug policy. In 2009 the Latin American Commission on Drugs and Democracy urged the United States to recognize that the drug war of the past thirty years had failed. The commission was headed not by the new left leadership, but by more conservative recent presidents: César Gaviria of Colombia (1990–1994), Ernesto Zedillo of Mexico (1994–2000), and Fernando Henrique Cardoso of Brazil (1995–2003).

"Prohibitionist policies based on eradication, interdiction and criminalization of consumption simply haven't worked," the leaders wrote. "Violence and the organized crime associated with the narcotics trade remain critical problems in our countries.... Today, we are further than ever from the goal of eradicating drugs."

The report blamed the drug war for high levels of political corruption and violence, particularly in Colombia and Mexico. The commission advocated that policy focus on demand as well as supply and that drug users be reclassified as patients rather than criminals. "Colombia is a clear example of the shortcomings of the repressive policies promoted at the global level by the United States," the report said. "For decades, Colombia implemented all conceivable measures to fight the drug trade in a massive effort whose benefits were not proportional to the vast amount of resources invested and the human costs involved."

Calls to decriminalize drug use in Latin America have begun to escalate. In 2002 and 2006, Brazil's legislature partially decriminalized possession for personal use, imposing mandatory treatment and community service instead of prison sentences. In 2009, while also battling the expansion of drug cartels, Mexico decriminalized the possession for personal use of small amounts of marijuana, cocaine, heroin, methamphetamines, and LSD. Days later, the Argentine Supreme Court ruled that it was unconstitutional to impose criminal sanctions for the personal possession of drugs. In 2013, Uruguay became the first Latin American country to legalize the marijuana trade.

While decriminalization held the promise of shrinking the Latin American drug market and eliminating some opportunities for local police corruption, the bigger problem for drug-exporting countries remained the seemingly insatiable demand in the main market, the United States. The U.S. approach, however, has focused on supply in Latin America, using deadly chemicals to clear thousands of acres of land, poisoning crops and people, only to have a balloon effect—when one area is squeezed, production reappears in another. The balloon effect is not being matched with a change in U.S. strategy. Despite the failure of Plan Colombia, the United States spent $2.4 billion on Plan Mérida in Mexico and Central America. Nonetheless, drug-related violence claimed some 70,000 lives from 2006 to 2013.

In 2013, the Global Commission on Drug Policy convened a conference in Mexico, where regional leaders reiterated their concerns that their countries have borne the brunt of efforts focused on production rather than consumption. The commission went a step further in 2014, calling for the decriminalization and regulation of, as former Brazilian President Fernando Henrique Cardoso put it, "as many of the drugs that are currently illegal as possible." Even the OAS released a report in 2013 calling for drug use to be treated as a public health issue and for regulatory rather than prohibitionist policies to be explored.

Latin America's new independence is based in part on reduced economic reliance on the United States. The United States is still the largest single investor in the region—$41.5 billion in 2012—but U.S. investment is declining while there has been phenomenal growth in investment by China. From 2005 to 2013, China invested $102.2 billion. The Chinese have invested heavily in oil, including Brazil's deepwater oil site, Libra, and a project in Venezuela's Orinoco region. Chinese investment has also gone into minerals, buying Peru's Las Bambas copper mine. The Chinese are also building hydroelectric facilities in Argentina and Brazil.

Trade between China and Latin America is also growing rapidly, with projections for it to shortly surpass trade with Europe and perhaps in the future even the United States. China had little trade with Latin America in 2000, but by 2012, trade between the regions amounted to $260 billion. While trade with the United States grew by 6.2 percent in 2012, that year the region's trade with China grew 8 percent. China has already become the biggest trade partner with Brazil, Chile, and Peru. Most trade has been for Latin American raw materials and commodities, including iron ore, copper, and soy. Some regional leaders were wary of Chinese manufacturing investments—for example, an automobile plant in Brazil—which it was feared would erode Latin American-owned manufacturing.

The importance of both China and Brazil can be seen in the formation of a new international organization, BRICS (Brazil, Russia, India, China, and South Africa), which showed the importance of these five large countries with growing population, economic development, and global presence. The term (originally BRIC) was coined by Goldman Sachs economist Jim O'Neill in 2001, and at first it merely represented a concept that O'Neill was considering. Gradually, however, the nations' leadership began to see a community of interests, and Russia sponsored the first summit of leaders in 2008. A formal BRICS organization was formed in 2009, with the intent of promoting trade among the countries.

In 2014, as the BRICS Summit opened in Brazil, the organization announced the formation of the BRICS Bank of Development, to be headquartered in Shanghai with initial capitalization of $100 billion. The Bank was greeted enthusiastically by Bolivian President Morales, who said the bank "will help Latin American countries to free themselves from the speculation and financial extortions of neoliberalism and neocolonialism." Argentine President Fernández voiced hope that the bank would promote development "without the hostile intention" of existing sources of international financing.

Latin American leaders had hoped that the election of Barack Obama as president of the United States in 2008 would alter the country's traditionally presumptuous attitude toward Latin America. Although Obama vowed to treat the region with more respect, there was little change in the U.S. approach—until December 2014. Obama astonished the region by announcing that the United States and Cuba, with the help of Pope Francis in arranging meetings, had agreed to restore diplomatic relations and begin discussions about normalizing trade and travel. In 2015, flags rose above reopened embassies, while the trade embargo and travel restrictions were still debated. The historic agreement raised the possibility of closing the last conflict of the Cold War.

12.5: Rise of the Middle Class—and the Vulnerable

In an historic change, Latin America's middle class grew by 50 percent from 2003 to 2009, to become 30 percent of the region's population. At the same time the percentage living in poverty dropped, from 44 to 30 percent. According to a 2013 study by the World Bank, the middle class and the poor now account for the same percentage of Latin America's population, while just ten years earlier the size of the poor sector was two and a half times the size of the middle class.

The concept of the middle class is slippery, sometimes based on education, assets, or income levels. The World Bank chose a different and useful definition: "the crucial notion of economic security (that is, a low probability of falling back into poverty)." The study notes that true middle-class status includes the ability to withstand economic shocks and contends that the most insecurity a household could withstand and still be considered middle class is a 10 percent likelihood of falling into poverty over a five-year period.

The study revealed that 68 percent of Latin Americans live below the middle-class level, and that group can be split into two categories: the 30.5 percent who are poor, earning no more than $4 a day, and the 37.5 percent who are vulnerable, living between poverty and the $10 daily minimum to be part of the Latin American middle class. A mere 2 percent of Latin Americans are in the upper class, earning at least $50 a day.

This change in the 2000s is in stark contrast to the lost decade of the 1980s for the entire region, and a lost decade in the 1990s for the middle class, which stagnated at around 21 percent of the population. Two components are responsible for the improvement: economic growth and government social policies. One of the most significant policies was distribution of targeted cash transfers, such as Brazil's *Benefício de Prestação Continuada* (Social Assistance Benefit), a social pension system, and the *Programa Bolsa Família* (Family Allowance), which subsidizes families who keep their children in school. Equally important was increased spending on education, health, nutrition, and basic infrastructure.

In short, the neoliberal policies of the 1990s—which, ironically, were supported by the World Bank—did nothing to improve economic mobility. In contrast, the programs of the new left governments of the 2000s helped the poor to move up. The United Nations Development Programme's Human Development Report for 2013 echoed the latter: "Investing in people's capabilities—through health, education and other public services—is not an appendage of the growth process but an integral part of it.... Without investment in people, returns from global markets tend to be limited."

The Human Development Report measures the Human Development Index, a combined measure of education, life expectancy, and income. Chile and Argentina are the only Latin American countries in the very high human development category. When income is removed, Cuba ranks even higher than Chile and Argentina on education and life expectancy. The change in rankings from 2007–2012 may also be indicative of results of new left programs: Ecuador moved up ten places in rank, and Venezuela moved up nine.

Latin America remained, however, the most unequal region in the world. There were a few improvements—inequality declined in Brazil, Mexico, and Argentina for the first time since data first was collected in the 1970s. Inequality persists despite the fact that Latin America is far richer than other poor areas of the globe, such as Africa. Latin America's income distribution is little different from the United States, but a comparison with select European countries offers a stark comparison, particularly in the share of income held by the bottom 40 percent.

Table 12.3 Latin American Inequality Data Share of Income by Quintile

	Poorest 20%	Second 20%	Third 20%	Fourth 20%	Richest 20%
Argentina	4.4	9.3	14.8	22.2	49.4
Bolivia	2.1	6.8	11.9	19.9	59.3
Brazil	2.9	7.1	12.4	19.0	58.6
Chile	4.3	7.9	11.7	18.4	57.7
Colombia	3.0	6.8	11.2	18.8	60.2
Costa Rica	3.9	8.0	12.4	19.9	55.9
Dominican Republic	4.7	8.6	13.2	20.8	52.8
Ecuador	4.3	8.2	13.0	20.7	53.8
El Salvador	3.7	8.8	13.7	20.7	53.1
Guatemala	3.1	6.9	11.4	18.5	60.3
Honduras	2.0	6.1	11.4	20.5	59.9
Mexico	4.7	8.7	13.1	19.9	53.7
Nicaragua	6.2	10.2	14.8	21.5	47.2
Panama	3.3	7.8	12.5	20.1	56.4
Paraguay	3.3	7.8	12.8	19.8	56.4
Peru	3.9	8.3	13.6	21.5	52.6
Uruguay	4.9	9.0	13.7	21.5	50.9
Venezuela	4.3	9.5	14.6	22.2	49.4
France	7.2	12.6	17.2	22.8	40.2
Germany	8.5	13.7	17.8	23.1	36.9
Norway	9.6	14.0	17.2	22.0	37.2
Sweden	9.1	14.0	17.6	22.7	36.6
United Kingdom	6.1	11.4	16.0	22.5	44.0
United States	5.4	10.7	15.7	22.4	45.8

Data from www.indexmundi.com, accessed October 2014.

12.6: Change and Continuity

As Latin America moved into the twenty-first century, there were tremendous changes but also remarkable continuities. The region repeated some earlier patterns, including a political turn to the left, renewed importance of the state, and attempts at a mixed economy after the failure once again of liberal economic policies. Underprivileged and underrepresented groups again clamored for inclusion, and charismatic leaders rose to power in several countries. The repeated patterns, however, occurred in a new setting, with new power centers and commitments to democratic practice. Perhaps the most disturbing continuity is the continued maldistribution of wealth, leaving most Latin Americans impoverished or vulnerable. The companions of poverty—hunger, disease, malnutrition, and illiteracy—still plague the region. That such deplorable conditions can still exist in the midst of plenty, of growth, of structural changes in the economy, indicates the continued limitations of the development models used in the region.

Half a century after the introduction of modernization theory and development plans, some argue, Latin America is in worse shape than it was before. Development programs frequently displaced people and disrupted traditional subsistence cultures that met people's basic needs and provided them with cultural well-being. Before 1960, when the majority of people still lived in the countryside, there were indeed subsistence cultures that provided a lifestyle that to North Americans might look impoverished but that might really be a more simple way of life that amply met community needs. These communities were frequently displaced by the spread of agro-industry and the so-called green revolution, which used many chemical inputs to increase yield and bring underutilized land into production. To the extent that development aid was aimed at modernizing Latin America, it facilitated the further expansion of commercial activity on the theory that subsistence equals poverty. The end result was often the production of a greater absolute poverty, both rural and urban, as well as the poverty that comes from destruction of community.

If, as Burns has argued elsewhere, the modernization of the nineteenth century brought with it the "poverty of progress," then the modernization of the twentieth century might be seen as the "devastation of development." How else can one view a situation in which the cure has often been worse than the supposed disease? For example, many countries that were not only self-sufficient in food but were actually food exporters ended up as food importers after the implementation of development programs that focused on exports, resulting in the creation and spread of hunger rather than its prevention.

Latin American theorists challenged mainstream economic development approaches in the 1970s and 1980s with the elaboration of dependency theory. Once again starting in the 1990s, Latin American scholars challenged the dominant paradigm. In *Encountering Development: The Making and Unmaking of the Third World*, Colombian Arturo Escobar shows how development programs created a discourse of problematic categories, such as peasants who needed to be modernized and mothers whose fertility needed to be controlled. The process of development created the problems and then failed to solve them. Yet, the discourse did not change.

"The rural development discourse repeats the same relation that has defined development discourse since its emergence: the fact that development is about growth, about capital, about technology, about becoming modern. Nothing else," Escobar argues. "… These statements were uttered pretty much in the same way in 1949 as in 1960 and in 1973, and today [1995] they are still repeated ad nauseam in many quarters. Such a poverty of imagination, one may think."

Escobar encourages a richer imagination, asking us to imagine a postdevelopment era, in a postmodern world. The debates about modernity and postmodernity are long and complex. To oversimplify, the modern was the project of the Enlightenment and industrialization, the optimistic belief in progress and universal truths. Postmodernity is the twentieth- and twenty-first-century condition of having seen the modern project fail profoundly in so many ways—wars, pollution, oppression, and destruction. In postmodernity, there is no one truth, but many truths, many ways of interacting with and viewing a decentered, fragmented world.

Some argue that postmodernity is irrelevant for Latin America, a region that is still not modern—or at least not modernized—in so many ways. Others quip that Latin America is the original land of the postmodern because it has always been fragmented, always been a mixture of past, present, and future. Chilean Martín Hopenhayn suggests that Latin Americans are "becoming postmoderns by osmosis in the midst of a still-pending modernization."

Argentine Néstor García Canclini offers the concept of *hybridity*—not the syncretism usually connected with religion, nor the mestizaje usually connected to race, but elements of the traditional and the modern, creating new forms, with a new hybrid emerging. This hybridity is the product of Latin America's distinctive interaction with the modern. "The pluralist perspective, which accepts fragmentation and multiple combinations among tradition, modernity, and postmodernity, is indispensable for considering the Latin American conjuncture at the end of the century," García Canclini writes.

What does that hybrid, postmodern, postdevelopment world look like? Perhaps it is exemplified by the Kayapo Indians of Brazil, who use video cameras in their fight to preserve a traditional way of life. Or by the Zapatistas, who evoke the figure of Emiliano Zapata from the Revolution of 1910 to preserve elements of an older indigenous culture—albeit modified by colonialism and neocolonialism—in a new project of sustainable economics that seeks autonomy within rather than conquest and control of the nation-state.

Latin American history has always been about the conflicts and consensus between individuals and groups, leaders and followers, the local and the international. What makes the twenty-first century different, perhaps, is the way in which these groups can come together now that technology is collapsing time and space. How are we to understand a world in which even elderly indigenous women in the highlands of Guatemala, who still dress in traditional woven *traje*, now use cell phones?

García Canclini, who gave us the concept of hybridity, offers a frame of analysis for considering the significance of the incorporation of modern technology and traditional goods in the confluence of citizenship and consumerism. He argues that the right of citizenship should be to decide how goods are produced, distributed, and used. "… [W]hen we recognize that when we consume we also think, select, and reelaborate social meaning, it becomes necessary to analyze how this mode of appropriation of goods and signs conditions more active forms of participation than those that are grouped under the label of consumption. In other words, we should ask ourselves if consumption does not entail doing something that sustains, nourishes, and to a certain extent, constitutes a new mode of being citizens."

It is disconcerting to think about citizens as consumers. But the idea of citizenship, he argues, must be reimagined under globalization. Although Latin Americans feel rooted in a local culture, it is a culture that is urban and cosmopolitan, encompassing the many nationalities and ethnicities of national migrants and international immigration. People travel with multiple passports or with no documents at all. "How can they believe themselves to be the citizens of only one country?" asks García Canclini.

If we now live in a world dominated by the market, why not reclaim what the market and the consumer mean? García Canclini contends that consumption is not just the elite preference to display class distinctions and live in luxury. Consumption is also at the heart of demands for wages, food, housing, health care, and education. Consumption is about meeting basic needs.

García Canclini's approach, however, is still located within the project of modernity, begun with the conquest and sustained in coloniality, the continuation of the subordination of Latin America to Western/Northern dominance. Another group of Latin American scholars have gathered in the twenty-first century around a new theoretical approach—modernity/coloniality/decoloniality (MCD), often referred to simply as decoloniality. The main intellectual sources of the group—Enrique Dussel, Aníbal Quijano, and Walter Mignolo—argue for the need to move beyond the assumptions of modernity, a way of thinking and of organizing society that was specific to European experience, and then exported with the assumption that their local historical experience should be universal. The assumption of universality entrenches European theory and intellectual approaches and negates the actual experience and intellectual approaches of the marginalized—people of the global south, people of color, indigenous, women, nonheterosexuals; MCD instead proposes pluriversal, not universal, ideas. The exaltation of modernity also obscures or ignores its darker side—slavery, poverty, and oppression. There would be no modernity without coloniality.

As Mignolo explains: "… [D]e-coloniality places itself in another, different arena: on the darker side of modernity. De-colonial projects dwell in the borders, are anchored in double consciousness, in mestiza consciousness. It is a colonial subaltern epistemology in and of the global and the variegated faces of the colonial wound inflicted by five hundred years of the historical foundation [of] modernity as a weapon of imperial/colonial global expansion of Western capitalism."

Modernity/coloniality/decoloniality poses a new challenge to dominant intellectual debates. MCD builds on earlier theories with roots in Latin America and Latina thought: liberation theology's emphasis on social justice; dependency theory's focus on the global south and the way that capitalism underdeveloped Latin America; Gloria Anzaldúa's focus on the borders between countries, people, and mentalities. Just as dependency theory argued that poverty in Latin America was the necessary product of northern capitalist dominance, decolonial theory posits that the economic, political, and social oppression of Latin America is the result of the imposition of modernity.

MCD scholars seek "to craft another space for the production of knowledge—an other way of thinking, *un paradigma otro*, the very possibility of talking about 'worlds and knowledges otherwise.' What this group suggests is that an other thought, an other knowledge (and another world, in the spirit of Porto Alegre's World Social Forum), are indeed possible," writes Arturo Escobar. But, as Escobar notes, the MCD project has yet to develop an economic theory that can be used to build a different future. The indigenous movements active in Latin America seek the native principle of "good living," which would provide support for everyone, respect nature, and not center on consumption.

That idea is echoed by Oswaldo de Rivero, a former Peruvian diplomat, who argues in *The Myth of Development* that basic needs and survivability must be the focus for Latin America in the future. "This reality is an invitation to discard the myth of development, abandon the search for El Dorado, and replace the elusive agenda of the wealth of nations with an agenda for the survival of nations."

The search for El Dorado is, indeed, where Latin America's complicated history began. The struggle for the majority of Latin Americans to meet their basic needs is not a new one. It has, in fact, shaped Latin American history for hundreds of years. It is a history characterized by international pressures and national responses; by diversity and complexity within each nation; by conflict and consensus between and among classes, races, ethnicities, and genders; and by wealth in the midst of plenty, with poor people inhabiting rich lands.

An Argentine Becomes Pope

In March 2013, Pope Benedict XVI stunned the Catholic world by announcing that at eighty-five he lacked the energy to continue to lead the church. His resignation—the first in 598 years—was historic. But the choice of his successor was even more momentous: Cardinal Jorge Mario Bergoglio of Argentina—the first Latin American, the first Jesuit, the first non-European in 1,272 years to be named to the papacy. Known in Argentina simply as Father Jorge, with a reputation for humility, he was the first to choose the name Pope Francis after St. Francis of Assisi, who dedicated his life to the poor.

With 40 percent of the world's Catholic population, Latin America is the most Catholic region in the world. But it remained to be seen whether Pope Francis would have an impact on the decline of the Latin American Catholic Church, where evangelical Christians have made significant inroads. For example, in Brazil, home of the world's largest Catholic population, the percentage of Catholics dropped from 92 percent in 1970 to 65 percent in 2010.

While many Latin Americans celebrated the selection of Pope Francis, questions were raised about his role as head of the Jesuit order in Argentina from 1973 to 1979, during the brutal military dictatorship (1976–1983).

He was accused of not challenging the regime and even of complicity in the 1976 abduction of two Jesuit priests, Orlando Yorio and Francisco Jalics. Jalics said Bergoglio was not to blame, and the e accusations were dismissed by Adolfo Pérez Esquivel, the 1980 Nobel Peace Prize winner who was tortured by the regime in 1977, and by Graciela Fernández Meijide, a former member of the National Commission on the Disappearance of Persons.

The new pope immediately gave up the luxury enjoyed by his predecessor—Pope Benedict XVI wore ermine-lined cloaks—and chose to live in a guest house at the Vatican rather than in the customary, spacious papal apartments. The symbolism was soon matched by Pope Francis's words: "Poverty in the world is a scandal." In November 2013, Pope Francis issued an apostolic exhortation, *Evangelii Guadium*, that attacked unfettered capitalism: "As long as the problems of the poor are not radically resolved by rejecting the absolute autonomy of markets and financial speculation and by attacking the structural causes of inequality, no solution will be found for the world's problems or, for that matter, to any problems."

Pope Francis courted controversy by indicating that it was not up to him to pass judgment on homosexuals, questioning celibacy, and calling for discussion about readmitting divorced Catholics to the sacraments. Although the pope never supported liberation theology, he seemed to rehabilitate the radical theology by inviting its founder, Gustavo Gutierrez, to Rome. The Vatican issued a statement that liberation theology can no longer "remain in the shadows to which it has been relegated for some years," and Pope Francis invited Brazilian liberation theologian Leonardo Boff—once silenced by Pope John Paul II—to contribute his ideas on eco-theology to a papal document on the environment. Francis also reinstated Nicaraguan priest Miguel d'Escoto, suspended in 1985 for serving in the Sandinista government. However, the Pope has reiterated his opposition to artificial birth control and to the ordination of women.

Pope Francis became a lightening rod for conservatives upset with his progressive stands and challenge to unfettered capitalism, while liberals applauded his defense of the poor. Pope Francis began his second year with a gesture of unity by having Pope Benedict join him in ceremonies canonizing conservative Pope John Paul II and liberal Pope John XXIII, who convened the Second Vatican Council, which modernized the church.

Questions for Discussion

1. What are the changes and continuities in Latin American political, economic, and social patterns in the twenty-first century?
2. Why have Latin Americans turned to left-leaning leaders in recent elections? How do these leaders and their programs compare to the revolutionary leaders of earlier decades?
3. What role does ethnicity play in recent movements for change in Latin America?
4. What are the reasons for Latin America's newfound independence from U.S. influence? How has that independence been acted upon?
5. How might the search for different theoretical paradigms affect the way Latin America is analyzed?

Latin America Through Art

When Spaniards arrived in the New World, they found people who were already illustrating their lives through art. This art ranged from the images of hands found in a cave in Patagonia to the richly colored murals on the eighth-century Temple Bonampak. For the next 500 years, the people of the Americas continued to represent and decorate their worlds, drawing on the indigenous, African, and European influences that have characterized Latin American society and culture.

In this section, we offer a small sample of Latin America's beautiful artwork, from the Conquest to the late twentieth century. The images show the ways in which Latin American culture has both changed and endured. They echo the historical realities of their time periods and offer a glimpse into Latin American society and culture.

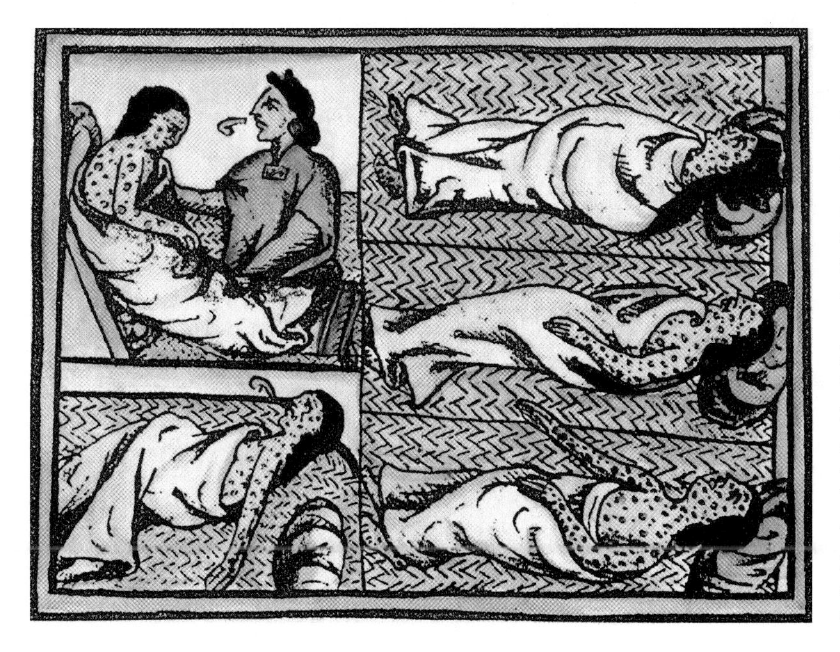

The Nahua people, under the guidance of Franciscan priest Bernardino de Sahagún, told the story of the preconquest Aztec empire and of the Conquest. Sahagún's sixteenth-century treatise, *General History of the Things of New Spain*, now known as the Florentine Codex, was a bilingual document that included this illustration of a Nahua healer attending to those suffering from smallpox. What does this illustration tell us about the devastating impact of European diseases on indigenous populations? *The Granger Collection, New York.*

Latin American society was the product of miscegenation, a process that confounded Spanish colonial elites. How are concerns about race mixture depicted in this eighteenth-century painting, one of the famous *casta* paintings? *"Human Races (Las Castas)," 18th century, oil on canvas, 1.04 × 1.48 m. Museo Nacional del Virreinato, Tepotzotlan, Mexico, Schalkwijk/Art Resource, NY.*

The details of everyday life in the colonies were captured in such paintings as this image of a Mexican market by an anonymous artist painting circa 1775. What does a close study of the scene reveal about the typical wares on sale and the roles of class and race represented in vendors and customers? *Oil on canvas. Museo Nacional de Historia, Castillo de Chapultepec, Mexico City, D.F., Mexico © Schalkwijk/ Art Resource, NY.*

The history of independence movements in Latin America typically created heroic leaders. Does Chilean painter Franco Gomez (1845–1880) contribute to the creation of national heroes with his painting of troops flanking Simón Bolívar, father of the Latin American independence movements, as he leads his troops across the Andes to liberate Peru? *SuperStock, Inc.*

Mexican artist Manuel Serrano (1814-1883) specialized in painting scenes of everyday life in nineteenth-century Mexico. In this 1855 painting, he shows a typical kitchen in a well-to-do household, staffed by servants. What can this seemingly peaceful domestic image also reveal about race and class distinctions of nineteenth-century Latin America? *Dea / G. Dagli orti/De Agostini/Getty Images*

Xul Solar was one of the key figures of the South American avant-garde movement of the 1920s as well as of the European modernist movement. Born Oscar Agustín Alejandro Schultz Solari (1887–1963), he changed his name to Xul Solar (Light of the Sun Reversed or Light from the Other Side), while living in Paris in 1916. He influenced writers such as Jorge Luis Borges as well as painters with his system of pictorial writing that he called "neocrillo" (Neo-Creole), which he intended to be understood throughout Latin America. Does this classic neocrillo 1925 painting, "Patria B," reveal new ideas of the 1920s in Latin America? *Christie's Images/SuperStock/Museo Xul Solar Fundación Pan Klub.*

Cándido Portinari (1903–1962) has been credited with developing a distinctly Brazilian style of painting. His topic for this dynamic 1935 painting is certainly a distinctly Brazilian phenomenon: the slavery of the coffee plantation. What does the painting show about the nature of the slaves' work, gender, and race? Does Portinari offer a romanticized or realistic view? *Mechika / Alamy*

Juan O'Gorman (1905–1982) was best known as an architect whose most famous work is the library at the Universidad Nacional Autónima de Mexico (UNAM). In 1932, he took the helm of Mexico City's department of building and construction for Mexico City. His interests in architecture, art (he became a renowned muralist), and the development of Mexico City are all on display in his 1949 painting, "El Ciudad de Mexico." How does this painting reflect the changes in Mexico through the Revolution and modernization efforts? *Mexico City, 1942 (tempera on masonite), O'Gorman, Juan (1905-82) / Museo de Arte Moderno, Mexico City, Mexico / Index / Bridgeman Images*

Frida Kahlo

No survey of Latin American art, however brief, would be complete without an image by Mexico's Frida Kahlo (1907–1954). Kahlo has become a pop icon, but more importantly, she has become known as one of Latin America's most important artists of the twentieth century, perhaps even eclipsing the fame of her husband, muralist Diego Rivera. Her stormy marriages to Rivera made him a central motif in many of Kahlo's works, and this one is no exception. He is at the center of the surrealistic "Love Embrace of the Universe, the Earth (Mexico), Myself, Diego and Señor Xolotl," a 1949 painting. How does this painting show precolonial influences, especially dichotomies such as life and death, night and day, moon and sun, man and woman?
Alamy Limited

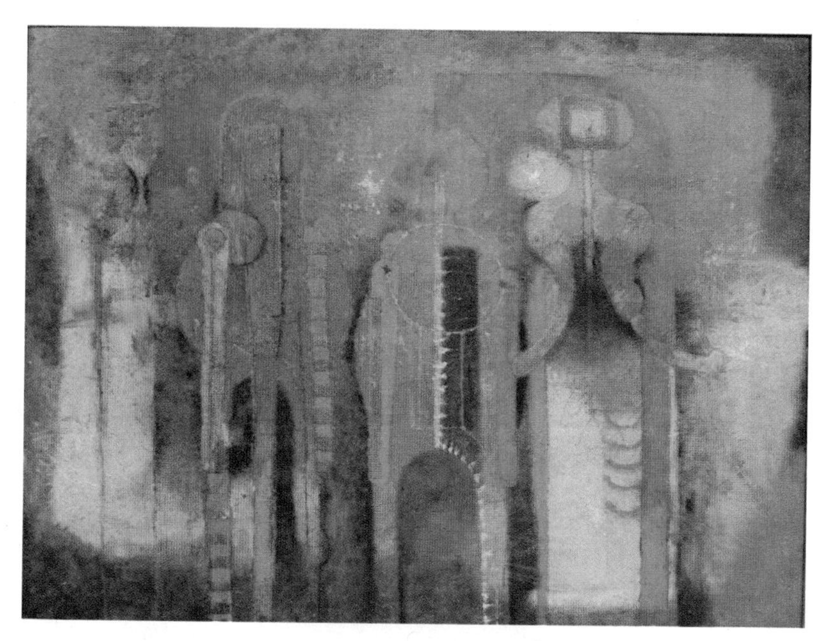

Rufino Tamayo (1899–1991) was a contemporary of Mexico's famous muralists—Diego Rivera, José Clemente Orozco, and David Alfaro Siquieros. Tamayo, however, rejected their emphasis on politics and muralism, choosing instead to focus on colors, textures, and abstract forms. This painting of abstract human shapes, "Tres Personajes," is a fine example of the style that so earned the artist enmity that he chose to live in the United States for a decade. Why was this style of painting so unusual for its time, and why would it be the target of attacks? *EMMANUEL DUNAND/AFP/Getty Images*

In the 1980s, Nicaraguan artists drew acclaim for their folkloric paintings of village life. This photograph shows an array of typical paintings on sale at the Mercado de Artesanias (National Artisans Market) in Masaya. These idealized images sold well among the tourists who flocked to Nicaragua to observe the Sandinista Revolution. What does their popularity reveal about national and international expectations about Central American culture? *Alamy Limited*

Oswaldo Guayasamin, an Ecuadoran painter of Quechua descent, became a renowned artist in the 1940s, shortly after his first exhibition. A human rights activist, particularly against the Pinochet regime in Chile, he devoted much of his art to a depiction of the suffering of oppressed Latin Americans. His work is divided into three distinct periods called: "Path of Sorrow," "Time of Wrath," and "Time of Endearment." Here he is shown in his Quito studio working on his last painting, entitled "El Grito" (The Cry). How does this painting express his protest against the dictatorships of the late twentieth century? *Martin BERNETTI/AFP PHOTO/ Newscom*

Another lovely work of art is this Ecuadoran tapestry, featuring figures in hats and ponchos gazing upon distant, snowy mountain peaks. Is this a timeless view of an unchanging Latin America? Or in art, as in history, do we find both continuity and change? *Fotos & Photos*/Photolibrary.com.

A Chronology of Significant Dates in Latin American History

1492	Columbus reaches the New World.
1494	The Treaty of Tordesillas divides the world between Spain and Portugal.
1500	Cabral discovers Brazil.
1503	Spain legalizes the encomienda in the New World; Casa de Contratación created.
1512	The Laws of Burgos regulate the treatment of the Indians.
1513	Balboa discovers the Pacific Ocean.
1521	Hernándo Cortés completes the conquest of the Aztec empire.
1524	Creation of the Council of the Indies.
1532	First permanent settlements in Brazil.
1535	Francisco Pizaro completes the conquest of the Incan empire; the first viceroy arrives in Mexico.
1542	The New Laws call for an end of the encomiendas.
1543	The first viceroy arrives in Peru.
1545	The Spaniards discover silver at Potosí.
1630–1654	The Dutch control as much as one-third of Brazil.
1695	The Luso-Brazilians discover gold in the Brazilian interior.
1763	The capital of the Viceroyalty of Brazil is moved to Rio de Janeiro.
1776	Creation of the Viceroyalty of La Plata.
1804	Haiti declares its independence.
1808	The royal family of Portugal arrives in Brazil.
1810	Padre Miguel Hidalgo initiates Mexico's struggle for independence.
1811	Paraguay and Venezuela declare their independence.
1816	Argentina declares its independence.

1818–1843 Jean-Pierre Boyer, a populist caudillo, rules Haiti.

1819 Brazil puts a steamship into service, the first in South America.

1821 Mexico, Peru, and Central America declare their independence.

1822–1823 Emperor Agustín I rules the Mexican empire.

1822 Prince Pedro declares Brazil's independence and receives the title of emperor.

1823 President James Monroe promulgates the Monroe Doctrine.

1824 The Battle of Ayacucho marks the final defeat of the Spaniards in South America.

1824–1838 The United Provinces of Central America in existence.

1825 Bolivia declares its independence.

1825–1828 The Cisplatine War between Brazil and Argentina to possess Uruguay results in a stalemate and Uruguayan independence.

1829–1852 The populist caudillo Juan Manuel de Rosas rules Argentina.

1830 The political union of Gran Colombia dissolves, leaving Colombia, Venezuela, and Ecuador to go their independent ways.

1838 The first railroad in Latin America is inaugurated in Cuba.

1839–1865 The populist caudillo Rafael Carrera governs Guatemala.

1846–1848 War of North American Invasion (Mexican–American War). The United States gains California, New Mexico, and Arizona from its victory.

1847–1903 The Cruzob rebellion in Yucatan and Mayan self-government.

1848–1855 The popular caudillo Manuel Belzú governs Bolivia.

1850 The United States and Great Britain sign the Clayton–Bulwer Treaty to check the expansion of each in Central America. Unionization of workers slowly begins in the largest Latin American nations.

1852 Chile inaugurates the first railroad in South America. Chile and Brazil initiate telegraphic systems.

1864–1867 Archduke Maximilian of Austria rules Mexico under French protection.

1865–1870 In the War of the Triple Alliance, Argentina, Brazil, and Uruguay fight and eventually defeat Paraguay.

1876 The first refrigerator ship carries beef from Buenos Aires to Europe.

1876–1911 Porfirio Díaz governs Mexico.

1879–1884 The War of the Pacific pits Chile against Peru and Bolivia.

1886 The University of Chile awards the first medical degree to a woman in Latin America.

1888 Brazil abolishes slavery.

1889 The military dethrones Emperor Pedro II of Brazil; Brazil becomes a republic.

1889–1890 The first Inter-American Conference meets in Washington, D.C.

1898 As a result of the Spanish–American War, Cuba gains its independence from Spain, and the United States takes possession of Puerto Rico.

1901 In the Hay–Pauncefote Treaty, Great Britain acknowledges U.S. supremacy in Central America.

1903 Panama gains its independence and signs a treaty with the United States for the construction of an interoceanic canal.

1903–1929 José Batlle dominates Uruguayan politics, bringing stability and economic growth as well as the middle class to power.

1909–1933 U.S. intervention and occupation of Nicaragua.

1910–1940 The Mexican Revolution.

1911 Emiliano Zapata advocates agrarian reform in his Plan of Ayala.

1912 Argentina adopts the Saenz Peña law, giving all male citizens the right to vote, without property or literacy requirements.

1914 The Panama Canal opens.

1915–1934 The United States occupies Haiti.

1916–1922 Hipólito Irigoyen governs Argentina as its first middle-class president.

1916–1924 The United States occupies the Dominican Republic.

1917 Promulgation of the Mexican constitution, the blueprint for the Revolution.

1919 Promulgation of the Uruguayan constitution, the blueprint for middle-class democracy.

1920–1924 Arturo Alessandri, a representative of middle-class interests, governs Chile.

1927–1933 Augusto César Sandino leads the guerrilla struggle in Nicaragua to expel the U.S. Marines.

1929 The world financial collapse reduces Latin American exports but encourages import-substitution industrialization. Ecuador grants the vote to women, the first in Latin America.

1932–1935 Bolivia and Paraguay fight the Chaco War.

1934–1940 The Mexican Revolution reaches its apogee under President Lázaro Cárdenas.

1937 Bolivia cancels foreign oil contracts and takes control of oil industry.

1938 Mexico nationalizes foreign oil companies.

1940 Promulgation of the Cuban constitution, a middle-class and nationalist blueprint for change.

1944–1954 The era of Guatemalan democracy.

1945 Gabriela Mistral, Chilean poet, is the first Latin American to receive the Nobel Prize in literature.

1952 Guatemala promulgates its land reform.

1952–1964 The Bolivian revolution.

1953 Bolivia puts into effect its land reform.

1954 The CIA overthrows President Jacobo Arbenz of Guatemala.

1959 Triumph of the Cuban revolution and the advent of Fidel Castro to power. Cuba issues its Agrarian Reform Law.

1960 For the first time, urban Latin Americans equal in number their rural counterparts.

1961 Washington breaks diplomatic relations with Cuba. The CIA sponsors the Bay of Pigs invasion in an attempt to overthrow Castro. President John F. Kennedy announces the Alliance for Progress.

1964 The Brazilian military deposes President João Goulart and establishes a dictatorship.

1965 The United States invades and occupies the Dominican Republic.

1967 Che Guevara is killed in Bolivia while attempting to spark a revolutionary uprising.

1968 The Mexican government squelches a significant student movement by firing on peaceful protestors at the Plaza of Tlatelolco in Mexico City. Hundreds are killed.

1970–1973 President Salvador Allende sets in motion profound reforms to peacefully and democratically change Chile.

1973 The Chilean military overthrows President Allende, who dies in the attack on the presidential palace.
The Uruguayan military terminates their nation's twentieth-century experiment with democracy.

1974–1976 Isabel Perón serves as president of Argentina, the first female chief of state in the Western Hemisphere.

1976–1983 Thousands are "disappeared" in the Dirty War in Argentina.

1977 Panama and the United States sign a treaty returning the Canal Zone to Panamanian control and putting the canal under Panamanian direction by 1999.

1979 Triumph of the Nicaraguan revolution. Young military reformers stage a coup d'etat in El Salvador.

1980 Sendero Luminoso initiates its armed struggle in Peru.

1981 Nicaragua promulgates its agrarian reform law. President Ronald Reagan begins the contra war against Nicaragua. Latin America enters a severe economic crisis.

1982 Argentina invades the Falkland Islands and is defeated by Great Britain. Mexico declares it cannot pay its foreign debt.

1983 The United States invades Grenada and overthrows government.

1984 Latin America's foreign debt reaches an unmanageable $350 billion.

1985 Brazil returns to civilian rule. Latin American population surpasses the 400 million mark.

1987 President Oscar Arias of Costa Rica wins the Nobel Peace prize.

1988 PRI (*Partido Revolucionario Institucional*) candidate Carlos Salinas de Gortari's defeat of Cuauhtemoc Cárdenas is decried as open fraud.
In Chile, Augusto Pinochet is ousted by plebiscite.

1989–1990 The United States invades and occupies Panama.

1990 Sandinistas lose presidential election to Violeta Chamorro, widow of the late newspaper editor Pedro Joaquin Chamorro, effectively ending the Sandinista Revolution.

1992 Quincentennial of European–American encounter.
Sendero Luminoso controls much of rural Peru.
Mexico's Carlos Salinas de Gortari amends the constitution to allow ejidos to be sold, rented, and mortgaged. Privatization of state firms begins.

1994 The North American Free Trade Agreement (NAFTA) links the markets of the United States, Mexico, and Canada. The day it takes effect, a rebellion breaks out in Chiapas, Mexico, led by the Zapatista Army of National Liberation.

1996 A peace agreement is signed in Guatemala.

1999 The Panama Canal zone is returned to Panamanian control.
Hugo Chávez is elected president of Venezuela, the first of a new wave of Latin American leaders who challenge neoliberalism.

2000 For the first time in seventy-one years, PRI allows a free election in Mexico, and the voters elect Vicente Fox of the National Action Party (*Partido de Acción Nacional*).
The Confederation of Indigenous Nationalities of Ecuador leads a national uprising that brings down Ecuador's government over plans to "dollarize" the economy.

2001 The United States launches Plan Colombia, a three-year agreement to provide $1.5 billion a year to Colombia to fight guerrilla groups.

2002 A coup ousts Chávez from the Venezuelan presidency for forty-eight hours. He is restored to office, and the rapid U.S. approval of his removal is widely decried by Latin American leaders.

2003 Luis Inácio "Lula" da Silva, former head of the metal workers union, is elected president in Brazil. He is reelected in 2006.
Néstor Kirchner is elected president in Argentina.

2004 Kirchner succeeds in renegotiating Argentina's debt and defying the International Monetary Fund.

2005 Petrosur, a petroleum company, and Telesur, a TV channel, are launched with the participation of several Latin American governments to provide oil and news to the region on Latin American terms.

Evo Morales is elected president of Bolivia, the first indigenous to head the country. He survives a recall effort in 2008 and is reelected president in 2009 and 2014.

2006 Michelle Bachelet is elected president of Chile, becoming South America's first elected female president.

Daniel Ortega is elected president of Nicaragua; he is reelected in 2011.

2007 Cristina Fernández de Kirchner becomes president of Argentina.

2008 International economic crisis begins. The United States launches Plan Mérida, modeled after Plan Colombia, to fight drug production in Mexico.

2009 Honduran president Manuel Zelaya is overthrown by a coup. Latin American leaders urge the United States to reconsider the war on drugs. FMLN candidate Mauricio Funes is elected president of El Salvador. José Mujica, a former leader of the Tupamaros, is elected president of Uruguay.

For the first time, the percentage of middle class in Latin America reaches 30 percent.

2010 Argentina is the first Latin American country to adopt a law at the national level legalizing same-sex marriage.

The Community of Latin American and Caribbean States is formed.

2012 While the United States remains Latin America's largest trade partner, the region's trade with China grows by 8 percent compared to 6 percent for the United States.

2013 Uruguay becomes the first Latin American country to legalize the sale of marijuana.

2014 Former FMLN guerrilla leader Salvador Sánchez Cerén is elected president of El Salvador.

Brazil hosts the World Cup.

Student protest leader Camila Vallejo is elected to the Chilean Congress.

U.S. President Barack Obama and Cuban President Raúl Castro agree to restore diplomatic relations.

2016 Brazil is scheduled to host the Summer Olympics.

Recommended Reading

Chapter 1

Alchon, Suzanne Austin. *A Pest in the Land: New World Epidemics in a Global Perspective*. Albuquerque: University of New Mexico Press, 2003.

Barton, Simon. *A History of Spain*. Basingstoke: Palgrave MacMillan, 2004.

Bethell, Leslie, ed. *Cambridge History of Latin America*, Vols. 1–11. Cambridge: Cambridge University Press, 1984.

Buisseret, David, and Steven G. Reinhardt, eds. *Creolization in the Americas*. College Station: Texas A&M Press, 2000.

Carrasco, David, Lindsay Jones, and Scott Sessions, eds. *Mesoamerica's Classic Heritage: From Teotihuacan to the Aztecs*. Boulder: University Press of Colorado, 2000.

Cook, Noble David. *Born to Die: Disease and New World Conquest, 1492–1650*. Cambridge: Cambridge University Press, 1998.

D'Altroy, Terence N. *The Incas*. Oxford: Blackwell Publishers, 2003.

De Blij, Harm. "American Regional Appellations: Farewell to 'Latin' America?" *AAG Newsletter*, September 2008, www.aag.org

Earle, Rebecca. *The Body of the Conquistador: Food, Race and the Colonial Experience in Spanish America, 1492–1700*. Cambridge: Cambridge University Press, 2012.

Gallup, John Luke, Alejandro Gaviria, and Eduardo Lora. *Is Geography Destiny? Lessons from Latin America*. Palo Alto: Stanford University Press, 2003.

Hall, Gwendolyn Midlo. *Slavery and African Ethnicities in the Americas: Restoring the Links*. Chapel Hill: University of North Carolina Press, 2005.

Henige, David. *Numbers from Nowhere: The American Indian Contact Population Debate*. Norman: University of Oklahoma Press, 1998.

Holloway, Thomas H., ed. *A Companion to Latin American History*. Oxford: Blackwell Publishing, 2008.

Latta, Alex, and Hannah Wittman. *Environment and Citizenship in Latin America: Natures, Subjects and Struggles*. New York: Berghahn Books, 2012.

Lewis, Martin W., and Karen E. Wigen. *The Myth of Continents: A Critique of Metageography*. Berkeley: University of California Press, 1997.

Miller, Mary Grace. *Rise and Fall of the Cosmic Race: The Cult of Mestizaje in Latin America*. Austin: University of Texas Press, 2004.

O'Callaghan, Joseph F. *Reconquest and Crusade in Medieval Spain*. Philadelphia: University of Pennsylvania Press, 2003.

Peloso, Vincent C. *Race and Ethnicity in Latin American History*. New York: Routledge, 2014.

Price, Marie D., and Catherine W. Cooper. "Competing Visions, Shifting Boundaries: The Construction of Latin America as a World Region," *Journal of Geography*, Vol. 106, No. 3. 113–122.

Rahier, Jean Muteba. *Blackness in the Andes: Ethnographic Vignettes of Cultural Politics in the Time of Multiculturalism*. New York: Palgrave Macmillan, 2014.

Sharer, Robert J., and Loa P. Traxler. *The Ancient Maya*. Stanford: Stanford University Press, 2006.

Sweet, James H. *Recreating Africa: Culture, Kinship, and Religion in the African–Portuguese*

World, 1441–1770. Chapel Hill: University of North Carolina Press, 2003.

Telles, Edward Eric, ed., and the Project on Ethnicity and Race in Latin America (PERLA). *Pigmentocracies: Ethnicity, Race, and Color in Latin America*. Chapel Hill: University of North Carolina Press, 2014.

Thornton, John. *Africa and Africans in the Making of the Atlantic World, 1400–1800*. Cambridge: Cambridge University Press, 1998.

Topik, Steven. "Historical Perspectives on Latin American Underdevelopment," *The History Teacher*, Vol. 20, No. 4 (1987), 545–560.

United Nations Environment Programme. *Global Environment Outlook: Latin America and the Caribbean*. Panama City, Panama: United Nations Environment Programme, 2010.

Chapter 2

Connell, William F. *After Moctezuma: Indigenous Politics and Self-government in Mexico City, 1524–1730*. Norman: University of Oklahoma Press, 2011.

Elliott, John Huxtable. *Empires of the Atlantic World: Britain and Spain in America, 1492–1830*. New Haven: Yale University Press, 2006.

Eltis, David. *The Rise of African Slavery in the Americas*. Cambridge: Cambridge University Press, 2000.

Fisher, Andrew B., and Matthew D. O'Hara, eds. *Imperial Subjects: Race and Identity in Colonial Latin America*. Durham: Duke University Press, 2009.

Griffiths, Nicholas, and Fernando Cervantes, eds. *Spiritual Encounters: Interactions Between Christianity and Native Religions in Colonial America*. Lincoln: University of Nebraska Press, 1999.

Hawthorne, Walter. *From Africa to Brazil: Culture, Identity, and an Atlantic Slave Trade, 1600–1830*. Cambridge: Cambridge University Press, 2010.

Kamen, Henry. *Golden Age Spain*. Basingstoke: Palgrave Macmillan, 2004.

Klein, Herbert S. *The Atlantic Slave Trade*. Cambridge: Cambridge University Press, 1999.

Landers, Jane G., and Barry M. Robinson, eds. *Slaves, Subjects and Subversives: Blacks in Colonial Latin America*. Albuquerque: University of New Mexico Press, 2006.

Lawrance, Jeremy. *Spanish Conquest, Protestant Prejudice: Las Casas and the Black Legend*. Nottingham: Critical, Cultural and Communications Press, 2009.

Lockhart, James, and Stuart B. Schwartz. *Early Latin America: A History of Colonial Spanish America and Brazil*. Cambridge: Cambridge University Press, 1983.

———. *The Nahuas After the Conquest: A Social and Cultural History of the Indians of Central Mexico, Sixteenth Through Eighteenth Centuries*. Stanford: Stanford University Press, 1992.

Matthew, Laura E., and Michel R. Oudijk, eds. *Indian Conquistadors: Indigenous Allies in the Conquest of Mesoamerica*. Norman: University of Oklahoma Press, 2007.

McEnroe, Sean. *From Colony to Nationhood in Mexico: Laying the Foundations, 1560–1840*. Cambridge: Cambridge University Press, 2012.

Naro, Nancy Priscilla. *A Slave's Place, A Master's World: Fashioning Dependency in Rural Brazil*. London: Continuum, 2000.

Nellis, Eric. *Shaping the New World: African Slavery in the Americas, 1500–1888*. Toronto: University of Toronto Press, 2013.

Nesvig, Martin Austin, ed. *Local Religion in Colonial Mexico*. Albuquerque: University of New Mexico Press, 2006.

Powers, Karen Vieira. *Women in the Crucible of Conquest: The Gendered Genesis of Spanish American Society, 1500–1600*. Albuquerque: University of New Mexico Press, 2005.

Proctor, Frank T. *Damned Notions of Liberty: Slavery, Culture, and Power in Colonial Mexico, 1640–1769*. Albuquerque: University of New Mexico Press, 2010.

Restall, Matthew, ed. *Beyond Black and Red: African-Native Relations in Colonial Latin America*. Albuquerque: University of New Mexico Press, 2005.

———. *Seven Myths of the Spanish Conquest*. Oxford: Oxford University Press, 2003.

Schwaller, John Frederick, ed. *The Church in Colonial Latin America*. Wilmington: Scholarly Resources Books, 2000.

Socolow, Susan Migden. *The Women of Colonial Latin America*. Cambridge: Cambridge University Press, 2000.

Sweet, James. *Recreating Africa: Culture, Kinship, and Religion in the African-Portuguese World, 1441–1770*. Chapel Hill: University of North Carolina Press, 2003.

Terraciano, Kevin. *The Mixtecs of Colonial Oaxaca: Ñudzahui History, Sixteenth Through Eighteenth Centuries*. Stanford: Stanford University Press, 2001.

Townsend, Camilla. *Malintzin's Choices: An Indian Woman in the Conquest of Mexico*. Albuquerque: University of New Mexico Press, 2006.

Twinam, Ann. *Purchasing Whiteness: Pardos, Mulattos, and the Quest for Social Mobility in the Spanish Indies*. Stanford: Stanford University Press, 2015.

Chapter 3

Archer, Christon I., ed. *The Wars of Independence in Spanish America*. Wilmington: Scholarly Resources, 2000.

Bauer, Ralph, and José Antonio Mazzotti, eds. *Creole Subjects in the Colonial Americas: Empires, Texts, Identities*. Chapel Hill: University of North Carolina Press, 2009.

Blanchard, Peter. *Under the Flags of Freedom: Slave Soldiers and the Wars of Independence in Spanish South America*. Pittsburgh: University of Pittsburgh Press, 2008.

Brading, David A. *The First America: The Spanish Monarchy, Creole Patriots and the Liberal State, 1492–1867*. Cambridge: Cambridge University Press, 1991.

Brown, Matthew. *Adventuring Through Spanish Colonies: Simón Bolívar, Foreign Mercenaries and the Birth of New Nations*. Liverpool: Liverpool University Press, 2006.

Burkholder, Mark A. *Spaniards in the Colonial Empire: Creoles vs. Peninsulars?* Malden: John Wiley & Sons, 2013.

Chasteen, John Charles. *Americanos: Latin America's Struggle for Independence*. Oxford: Oxford University Press, 2009.

Esdaile, Charles J. *Spain in the Liberal Age: From Constitution to Civil War, 1808–1939*. Oxford: Blackwell Publishers, 2000.

Geggus, David Patrick, and Norman Fiering. *The World of the Haitian Revolution*. Bloomington: Indiana University Press, 2009.

Graham, Richard. *Independence in Latin America: Contrasts and Comparisons*. 3rd ed. Austin: University of Texas Press, 2013.

Guardino, Peter F. *The Time of Liberty: Popular Political Culture in Oaxaca, 1750–1850*. Durham: Duke University Press, 2005.

Henderson, Timothy J. *The Mexican Wars for Independence*. New York: Hill and Wang, 2009.

Kamen, Henry. *Philip V of Spain: The King Who Reigned Twice*. New Haven: Yale University Press, 2001.

Kinsbruner, Jay. *Independence in Spanish America: Civil Wars, Revolutions, and Underdevelopment*. Albuquerque: University of New Mexico Press, 2000.

Langley, Lester D. *Simón Bolívar: Venezuelan Rebel, American Revolutionary*. Lanham: Rowman & Littlefield Publishers, 2009.

Lynch, John. *Simón Bolívar: A Life*. New Haven: Yale University Press, 2006.

———. *Latin America Between Colony and Nation: Selected Essays*. New York: Palgrave Macmillan, 2001.

McFarlane, Anthony. *War and Independence in Spanish America*. New York: Routledge, 2014.

Murray, Pamela S. *For Glory and Bolívar: The Remarkable Life of Manuela Sáenz*. Austin: University of Texas Press, 2010.

Phelan, John Leddy. *The People and the King: The Comunero Revolution in Colombia, 1781*. Madison: University of Wisconsin Press, 1978.

Racine, Karen. *Francisco de Miranda: A Transatlantic Life in the Age of Revolution*. Wilmington: Scholarly Resources, 2003.

Rodriguez O., Jaime E. *The Independence of Spanish America*. Cambridge: Cambridge University Press, 1998.

———. *"We Are Now the True Spaniards": Sovereignty, Revolution, Independence, and the Emergence of the Federal Republic of Mexico, 1808–1824*. Stanford: Stanford University Press, 2012.

Santoni, Pedro, ed. *Daily Lives of Civilians in Wartime Latin America: From the Wars of Independence to the Central American Wars*. Westport: Greenwood Press, 2008.

Stavig, Ward. *The World of Tupac Amaru: Conflict, Community and Identity in Colonial Peru*. Lincoln: University of Nebraska Press, 1999.

Van Young, Eric. *The Other Rebellion: Popular Violence, Ideology, and the Mexican Struggle for Independence, 1810–1821*. Stanford: Stanford University Press, 2001.

Zeuske, Michael. *Simón Bolívar: History and Myth*. Princeton: Markus Wiener Publishers, 2013.

Chapter 4

Bazant, Jan. *Alienation of Church Wealth in Mexico: Social and Economic Aspects of the Liberal Revolution, 1856–1875*. Cambridge: Cambridge University Press, 1971.

Bushnell, David. *The Emergence of Latin America in the Nineteenth Century*. Oxford: Oxford University Press, 1988.

Centeno, Miguel Angel, and Agustín Ferraro, eds. *State and Nation Making in Latin America and Spain: Republics of the Possible*. Cambridge: Cambridge University Press, 2013.

Chasteen, John Charles. *Heroes on Horseback: A Life and Times of the Last Gaucho Caudillos*. Albuquerque: University of New Mexico Press, 1995.

Fowler, Will. *Santa Anna of Mexico*. Lincoln: University of Nebraska Press, 2007.

———. *Malcontents, Rebels, and Pronunciados: The Politics of Insurrection in Nineteenth-Century Mexico*. Lincoln: University of Nebraska Press, 2012.

Guardino, Peter F. *Peasants, Politics, and the Formation of Mexico's National State: Guerrero, 1800–1857*. Stanford: Stanford University Press, 1996.

———. *The Time of Liberty: Popular Political Culture in Oaxaca, 1750–1850*. Durham: Duke University Press, 2005.

Hamill, Hugh, ed. *Caudillos: Dictators in Spanish America*. Norman: University of Oklahoma Press, 1992.

Lynch, John. *Argentine Caudillo: Juan Manuel de Rosas*. Wilmington: Scholarly Resources, 2001.

———. *Caudillos in Spanish America, 1808–1850*. Oxford: Oxford University Press, 1992.

Mallon, Florencia E. *Peasant and Nation: The Making of Postcolonial Mexico and Peru*. Berkeley: University of California Press, 1995.

Peloso, Vincent C., and Barbara A. Tenenbaum, eds. *Liberals, Politics, and Power: State Formation in Nineteenth-Century Latin America*. Athens: University of Georgia Press, 1996.

Salvatore, Ricardo. *Wandering Paysanos: State Order and Subaltern Experience in Buenos Aires During the Rosas Era*. Durham: Duke University Press, 2003.

Sanders, James E. " 'A Mob of Women' Confront Post-Colonial Republican Politics: How Class, Race, and Partisan Ideology Affected Gendered Political Space in Nineteenth-Century Southwestern Colombia." *Journal of Women's History* 20, No. 1 (Spring 2008): 63–89.

Sexton, Jay. *The Monroe Doctrine: Empire and Nation in Nineteenth-Century America*. New York: Hill and Wang, 2011.

Shumway, Jeffrey M. "Juan Manuel de Rosas: Authoritarian Caudillo and Primitive Populist," *History Compass*, Vol. 2 (2004) LA 113, 1–14.

Slatta, Richard W. *Gauchos and the Vanishing Frontier*. Lincoln: University of Nebraska Press, 1992.

Thiessen-Reily, Heather K. "Caudillo Nationalism in Bolivia," in Don H. Doyle and Marco Antonio Pamplona, eds., *Nationalism in the New World*. Athens: University of Georgia Press, 2006.

Voss, Stuart F. *Latin America in the Middle Period, 1750–1920*. Wilmington: Scholarly Resources, 2002.

Walker, Charles F. *Smoldering Ashes: Cuzco and the Creation of Republican Peru, 1780–1840*. Durham: Duke University Press, 1999.

Wasserman, Mark. *Everyday Life and Politics in Nineteenth Century Mexico: Men, Women, and War*. Albuquerque: University of New Mexico Press, 2000.

Woodward, Ralph Lee. *Rafael Carrera and the Emergence of the Republic of Guatemala, 1821–1871*. Athens: University of Georgia Press, 1993.

Chapter 5

Bauer, Arnold J. *Chilean Rural Society from Spanish Conquest to 1930*. Cambridge: Cambridge University Press, 1975.

Bulmer-Thomas, Victor. *The Economic History of Latin America Since Independence*. 2nd ed. New York: Cambridge University Press, 2003.

Burns, E. Bradford. *The Poverty of Progress: Latin America in the Nineteenth Century*. Berkeley: University of California Press, 1980.

Charlip, Julie A. *Cultivating Coffee: The Farmers of Carazo, Nicaragua, 1880–1930*. Athens: Ohio University Press, 2003.

Coatsworth, John H. *Growth Against Development: The Economic Impact of the Railroads in Porfirian Mexico*. DeKalb: Northern Illinois University Press, 1981.

———, and Alan M. Taylor. *Latin America and the World Economy Since 1800*. Cambridge: Harvard University Press, 1998.

Da Cunha, Euclides. *Rebellion in the Backlands* [1902]. Chicago: University of Chicago Press, 1964.

Dean, Warren. *The Industrialization of São Paulo, 1880–1945*. Austin: University of Texas Press, 1969.

Foote, Nicola, and Rene D. Harder Horst, eds. *Military Struggle and Identity Formation in Latin America: Race, Nation, and Community During the Liberal Period*. Gainesville: University Press of Florida, 2010.

Garrigan, Shelley E. *Collecting Mexico: Museums, Monuments, and the Creation of National Identity*. Minneapolis: University of Minnesota Press, 2012.

Gilson, Gregory D., and Irving W. Levinson, eds. *Latin American Positivism: New Historical and Philosophical Essays*. Lanham: Lexington Books, 2013.

Hecht, Susanna B. *The scramble for the Amazon and the "Lost paradise" of Euclides da Cunha*. Chicago: University of Chicago Press, 2013.

Holloway, Thomas H. *Immigrants on the Land: Coffee and Society in São Paulo, 1886–1934*. Chapel Hill: University of North Carolina Press, 1980.

Langer, Erick D. *Economic Change and Rural Resistance in Southern Bolivia, 1880–1930*. Stanford: Stanford University Press, 1989.

Lauria Santiago, Aldo. *An Agrarian Republic: Commercial Agriculture and the Politics of Peasant Communities in El Salvador, 1823–1914*. Pittsburgh: University of Pittsburgh Press, 1999.

Levine, Robert M. *Vale of Tears: Revisiting the Canudos Massacre in Northeastern Brazil, 1893–1897*. Berkeley: University of California Press, 1992.

Levy, Juliette. *The Making of a Market: Credit, Henequen, and Notaries in Yucatán, 1850–1900*. University Park: Pennsylvania State University Press, 2012.

Matthews, Michael. *The Civilizing Machine: A Cultural History of Mexican Railroads, 1876–1910*. Lincoln: University of Nebraska Press, 2013.

McCreery, David. *Rural Guatemala: 1760–1940*. Stanford: Stanford University Press, 1994.

Orlove, Benjamin, ed. *The Allure of the Foreign: Imported Goods in Postcolonial Latin America*. Ann Arbor: University of Michigan Press, 1997.

Pineo, Ronn, and James A. Baer, eds. *Cities of Hope: People, Protests, and Progress in Urbanizing*

Latin America, 1870–1930. Boulder: Westview Press, 1998.

Reed, Nelson A. *The Caste War of the Yucatan*. Stanford: Stanford University Press, 2001.

Rugeley, Terry. *Rebellion Now and Forever: Mayas, Hispanics, and Caste War Violence in Yucatán, 1800–1880*. Stanford: Stanford University Press, 2009.

Topik, Steven C., and Allen Wells, eds. *The Second Conquest of Latin America: Coffee, Henequen, and Oil During the Export Boom, 1850–1930*. Austin: University of Texas Press, 1998.

Van Hoy, Teresa Miriam. *A Social History of Mexico's Railroads: Peons, Prisoners, and Priests*. Lanham: Rowman & Littlefield, 2008.

Chapter 6

Barr-Melej, Patrick. *Reforming Chile: Cultural Politics, Nationalism, and the Rise of the Middle Class*. Chapel Hill: University of North Carolina Press, 2001.

Bergquist, Charles. *Labor in Latin America: Comparative Essays on Chile, Argentina, Venezuela, and Colombia*. Stanford: Stanford University Press, 1986.

Blum, Ann Shelby. *Domestic Economies: Families, Work, and Welfare in Mexico City, 1884–1943*. Lincoln: University of Nebraska Press, 2009.

Coerver, Don M., and Linda B. Hall. *Tangled Destinies: Latin America and the United States*. Albuquerque: University of New Mexico Press, 1999.

DeShazo, Peter. *Urban Workers and Labor Unions in Chile, 1902–1927*. Madison: University of Wisconsin Press, 1983.

Ehrick, Christine. *The Shield of the Weak: Feminism and the State in Uruguay, 1903–1933*. Albuquerque: University of New Mexico Press, 2005.

Farnsworth-Alvear, Ann. *Dulcinea in the Factory: Myths, Morals, Men, and Women in Colombia's Industrial Experiment, 1905–1960*. Durham: Duke University Press, 2000.

French, John D. *The Brazilian Workers' ABC: Class Conflict and Alliances in Modern São Paulo*. Chapel Hill: University of North Carolina Press, 1992.

Hutchison, Elizabeth A. *Labors Appropriate to Their Sex: Gender, Labor, and Politics in Urban Chile, 1900–1930*. Durham: Duke University Press, 2001.

Karush, Matthew B. *Workers or Citizens: Democracy and Identity in Rosario, Argentina (1912–1930)*. Albuquerque: University of New Mexico Press, 2002.

Klubock, Thomas Miller. *Contested Communities: Class, Gender, and Politics in Chile's El Teniente Copper Mine, 1904–1951*. Durham: Duke University Press, 1998.

Langley, Lester D. *The Banana Wars: United States Intervention in the Caribbean, 1898–1934*. Wilmington: Scholarly Resources Books, 2002.

Leonard, Thomas M., ed. *United States-Latin American Relations, 1850–1903: Establishing a Relationship*. Tuscaloosa: University of Alabama Press, 1999.

Lesser, Jeff. *Immigration, Ethnicity, and National Identity in Brazil, 1808 to the Present*. Cambridge: Cambridge University Press, 2013.

McCartney, Paul T. *Power and Progress: American National Identity, the War of 1898, and the Rise of American Imperialism*. Baton Rouge: Louisiana State University Press, 2006.

McCreery, David J. *The Sweat of Their Brow: A History of Work in Latin America*. Armonk: M.E. Sharpe Inc., 2000.

Nouwen, Mollie Lewis. *Oy, My Buenos Aires: Jewish Immigrants and the Creation of Argentine National Identity*. Albuquerque: University of New Mexico Press, 2013.

Owensby, Brian P. *Intimate Ironies: Modernity and the Making of Middle-Class Lives in Brazil*. Stanford: Stanford University Press, 1999.

Parker, David S. *The Idea of the Middle Class: White Collar Workers and Peruvian Society, 1900–1950*. University Park: Pennsylvania State University Press, 1998.

Pérez Jr., Louis A. *The War of 1898: The United States and Cuba in History and Historiography*.

Chapel Hill: University of North Carolina Press, 1998.

Porter, Susie S. *Working Women in Mexico City: Public Discourses and Material Conditions, 1879–1930*. Tucson: University of Arizona Press, 2003.

Silva Dias, Maria Odila Leite Da. Power and Everyday Life: The Lives of Working Women in Nineteenth-Century Brazil. New Brunswick: Rutgers University Press, 1995.

Smith, Peter H. *Talons of the Eagle: Latin America, the United States, and the World*. New York: Oxford University Press, 2008.

Spalding, Hobart. *Organized Labor in Latin America: Historical Case Studies of Workers in Dependent Societies*. New York: New York University Press, 1977.

Chapter 7

Becker, Marjorie. *Setting the Virgin on Fire: Lázaro Cárdenas, Michoacán Peasants, and the Redemption of the Mexican Revolution*. Berkeley: University of California Press, 1995.

Benjamin, Thomas. *La Revolución: Mexico's Great Revolution as Memory, Myth, & History*. Austin: University of Texas Press, 2000.

Benjamin, Thomas, and Mark Wasserman, eds. *Provinces of the Revolution: Essays on Regional Mexican History, 1910–1929*. Albuquerque: University of New Mexico Press, 1990.

Bortz, Jeff. *Revolution within the Revolution: Cotton Textile Workers and the Mexican Labor Regime, 1910–1923*. Stanford: Stanford University Press, 2008.

Brunk, Samuel. *¡Emiliano Zapata! Revolution and Betrayal in Mexico*. Albuquerque: University of New Mexico Press, 1995.

Gonzales, Michael J. *The Mexican Revolution, 1910–1940*. Albuquerque: University of New Mexico Press, 2002.

Hart, John Mason. *Revolutionary Mexico: The Coming and Process of the Mexican Revolution*. Berkeley: University of California Press, 1987.

Hart, Paul. *Bitter Harvest: The Social Transformation of Morelos, Mexico, and the Origins of the Zapatista Revolution, 1840–1910*. Albuquerque: University of New Mexico Press, 2005.

Joseph, Gilbert M., and Jürgen Buchenau. *Mexico's Once and Future Revolution: Social Upheaval and the Challenge of Rule Since the Late Nineteenth Century*. Durham: Duke University Press, 2013.

Katz, Friedrich. *The Life and Times of Pancho Villa*. Stanford: Stanford University Press, 1998.

Knight, Alan. *The Mexican Revolution, v. 1, Porfirians, Liberals and Peasants; v. 2, Counterrevolution and Reconstruction*. Cambridge: Cambridge University Press, 1986.

Lear, John. *Workers, Neighbors, and Citizens: The Revolution in Mexico City*. Lincoln: University of Nebraska Press, 2001.

Mitchell, Stephanie E., and Patience A. Schell, eds. *The Women's Revolution in Mexico, 1910–1953*. Lanham: Rowman & Littlefield, 2007.

Mraz, John. *Photographing the Mexican Revolution: Commitments, Testimonies, Icons*. Austin: University of Texas Press, 2012.

Newman, Elizabeth Terese. *Biography of a Hacienda: Work and Revolution in Rural Mexico*. Tucson: University of Arizona Press, 2014.

Olcott, Jocelyn, Mary K. Vaughan, and Gabriela Cano, eds. *Sex in Revolution: Gender, Politics, and Power in Modern Mexico*. Durham: Duke University Press, 2006.

Randall, Laura, ed. *Reforming Mexico's Agrarian Reform*. New York: M.E. Sharpe, 1996.

Richmond, Douglas W., and Sam W. Haynes, eds. *The Mexican Revolution: Conflict and Consolidation, 1910–1940*. College Station: Texas A&M University Press, 2013.

Salas, Elizabeth. *Soldaderas in the Mexican Military: Myth and History*. Austin: University of Texas Press, 1990.

Santiago, Myrna I. *The Ecology of Oil: Environment, Labor, and the Mexican Revolution, 1900–1938*. Cambridge: Cambridge University Press, 2006.

Smith, Stephanie J. *Gender and the Mexican Revolution: Yucatán Women & the Realities of Patriarchy*.

Chapel Hill: University of North Carolina Press, 2009.

Snodgrass, Michael. *Deference and Defiance in Monterrey: Workers, Paternalism, and Revolution in Mexico, 1890–1950*. Cambridge: Cambridge University Press, 2003.

Vaughan, Mary K., and Stephen E. Lewis, eds. *The Eagle and the Virgin: Nation and Cultural Revolution in Mexico, 1920–1940*. Durham: Duke University Press, 2006.

Womack Jr., John. *Zapata and Mexican Revolution*. New York: Random House, 1968.

Wood, Andrew Grant. *Revolution in the Street: Women, Workers, and Urban Protest in Veracruz, 1870–1927*. Wilmington: SR Books, 2001.

Chapter 8

Albert, Bill, and Paul Henderson. *South America and the First World War: The Impact of the War on Brazil, Argentina, Peru, and Chile*. Cambridge: Cambridge University Press, 1988.

Bethel, Leslie, and Ian Roxborough, eds. *Latin America between the Second World War and the Cold War, 1944–1948*. Cambridge: Cambridge University Press, 1992.

Brands, Hal. *Latin America's Cold War*. Cambridge: Harvard University Press, 2010.

Bulmer-Thomas, Victor. *The Economic History of Latin America Since Independence*. 2nd ed. Cambridge: Cambridge University Press, 2003.

Carillo, Justo. *Cuba 1933: Students, Yankees, and Soldiers*. New Brunswick: Transaction Publishers, 1994.

Coniff, Michael L. *Populism in Latin America*. Tuscaloosa: University of Alabama Press, 1999.

Crawley, Andrew. *Somoza and Roosevelt: Good Neighbour Diplomacy in Nicaragua, 1933–1945*. Oxford: Oxford University Press, 2007.

Derby, Lauren Hutchinson. *The Dictator's Seduction: Politics and the Popular Imagination in the Era of Trujillo*. Durham: Duke University Press, 2009.

Drinot, Paulo, and Alan Knight, eds. *The Great Depression in Latin America*. Durham: Duke University Press, 2014.

Elena, Eduardo. *Dignifying Argentina: Peronism, Citizenship, and Mass Consumption*. Pittsburgh: University of Pittsburgh Press, 2011.

Forster, Cindy. *The Time of Freedom: Campesino Workers in Guatemala's October Revolution*. Pittsburgh: University of Pittsburgh Press, 2001.

Fraser, Nicholas, and Marysa Navarro. *Evita*. New York: W. W. Norton & Co., 1996.

Garrard-Burnett, Virginia, Mark Atwood Lawrence, and Julio E. Moreno, eds. *Beyond the Eagle's Shadow: New Histories of Latin America's Cold War*. Albuquerque: University of New Mexico Press, 2013.

Gleijeses, Piero. *Shattered Hope: The Guatemalan Revolution and the United States, 1944–1954*. Princeton: Princeton University Press, 1991.

Gould, Jeffrey L., and Aldo A. Lauria-Santiago. *To Rise in Darkness: Revolution, Repression, and Memory in El Salvador, 1920–1932*. Durham: Duke University Press, 2008.

Grieb, Kenneth J. *Guatemalan Caudillo, the Regime of Jorge Ubico: Guatemala, 1931–1944*. Athens: Ohio University Press, 1979.

Grow, Michael. *U.S. Presidents and Latin American Interventions: Pursuing Regime Change in the Cold War*. Lawrence: University Press of Kansas, 2008.

Handy, Jim. *Revolution in the Countryside: Rural Conflict and Agrarian Reform in Guatemala, 1944–1954*. Chapel Hill: University of North Carolina Press, 1994.

Immerman, Richard H. *The CIA in Guatemala: The Foreign Policy of Intervention*. Austin: University of Texas Press, 1982.

Karush, Matthew B., and Oscar Chamosa. *The New Cultural History of Peronism: Power and Identity in Mid-Twentieth-Century Argentina*. Durham: Duke University Press, 2010.

Levine, Robert M. *Father of the Poor?: Vargas and His Era*. Cambridge: Cambridge University Press, 1998.

Plotkin, Mariano Ben. *Mañana es San Perón: A Cultural History of Perón's Argentina*. Wilmington: Scholarly Resources Books, 2003.

Rock, David, ed. *Latin America in the 1940s: War and Postwar Transition*. Berkeley: University of California Press, 1994.

Roorda, Eric. *The Dictator Next Door: The Good Neighbor Policy and the Trujillo Regime in the Dominican Republic, 1930–1945*. Durham: Duke University Press, 1998.

Rose, R. S. *One of the Forgotten Things: Getúlio Vargas and Brazilian Social Control, 1930–1954*. Westport: Greenwood Press, 2000.

Schlesinger, Stephen C., and Stephen Kinzer. *Bitter Fruit: The Untold Story of the American Coup in Guatemala*. Garden City: Doubleday, 1982.

Smith, Timothy J., and Abigail E. Adams, eds. *After the Coup: An Ethnographic Reframing of Guatemala, 1954*. Urbana: University of Illinois Press, 2011.

Walter, Knut. *The Regime of Anastasio Somoza, 1936–1956*. Chapel Hill: University of North Carolina Press, 1993.

Chapter 9

Besancenot, Olivier, and Michael Löwy. *Che Guevara: His Revolutionary Legacy*. New York: Monthly Review Press, 2009.

Black, George. *Triumph of the People: The Sandinista Revolution in Nicaragua*. London: Zed Press, 1981.

Bohning, Don. *The Castro Obsession: U.S. Covert Operations Against Cuba, 1959–1965*. Washington: Potomac Books, 2005.

Booth, John A. *The End and the Beginning: The Nicaraguan Revolution*. Boulder: Westview Press, 1982.

Brenner, Phillip, et al., eds. *A Contemporary Cuba reader: Reinventing the Revolution*. Lanham: Rowman and Littlefield, 2008.

Castro, Daniel, ed. *Revolution and Revolutionaries: Guerrilla Movements in Latin America*. Wilmington: SR Books, 1999.

Chilcote, Ronald H., ed. *Development in Theory and Practice: Latin American Perspectives*. Lanham: Rowman & Littlefield, 2003.

Field, Thomas C. Jr. *From Development to Dictatorship: Bolivia and the Alliance for Progress in the Kennedy Era*. Ithaca: Cornell University Press, 2014.

Grindle, Merilee, and Pilar Domingo, eds. *Proclaiming Revolution: Bolivia in Comparative Perspective*. London: Institute of Latin American Studies, and Cambridge: David Rockefeller Center for Latin American Studies, Harvard University, 2003.

Guevara, Ernesto, Brian Loveman, and Thomas M. Davies. *Guerrilla Warfare*, 3rd ed., with case studies by Loveman and Davies. Wilmington: SR Books, 1997.

Harmer, Tanya. *Allende's Chile and the Inter-American Cold War*. Chapel Hill: University of North Carolina Press, 2011.

Kapcia, Antoni. *Cuba in Revolution: A History Since the Fifties*. London: Reaktion Books, 2008.

———. *Leadership in the Cuban Revolution: The Unseen Story*. London: Zed Books, 2014.

Kunzle, David. *The Murals of Revolutionary Nicaragua, 1979–1992*. Berkeley: University of California Press, 1995.

Montoya, Rosario. *Gendered Scenarios of Revolution: Making New Men and New Women in Nicaragua, 1975–2000*. Tucson: University of Arizona Press, 2012.

Nelson, Lowry. *Cuba: The Measure of a Revolution*. Minneapolis: University of Minnesota Press, 1972.

O'Brien, Philip J. *Allende's Chile*. New York: Praeger, 1976.

Pérez-Stable, Marifeli. *The Cuban Revolution: Origins, Course, and Legacy*. 2nd ed. Oxford: Oxford University Press, 1999.

Qureshi, Lubna Z. *Nixon, Kissinger, and Allende: U.S. Involvement in the 1973 Coup in Chile*. Lanham: Lexington Books, 2009.

Roman, Peter. *People's Power: Cuba's Experience with Representative Government*. Boulder: Westview Press, 1999.

Schoultz, Lars. *That Infernal Little Cuban Republic: The United States and the Cuban Revolution*.

Chapel Hill: University of North Carolina Press, 2009.

Selbin, Eric. *Modern Latin American Revolutions.* 2nd ed. Boulder: Westview Press, 1998.

Smith, Wayne S. *The Closest of Enemies: A personal and Diplomatic Account of U.S.-Cuban Relations Since 1957.* New York: W. W. Norton, 1987.

Taffet, Jeffrey F. *Foreign Aid as Foreign Policy: The Alliance for Progress in Latin America.* New York: Routledge, 2007.

Veltmeyer, Henry, and Mark Rushton. *The Cuban Revolution as Socialist Human Development.* Leiden: Brill, 2012.

Walker, Thomas W., and Christine J. Wade. *Nicaragua: Living in the Shadow of the Eagle.* 5th ed. Boulder: Westview Press, 2011.

Wright, Thomas C. *Latin America in the Era of the Cuban Revolution.* Westport: Praeger, 2001.

Chapter 10

Angell, Alan. *Democracy After Pinochet: Politics, Parties and Elections in Chile.* London: Institute for the Study of the Americas, 2007.

Arditti, Rita. *Searching for Life: The Grandmothers of the Plaza de Mayo and the Disappeared Children of Argentina.* Berkeley: University of California Press, 1999.

Armstrong, Robert, and Janet Shenk. *El Salvador: The Face of Revolution.* Boston: South End Press, 1982.

Bouvard, Marguerite Guzman. *Revolutionizing Motherhood: The Mothers of the Plaza de Mayo.* Wilmington: Scholarly Resources, 1994.

Canak, William L. *Lost Promises: Debt, Austerity, and Development in Latin America.* Boulder: Westview Press, 1989.

Chesnut, R. Andrew. *Competitive Spirits: Latin America's New Religious Economy.* Oxford: Oxford University Press, 2003.

Coatsworth, John H. *Central America and the United States: The Clients and the Colossus.* New York: Twayne Publishers, 1994.

Collier, David, ed. *The New Authoritarianism in Latin America.* Princeton: Princeton University Press, 1979.

Cruden, Alex. *Genocide and Persecution: El Salvador and Guatemala.* Detroit: Greenhaven Press, 2013.

Dinges, John. *The Condor Years: How Pinochet and His Allies Brought Terrorism to Three Continents.* New York: New Press, 2004.

Escobar, Arturo, and Sonia E. Alvarez, eds. *The Making of Social Movements in Latin America: Identity, Strategy, and Democracy.* Boulder: Westview Press, 1992.

Esparza, Marcia, Henry R. Huttenbach, and Daniel Feierstein, eds. *State Violence and Genocide in Latin America: The Cold War Years.* London: Routledge, 2010.

Grandin, Greg. *The Last Colonial Massacre: Latin America in the Cold War.* Updated edition. Chicago: University of Chicago Press, 2011.

Handelman, Howard, and Werner Baer, eds. *Paying the Costs of Austerity in Latin America.* Boulder: Westview Press, 1989.

Jelin, Elizabeth, J. Ann Zammit, and Marilyn Thomson, eds. *Women and Social Change in Latin America.* London: Zed Books, 1990.

Jonas, Susanne. *The Battle for Guatemala: Rebels, Death Squads, and U.S. Power.* Boulder: Westview Press, 1991.

LaFaber, Walter. *Inevitable Revolutions: The United States in Central America*, 2nd ed. New York: W. W. Norton, 1993.

Lernoux, Penny. *Cry of the People: The Struggle for Human Rights in Latin America—The Catholic Church in Conflict With U.S. Policy.* New York: Penguin Books, 1982.

Lewis, Paul H. *Guerrillas and Generals: The 'Dirty War' in Argentina.* Westport: Praeger Publishers, 2002.

Menjívar, Cecilia, and Nestor Rodriguez, eds. *When States Kill: Latin America, the U.S., and Technologies of Terror.* Austin: University of Texas Press, 2005.

Montgomery, Tommie Sue. *Revolution in El Salvador: From Civil Strife to Civil Peace*. Boulder: Westview Press, 1995.

Munck, Ronaldo. *Latin America: The Transition to Democracy*. London: Zed Books, 1989.

Ondetti, Gabriel. *Land, Protest, and Politics: The Landless Movement and the Struggle for Agrarian Reform in Brazil*. University Park: Pennsylvania State University Press, 2008.

Rabe, Stephen G. *The Killing Zone: The United States Wages Cold War in Latin America*. Oxford: Oxford University Press, 2012.

Schirmer, Jennifer G. *The Guatemalan Military Project: A Violence Called Democracy*. Philadelphia: University of Pennsylvania Press, 1998.

Walker, Thomas W., and Ariel C. Armony. *Repression, Resistance, and Democratic Transition in Central America*. Wilmington: Scholarly Resources, 2000.

Chapter 11

Audley, John J., Demetrios G. Papademetriou, Sandra Polaski, and Scott Vaughan. *NAFTA's Promise and Reality: Lessons from Mexico for the Hemisphere*. Washington: Carnegie Endowment for International Peace, 2004.

Bennett, Vivienne, Sonia Davila-Poblete, and Maria Nieves Rico, eds. *Opposing Currents: The Politics of Water and Gender in Latin America*. Pittsburgh: University of Pittsburgh Press, 2005.

Brenner, Philip, Marguerite Rose Jiménez, John M. Kirk, and William M. Leogrande, eds. *A Contemporary Cuba Reader: The Revolution Under Raúl Castro*. Lanham: Rowman & Littlefield, 2015.

Brittain, James J. *Revolutionary Social Change in Colombia: The Origin and Direction of the FARC-EP*. London: Pluto Press, 2010.

Campbell, Bruce. *Viva la historieta: Mexican Comics, NAFTA, and the Politics of Globalization*. Jackson: University Press of Mississippi, 2009.

Carpenter, Ted. *Bad Neighbor Policy: Washington's Futile War on Drugs in Latin America*. New York: Palgrave Macmillan, 2003.

Collier, George A. *Basta!: Land and the Zapatista Rebellion in Chiapas*. Oakland: Food First Books, 2005.

Conant, Jeff. *A Poetics of Resistance: The Revolutionary Public Relations of the Zapatista Insurgency*. Edinburgh and Oakland: AK Press, 2010.

di Piramo, Daniela. *Political Leadership in Zapatista Mexico: Marcos, Celebrity, and Charismatic Authority*. Boulder: FirstForumPress, 2010.

Gollnick, Brian. *Reinventing the Lacandón: Subaltern Representations in the Rain Forest of Chiapas*. Tucson: University of Arizona Press, 2008.

Gott, Richard. *Hugo Chávez and the Bolivarian Revolution*. 2nd ed. London: Verso, 2011.

Grandin, Greg. *Who is Rigoberta Menchú?* London: Verso, 2011.

Harris, Richard L. *Globalization and Development in Latin America*. Whitby, ON: De Sitter, 2005.

Henck, Nick. *Subcommander Marcos: The Man and the Mask*. Durham: Duke University Press, 2007.

Klein, Hilary. *Compañeras: Zapatista Women's Stories*. New York: Seven Stories Press, 2015.

Krull, Catherine, and Louis A. Pérez, Jr., eds. *Cuba in a Global Context: International Relations, Internationalism, and Transnationalism*. Gainesville: University Press of Florida, 2014.

Lamrani, Salim. *Cuba, the Media, and the Challenge of Impartiality*. New York: Monthly Review Press, 2014.

Leech, Garry M. *The FARC: The Longest Insurgency*. London: Zed Books, 2011.

Livingstone, Grace. *Inside Colombia: Drugs, Democracy and War*. New Brunswick: Rutgers University Press, 2004.

Murillo, Mario A. *Colombia and the United States: War, Unrest and Destabilization*. New York: Seven Stories Press, 2014.

O'Toole, Gavin. *The Reinvention of Mexico: National Ideology in a Neoliberal Era*. Liverpool: Liverpool University Press, 2010.

Prevost, Gary, and Carlos Oliva Campos, eds. *The Bush Doctrine and Latin America*. New York: Palgrave Macmillan, 2007.

Rus, Jan, Rosalva Aída Hernández Castillo, and Shannan L. Mattiace. *Mayan Lives, Mayan Utopias: The Indigenous Peoples of Chiapas and the Zapatista Rebellion*. Lanham: Rowman & Littlefield, 2003.

Safford, Frank, and Marco Palacios. *Colombia: Fragmented Land, Divided Society*. Oxford: Oxford University Press, 2002.

Weisbrot, Mark, Stephan Lefebvre, and Joseph Sammut. *Did NAFTA Help Mexico? An Assessment After 20 Years*. Washington: Center for Economic and Policy Research, 2014.

Chapter 12

Ali, Tariq. *Pirates of the Caribbean: Axis of Hope*. London: Verso, 2006.

Anzaldúa, Gloria. *Borderlands/La Frontera: The New Mestiza*. San Francisco: Spinsters/Aunt Lute, 1987.

Babbitt, Susan E. *José Martí, Ernesto "Che" Guevara, and Global Development Ethics: The Battle for Ideas*. New York: Palgrave Macmillan, 2014.

Ballvé, Teo, and Vijay Prashad, eds. *Dispatches from Latin America: On the Frontlines Against Neoliberalism*. Cambridge: South End Press, 2006.

Burbach, Roger, Michael Fox, and Federico Fuentes. *Latin America's Turbulent Transitions: The Future of Twenty-First Century Socialism*. London: Zed Books, 2013.

De Rivero, Oswaldo. *The Myth of Development: Non-Viable Economies of the 21st Century*. London: Zed Books, 2003.

Ellner, Steve, ed. *Latin America's Radical Left: Challenges and Complexities of political Power in the Twenty-First century*. Lanham: Rowman & Littlefield, 2014.

Escobar, Arturo. *Encountering Development: The Making and Unmaking of the Third World*. Princeton: Princeton University Press, 1995.

———. *Territories of Difference: Place, Movements, Life, Redes*. Durham: Duke University Press, 2008.

García Canclini, Néstor. *Consumers and Citizens: Globalization and Multicultural Conflicts*. Minneapolis: University of Minnesota Press, 2001.

———. *Hybrid Cultures: Strategies for Entering and Leaving Modernity*. Minneapolis, MN: University of Minnesota Press, 1995.

Ferreira, Francisco H.G. *Economic Mobility and the Rise of the Latin American Middle Class*. Washington: World Bank, 2013.

Goodale, Mark, and Nancy Grey Postero. *Neoliberalism, Interrupted: Social Change and Contested Governance in Contemporary Latin America*. Stanford: Stanford University Press, 2013.

Gootenberg, Paul, and Luis Reygadas, eds. *Indelible Inequalities in Latin America: Insights from History, Politics, and Culture*. Durham: Duke University Press, 2010.

Guardiola-Rivera, Oscar. *What if Latin America Ruled the World?: How the South will Take the North Through the 21st century*. New York: Bloomsbury Press, 2010.

Gustafson, Bret Darin. *New Languages of the State: Indigenous Resurgence and the Politics of Knowledge in Bolivia*. Durham: Duke University Press, 2009.

Harten, Sven. *The rise of Evo Morales and the MAS*. London: Zed Books, 2011.

Hernández, Tanya Katerí. *Racial Subordination in Latin America: The Role of the State, Customary Law, and the New Civil Rights Response*. Cambridge: Cambridge University Press, 2013.

Jaquette, Jane S. *Feminist Agendas and Democracy in Latin America*. Durham: Duke University Press, 2009.

Kozloff, Nikolas. *Hugo Chávez: Oil, Politics, and the Challenge to the United States*. New York: Palgrave Macmillan, 2006.

Lievesley, Geraldine, and Steve Ludlam, eds. *Reclaiming Latin America: Experiments in Radical Social Democracy*. London: Zed Books, 2009.

Macdonald, Laura, and Anne Ruckert, eds. *Post-neoliberalism in the Americas: Beyond the Washington Consensus*. New York: Palgrave Macmillan, 2009.

Mignolo, Walter D. *The Darker Side of Western Modernity: Global Futures, Decolonial Options*. Durham: Duke University Press, 2011.

———, and Arturo Escobar, eds. *Globalization and the Decolonial Option*. London: Routledge, 2010.

Pierceson, Jason, Adriana Piatti-Crocker, and Shawn Shulenberg, eds. *Same-Sex Marriage in Latin America: Promise and Resistance*. Lanham: Lexington Books, 2013.

Postero, Nancy Grey. *Now We Are Citizens: Indigenous Politics in Postmulticultural Bolivia*. Stanford: Stanford University Press, 2007.

Prevost, Gary, Carlos Oliva Campos, and Harry E. Vanden. *Social Movements and Leftist Governments in Latin America: Confrontation or Co-optation?* London: Zed Books, 2012.

Stahler-Sholk, Richard, Harry E. Vanden, and Glen David Kuecker, eds. *Latin American Social Movements in the Twenty-First Century*. Lanham: Rowman & Littlefield, 2008.

Strauss, Julia C. and Ariel C. Armony, eds. *From the Great Wall to the New World: China and Latin America in the 21st Century*. Cambridge: Cambridge University Press, 2012.

Sue, Christina A. *Land of the Cosmic Race: Race Mixture, Racism, and Blackness in Mexico*. Oxford: Oxford University Press, 2013.

Webber, Jeffery R., and Barry Carr, eds. *The New Latin American Left: Cracks in the Empire*. Lanham: Rowman & Littlefield Publishers, 2013.

Weitzman, Hal. *Latin Lessons: How South America Stopped Listening to the United States and Started Prospering*. Hoboken: Wiley, 2012.

A Glossary of Spanish, Portuguese, and Indigenous Terms

Adelantado An individual in colonial Spanish America authorized by the crown to explore, conquer, and hold new territory. He pushed back the frontier and extended Spanish claims and control of the New World.

Alcabala Sales tax levied by Spain and particularly hated by colonists.

Alcaldes mayores In colonial Spanish America, appointed officials who held administrative and judicial responsibility on local or district level.

Aldeia An indigenous village or settlement in Portuguese America administered by the religious orders until the mid-eighteenth century and then by secular officials thereafter.

Altepetl City-state in the Aztec empire, the basic provincial unit.

Arpilleras Patchwork tapestries created in Chile to tell stories of the repression during the dictatorship of Augusto Pinochet, 1973–1989.

Audiencia The highest royal court and consultative council in colonial Spanish America.

Ayllu A communal unit in the Incan empire that worked the land in common, part for themselves and part for the Incan ruler and priestly elite.

Balseros Cubans who left the country on flimsy rafts in a mass exodus in 1995.

Bandeirante Particularly active during the 1650–1750 period, these individuals penetrated the interior of Brazil to explore, to capture indigenous slaves, or to search for gold.

Cabildo The municipal government in Spanish America. Sometimes called *ayuntamiento.*

Cabildo abierto The municipal council in Spanish America, which expanded under special circumstances to include most of the principal citizens of the municipality.

Cacique An indigenous chief in the Caribbean region. The Spaniards adopted the word to refer to indigenous leaders throughout their colonies.

Calidad Literally, quality. The nineteenth-century view of elites regarding respectability, earned by family background, the organization and location of one's household, formal training and education, occupation, economic resources, and perceived color.

Campesino Literally, a person from the country. It is frequently translated as peasant, a problematic English term that evokes medieval institutions of servile relationships, such as serfdom, and that implies a particular culture built around subsistence farming.

Capitulación A contract between monarch and *adelantado* stating the duties and rewards of the latter.

Casa de Contratación The Board of Trade established in Spain in 1503 to organize, regulate, and develop trade with the New World.

Casta Person of mixed race. Frequently used to mean dark-skinned people.

Caudillo (Portuguese, *caudilho*) A strong leader who wields complete power over subordinates. Usually refers to nineteenth-century leaders whose power was personalistic, not dependent on institutions.

Chinampa Floating gardens used in the Aztec empire.

Científico A high administrator in the government of President Porfirio Díaz of Mexico (1876–1911), infused with Positivist ideas, who believed national problems could be solved by scientific solutions. Such men were prominent during the last two decades of his administration.

Coatequitl Draft rotary labor system in Aztec empire.

Cochineal A red dye made from the insects feeding on nopal cactus.

Composición A Spanish legal device for claiming land through surveys.

Congregación The Spanish policy of concentrating indigenous people into villages.

Conquistador Conqueror, a term used to refer to all those who served in the conquest.

Consejo de las Indias The Council of the Indies established in Spain in 1524 to advise the monarch on American affairs.

Consulado In colonial Spanish America, a guild of merchants acting as a sort of chamber of commerce.

Contra Short for *contrarevolucionario*, counterrevolutionary. The term was used to refer to U.S.-trained forces fighting to overthrow the Sandinista government in Nicaragua in the 1980s.

Corregidor An official in colonial Spanish America who was assigned to Spanish as well as indigenous communities as tax collector, police officer, magistrate, and administrator.

Criollo Anglicized as creole. A white born in the Spanish American empire.

Cumbe A settlement of runaway slaves in Spanish America.

Denuncia Under Spanish law, the process of claiming land that does not have legally recognized owners.

Desaparacido Disappeared person. A term coined to refer to people kidnapped by the military regime during Argentina's dictatorship, 1976–1983.

Ejido The common land held by communities and used for agriculture in Mexico. Also the municipal lands available for common use in most regions of Latin America during the nineteenth and early twentieth centuries.

Encomendero The person who received an *encomienda.*

Encomienda A tribute institution used in Spanish America in the sixteenth century. The Spaniard received indigenous workers as an entrustment, *encomienda*, to protect and to Christianize, but in return he could demand tribute including labor.

Fazenda A large estate or plantation in Brazil.

Fazendeiro The owner of a large estate or plantation in Brazil.

Fuero militar A special military privilege in Spanish America that exempted officers from civil legal jurisdiction.

Gaucho The cowboy of the Pampas.

Gente alta Elites.

Gente baja Lower classes.

Gente decente Literally, decent people. Refers to the elites, or *gente alta*.

Gente de pueblo Common people, or lower classes. Also *gente baja*.

Guerrilla Literally *small war*. As a noun, a member of an irregular force fighting against larger, regular forces. As an adjective, it describes a war waged by irregular forces or in a nontraditional manner, including ambushes and hit-and-run tactics. The word has its origins in the war waged by the Spanish against French occupation (1808–1814).

Hacendado The owner of a large estate or plantation in Spanish America.

Hacienda A large estate in Spanish America.

Hidalgo Nobleman. Literally son of someone (*hijo de algo*).

Jefe Chief or leader; boss. In Spanish America, it is often used as synonymous with *caudillo.*

Kuraka Indigenous leader among Andean peoples.

Ladino A person of mixed European and indigenous ancestry, or an indigenous person who adopts a Hispanic lifestyle. Used in place of mestizo in much of Central America. Originally a free, acculturated African from Spain.

Latifundia The system of large landholdings in Latin America.

Leva Military draft, sometimes taking men directly from the street.

Llaneros Cowboys of the plains in New Granada.

Mandamiento A forced labor system.

Maquiladora An assembly plant, originally on the Mexican side of the border though now found in other parts of Latin America. Also called *maquila*, the plant assembles goods from imported materials for reexport.

Matanza Massacre.

Mazombo In Portuguese America, a white born in the New World.

Mestizaje Race mixture. (*mestiçagem* in Portuguese).

Mestizo A person of mixed parentage. Usually it refers to a European–indigenous mixture.

Meseta Highland plateau in Spain.

Mita A forced labor system in which the indigenous were required to labor for Spanish settlers, taken from the Quechua and Aymara tradition.

Mulatto A person of European and African ancestry. The word is thought to have its roots in *mula*, or *mule*, a product of the donkey and horse. The term tends to have pejorative connotations in the United States, but it is commonly used in Latin America.

Obedezco pero no cumplo Literally, "I obey but I don't fulfill." In this way, colonial officials recognized the royal authority but bent the laws to suit local circumstances.

Obraje Workshop. Generally refers to small manufacturing enterprises.

Oidor A judge on the *audiencias* of Spanish America.

Palenque A settlement of runaway slaves in Spanish America.

Patria chica Literally, the small country, it refers to the immediate region with which people identify rather than the nation.

Patrón In Spanish America, the owner or boss or one in a superior position.

Peninsular In Spanish America, a white born in Europe who later came to the New World.

Pleybeyos Plebeians. The lower classes, considered coarse and common by elites.

Porteño An inhabitant of the city of Buenos Aires.

Pueblo A town, but it can also mean "people."

Quilombo A settlement of runaway slaves in Portuguese America.

Quipu Knotted cord on which Inca empire officials recorded information.

Reconquista Spanish for reconquest, it refers to the Christian struggle from 732–1492 to retake the Iberian peninsula from Muslim conquerors.

Reinol (plural, *reinóis*) In Portuguese America, a white born in Europe who later came to the New World.

Relaçao The high court in Portuguese America.

Repartimiento A labor institution in colonial Spanish America in which a royal judge made a temporary allotment of indigenous workers for a given task.

Reparto Forced sale of goods, particularly to the indigenous, during the colonial period.

Repúbica de españoles Colonial Republic of the Spanish, an attempt to keep the Spanish world separate from the indigenous.

República de indios Colonial Republic of Indians, a Spanish attempt to protect the indigenous communities from elements of the Spanish world.

Senado da Câmara In Brazil, the municipal government, in particular the town council.

Sesmaria A land grant in colonial Brazil.

Soldadera During the Mexican Revolution, a woman who was attached to a soldier. The *soldaderas* cooked for the soldiers, tended the ill and wounded, and fought.

Tienda de raya The store on an hacienda that sold resident workers their supplies, frequently at inflated prices.

Tratante A merchant who dealt in locally produced goods in colonial Latin America.

Vecino Literally neighbor, or a resident of a particular place.

Visita In both the Spanish and Portuguese American empires, an on-the-spot administrative investigation of a public employee ordered by the monarch.

Zambo A person of mixed indigenous and African parentage.

A Glossary of Concepts and Terms

Scattered throughout this text are a series of concepts and terms, some of which are defined—"reform" and "revolution," for example—and some of which are not—"capitalism," "socialism," and "Enlightenment," for example. The purpose of this glossary is to provide brief working definitions for the concepts and terms frequently encountered in the text. Definitions vary widely. They can be slippery. We have attempted to define the words in accordance with their use in the text, but we realize these definitions will neither satisfy everyone nor be universally applicable.

Capitalism An economic system characterized by private ownership and investment, economic competition, wage labor, and profit incentive.

Centralism A high concentration of political power in the federal government.

Communism Communism really denotes a future society, one yet to be achieved. Societies often termed *Communist* are really at best in a transitional phase and whose goal is a form of community living free from hierarchical controls and enjoying common property. This book uses the term within a contemporary context to mean a government ruling in the name of the workers and peasants to best serve their ends and an economic system controlled by that government. The government owns the means of production in the name of the workers and peasants. This text distinguishes socialism from communism by the degree of state ownership, by the degree of democracy, and the existence of a plurality of political parties.

Conservatism The term is generally associated with nineteenth-century political parties whose disposition was to preserve things much as they were, exercising caution in the acceptance of change.

Decoloniality A theoretical approach developed by Latin American intellectuals in the early twenty-first century that challenges the concept of modernity as enforced by Europe beginning with the conquest and continued via neocolonial arrangements dominated by the United States. Decoloniality stresses the darker side of modernity that requires domination of subject peoples and denies the validity of their own ways of living and epistemological approaches.

Democracy A system of government in which all or most of the citizenry participate in the decision-making process. Western democracy stresses equality of all citizens before the law, a government responsive to the majority, regular elections, civil liberties, and plural political parties.

Dependency Dependency describes a situation in which the economic well-being, or lack of it, of one nation, colony, or area results from the consequences of decisions made elsewhere. Latin America was first dependent on the Iberian motherlands, then in the nineteenth century on England, and in the twentieth on the United States, whose decisions and policies directly influenced, or influence, its economic prosperity or poverty. Obviously to the degree a nation is dependent, it will lack "independence" of action.

Development The maximum use of a nation's potential for the greatest benefit of the largest number of inhabitants.

Developmentalism The belief that adoption of programs aimed at economic growth would bring prosperity and the growth of a stabilizing middle class.

Elites Those persons who occupy the highest or most eminent positions in society.

Enclave economy An economic activity that depends on little from the host country other than low-wage labor, rather than depending on local material imports or interactions with local businesses. A class example is the banana industry, in which local workers cut bunches of bananas that were immediately loaded onto foreign-owned ships and sent out of the country.

Enlightenment Broadly identified with eighteenth-century Europe, the Enlightenment introduced a series of ideas associated with the forms of democracy and capitalism of the nineteenth century. The Enlightenment thinkers believed in human social evolution and thus a kind of philosophy of progress and perfectability. The ideas of the Enlightenment exerted a profound influence on the writing of the U.S. Constitution and on the ideology of the Latin American elites.

Federalism In a federal political system, political power is divided and/or shared between a central government and regional or local governments.

Feudalism Strictly speaking, this term refers to a form of social organization prevalent in Europe from the time of the dissolution of Charlemagne's empire until the rise of the absolute monarchies, roughly from the ninth to the fifteenth centuries. Its general characteristics were strict class division; private jurisdiction based on local custom; and a landowning system in which the owner, the lord, allowed the serf to work land in return for services and/or payments. By extension, the term sometimes is used in Latin America to designate a system in which a few own the land and control the lives of the many who work the land for them. Those few enjoy comfortable lives, while the workers live in misery largely dependent on the whims of the landowners.

Globalization The increased interconnectedness of communications, economics, and cultural exchange of the late twentieth and early twenty-first centuries. The term frequently refers to economic changes interlinked with neoliberal trade policies.

Growth Growth indicates numerical accumulation in a country or region's economy and generally does not reveal who, if anyone, benefits from it.

Hegemony Refers both to dominance, as in "the United States is the hegemonic power in the hemisphere," and to an acceptance by the majority of the role of the dominant power. A government has hegemony when the majority of the citizens recognize the government's rights to rule.

Hybridity A mixture of premodern, modern, and postmodern ways of life characteristic of Latin America in the late twentieth and early twenty-first centuries. The term was coined by Argentine writer Néstor García Canclini.

Institutions This book's most frequently used term and its most difficult to define, "institutions" represent the recognized usages governing relations between people, an entire complex of such usages and the principles governing it, and the formal organizations supporting such a complex. Perhaps Webster's unabridged dictionary offers a more satisfactory and comprehensive definition: "A significant and persistent element (as a practice, a relationship, an organization) in the life of a culture that centers on a fundamental human need, activity, or value, occupies an enduring and cardinal position within a society, and is usually maintained and stabilized through social regulatory agencies." Examples range from patriarchal families to the military, from village social structure to land division.

Liberalism This term is generally associated with nineteenth-century political parties whose disposition was to relax governmental control, to expand individual freedom, and to innovate. Economically, liberalism advocated free trade.

Liberation theology An interpretation of Christianity, based primarily in the Gospel of Luke, that cast Jesus as a liberator of the people. The term was coined by Peruvian theologian Gustavo Gutierrez in 1973.

Luso-Brazilian This term encompasses both Portugal and Brazil. In Roman times, the area we now associate with Portugal bore the name *Lusitania*, the adjective being *Luso*.

Mercantilism A term coined in the eighteenth century, it is a belief that the nation's economic welfare can best be ensured by governmental regulation of a nationalist character. Wealth was measured by the accumulation of bullion, and colonial economies were controlled through trade restrictions and government monopolies. Mercantilism characterized the economic systems of Spain and Portugal.

Metropolis This term refers to that country exerting direct or indirect control over another. For Spanish America, the metropolis in the colonial period was Spain; for Brazil, Portugal. During the nineteenth century, the metropolis for Latin America was England; in the twentieth century, it was the United States.

Modernization In Latin America, modernization consisted largely of copying and adopting, rarely adapting, the styles, ideas, technology, and patterns of Northern Europe in the nineteenth century and the United States in the twentieth.

Nationalism This term refers to a group consciousness that attributes great value to the nation-state, to which total loyalty is pledged. Members of the group agree to maintain the unity, independence, and sovereignty of the nation-state as well as to pursue certain broad and mutually acceptable goals.

Nation-state This term, implying more than the area encompassed by the geographic boundaries of a country, signifies that a central authority effectively exercises political power over that entire area.

Neofeudalism *Neo*, from the Greek, signifies "new" or "recent." See the entry "feudalism."

Neoliberalism Market-oriented reforms begun in the 1990s that echoed the liberalism of the nineteenth century. Such reforms included privatization of state businesses and public services along with elimination of tariff barriers.

Oligarchy These privileged few rule for their own benefit, demonstrating little or no responsibility toward the many.

Patriarchal, Patriarchy This term refers to a type of family arrangement, or government, in which the father or an older male rules.

Patrimonialism A system in which the landowner exerts authority over his followers as one aspect of property ownership. Those living on his land fall under his control. He rules the estate at will and controls all contact with the outside world. The term describes the hacienda system.

Physiocrat Doctrine This concept originating in the eighteenth century urges society to survey scientifically its resources and, once knowing them, to exploit them. Maximum profit results from the exploitation and international sale of those resources.

Populism Political movements or governments that seem at least outwardly opposed to the status quo are in some cases termed "populist." They advocate a system appealing to and supported by large numbers of the ordinary citizens, generally the urban working class. Usually change is directed from the top down and is dependent on a charismatic leader. In practice, they often provide temporary relief or benefits without actually reforming basic social structures.

Positivism This nineteenth-century ideology originated in France. Its principal philosopher was Auguste Comte. Positivism affirmed the inevitability of social innovation and progress. According to Comte, that progress was attainable through the acceptance of scientific social laws codified by Positivism.

Postmodernism A set of theories that posit postmodernity as the twentieth- and twenty-first-century condition of failed "modern" projects based on Enlightenment ideals and the belief in progress. Postmodern theorists argue against the overarching narrative of modern theories and instead see many truths and views in a decentered, fragmented world.

Reform To reform is to gradually change or modify established economic, political, or social institutions.

Revolution The sudden, forceful, and violent overturn of a previously stable society and the substitution of other institutions for those discredited.

Socialism A democratic society in which the community owns or controls the major means of production, administering them for the benefit of all.

Index

B

C

F

G

H

I

P

T

U

V

Credits

Chapter 1 P.01: Thomas Gage. The English-American: A New Survey of the West Indies, 1648. First published in 1928. Reprinted in 2005 by Routledge-Curzon, p. 84; Amerigo Vespucci's Letters from Cape Verde, 1501, and Libson, 1502, quoted in Amerigo Vespucci Pilot Major, Routledge, 1967, 249 pages; Latin America, An Interpretive History, by E. Bradford Burns, Pearson Education, 1972; p.07: Euclides da Cunha, Rebellion in the backlands. Translated by Samuel Putnam. Chicago: University of Chicago Press; 1944; Stephen Minta. García Márquez: writer of Colombia. NY: Harper & Row, 1987, p. 4; p.10: Based on population statistics from The World Bank, 2013, http://wdi.worldbank.org. and agri-cultural area statistics from United Nations Food and Agriculture Organization, 2012, http://faostat.fao.org. Agricultural area comprises areas of temporary crops, permanent crops, and permanent meadows and pastures; p.13: Popol Vuh: The Sacred Book of the Ancient Quiché Maya. English version by Delia Goetz and Sylvanus G. Morley, from the translation of Adrián Recinos. Norman: University of Oklahoma Press, 1950, p. 167; p.15: Translated from Pedro de Cieza de León, La Crónica del Perú. Madrid: Calpe, 1922. Originally published in Seville 1553. https://archive.org/details/lacrnicadelper00ciez; Translated from Pedro de Cieza de León, La Crónica del Perú. Madrid: Calpe, 1922. Originally published in Seville 1553. https://archive.org/details/lacrnicadelper00ciez; Bernal Diaz del Castillo, The discovery and conquest of Mexico, 1517-1521. (NY: George Routledge & Sons, Ltd., 1928). Universal Digital Library, p 220; p.16: Adapted from Howard Spodek, The World's History Combined, 3rd ed. © 2006. Electronically reproduced by permission of Pearson Education, Inc., Upper Saddle River, NJ; p.17: Arthur J. O. Anderson, "Aztec Hymns of Life and Love," p. 36 revised translation of Florentine Codex, New Scholar, 1982, 6:33:179; p.18: E. Bradford Burns, A History of Brazil. New York: Columbia University Press, 1993, p. 18; Miguel León Portilla, The Broken Spears: The Aztec Account of the Conquest of Mexico. Boston: Beacon Press, first published 1962; updated editions 1992, 2006; p.19: Kenneth Baxter Wolf, ed. and trans. Conquerors and Chroniclers of Early Medieval Spain. Liverpool University Press, 1999., p. 33; p.20: Based on http://www.fordham.edu/halsall/maps/1492spain.jpg; p.23: Adapted from Howard Spodek, The World's History Combined, 3rd ed. © 2006. Electronically reproduced by permission of Pearson Education, Inc., Upper Saddle River, NJ; p.25: Giacinto Brugiotti da Vetralla, quoted in Cheikh Anta Diop, Precolonial Black Africa. Chicago Review Press, 1988; p.26: Olaudah Equiano, The Interesting Narrative of the Life of Olaudah Equiano, Or Gustavus Vassa, The African. Written By Himself. https://www.gutenberg.org/files/15399/15399-h/15399-h.htm; p.26: Magnus Mörner, Race Mixture in the History of Latin America (Boston, MA: Little, Brown, 1967), 58; p.27: Magnus Mörner, Race Mixture in the History of Latin America (Boston, MA: Little, Brown, 1967), 58–59; p.28: Amado's preface to Haroldo Costa, Fala Crioulo. Rio de Janeiro: Record, 1982. p. 15; Juan Felipe Herrera, 187 Reasons why Mexicanos Can't Cross the Border. City Lights Publishers, 2007; David Buisseret and Steven G. Reinhardt, eds. Creolization in the Americas. Arlington: University of Texas Press, 2000, p. 3; Count Joseph Arthur Gobineau and Gustave Le Bon, quoted in The Spectacle of the Races: Scientists, Institutions, and the Race Question in Brazil, 1870-1930, Hill and Wang, 199,224 pages; Creolization in the Americas, by David Buisseret, Texas A&M University Press, 2000; p.30: Arturo Escobar, "Revisioning Latin American and Caribbean Studies: A Geopolitics of Knowledge Approach." LASA Forum 37, No. 2 (Spring 2006): 11-13, p. 11.

Chapter 2 P.36: Hernando de Castro, quoted in James Lockhart, Enrique Otte, Letters and People of the Spanish Indies: Sixteenth Century. Cambridge University Press, 1976; p.37: Bernal Diaz del Castillo, The discovery and conquest of Mexico, 1517-1521. (NY: George Routledge & Sons, Ltd., 1928). Universal Digital Library. P8; Paolo Bernardini and Norman Fiering, eds. The Jews and the Expansion of Europe to the West, 1450 to 1800. Berghahn Books, 2001. p.9; p.39: James Lockhart and Enrique Otte, Letters and People of the Spanish Indies: Sixteenth Century. Cambridge: Cambridge University Press, 1976, p. 12; p.40: Leslie B. Rout Jr., The African Experience in Spanish America, 1502 to the Present Day. Cambridge: Cambridge University Press, 1976.

p. 34; Leslie B. Rout, The African Experience in Spanish America (CUP Archive, 1976);"p.44: E. Bradford Burns, A History of Brazil. New York: Columbia University Press, 1993, p. 84; A History of Brazil, Third Edition, by E. Bradford Burns, Columbia University Press, 1993, 554 pages; p.46: Craig, Albert M.; Graham, William A.; Kagan, Donald; Ozment, Steven M.; Turner, Frank M. Heritage of World Civilizations Combined, 7th ed. © 2006. Electronically reproduced by permission of Pearson Education, Inc., Upper Saddle River, NJ; p.48: William Stanley Rycroft, Religion and Faith in Latin America. Westminster Press, 1958, p. 97; p.50: Justo L. González, Ondina E. González. Christianity in Latin America: A History. Cambridge: Cambridge University Press, 2008, p. 30; p.52: Philip II, quoted in Leland Dewitt Baldwin, The Story of the Americas: The Discovery, Settlement, and Development of the New World. Simon and Schuster, 1943; p.53: Harold Bloom, Sandra Cisneros's The House on Mango Street (Infobase Publishing, 2010); Our Lady of Guadalupe: The Origins and Sources of a Mexican National Symbol, 1531-1797, by Stafford Poole, University of Arizona Press, 1995, 325 pages; p.54: Alfred W. Crosby Jr., The Columbian Exchange: Biological and Cultural Consequences of 1492. Westport, CT: Praeger, 2003, p. 37; German Missonary, 1699, as translated in Colonial Latin America, by Donald Mabry, Llumina Press, 2002, 296 pages; p.58: James Lockhart and Enrique Otte, Letters and People of the Spanish Indies: Sixteenth Century. Cambridge: Cambridge University Press, 1976, p. 15.

Chapter 3 P.62: D. A. Brading, The First America: The Spanish Monarchy, Creole Patriots and the Liberal State. Cambridge: Cambridge University Press, 1991, p. 317; Ralph Bauer, José Antonio Mazzotti, eds. Creole Subjects in the Colonial Americas: Empires, Texts, Identities. University of North Carolina Press, 2009. p. 4); p.63: Ralph Bauer, José Antonio Mazzotti, eds. Creole Subjects in the Colonial Americas: Empires, Texts, Identities. University of North Carolina Press, 2009. p. 1; Ralph Bauer, José Antonio Mazzotti, eds. Creole Subjects in the Colonial Americas: Empires, Texts, Identities. University of North Carolina Press, 2009. p. 4); Jaime Rodriguez O. "We Are Now the True Spaniards": Sovereignty, Revolution, Independence, and the Emergence of the Federal Republic of Mexico, 1808-1824. Stanford University Press, 2012, p. 22; Juan Vicente de Güemes Pacheco y Padilla, quoted in Jaime E. Rodríguez O., The Independence of Spanish America. Cambridge University Press, 1998; p.64: John Lynch; p.66: D. A. Brading, The First America: The Spanish Monarchy, Creole Patriots and the Liberal State. Cambridge: Cambridge University Press, 1991, p. 481; p.66: E. Bradford Burns, ed. A Documentary History of Brazil. NY: Alfred A. Knopf, 1966, p. 134; Ruth Mackay. "Lazy, Improvident People": Myth and Reality in the Writing of Spanish History. Ithaca: Cornell University Press, 2006. p. 230; p.69: Translated from Manuel de Salas, José de Cos Iriberri, Real Consulado de Comercio de Chile, "Carta 1799 May. 8, Santiago a SS. de la Junta del Consulado," 1799, in Industria durante el coloniaje: documentos varios, pp 467-473; Anselmo de La Cruz; p.70: Robert Morris: Financier of the American Revolution, by Charles Rappleye, Simon and Schuster, 2010, 640 pages; 18th century song; p.71: Charles de Secondat baron de Montesquieu, Thomas Nugent, Jean Le Rond d' Alembert. The Spirit of Laws, Volume 1. Colonial Press, 1900, p. 206; Jean-Jacques Rousseau, The Social Contract, Cosimo, Inc., 2008; Jeremy Bentham, address to the national convention of France, 1793, The Works of Jeremy Bentham Volume 4. Edinburgh: William Tait, 1844, p. 408; p.72: Thomas Paine, Common Sense, p. 16. London: H.D. Symonds, 1792; William Walton, Present State of the Spanish Colonies, Vol. II. London: Longman Hurst, Rees, Orme, and Brown, 1810, p. 338; William Walton, Present State of the Spanish Colonies, Vol. II. London: Longman Hurst, Rees, Orme, and Brown, 1810, pp 345, 346-347; p.74: Leslie Bethell, ed. The Independence of Latin America, Cambridge: Cambridge University Press, 1987, p. 44; p.75: A Voyage to Cochin, China, T. Cadell, 1806; Beyond Slavery: The Multilayered Legacy of Africans in Latin America and the Caribbean, by Darien J. Davis, Rowman & Littlefield, 2007, 289 pages.; p.77: Gazeta de Caracas, III, No. 123, (Nov. 9, 1810), as translated in The Independence of Spanish America, by Jaime E. Rodriguez, Cambridge University Press, 1998, 274 pages; p.79: Cornelio Saavedra, 1809; p.82: Miguel Hidalgo, Quoted in James Aloysius Magner, James T. Hurley, Louise F. Spaeth, Latin America Pattern. Catholic student's mission crusade, 1945; p.85: Craig, Albert M.; Graham, William A.; Kagan, Donald; Ozment, Steven M.; Turner, Frank M. Heritage of World Civilizations Combined, 7th ed. © 2006. Electronically reproduced by permission of Pearson Education, Inc., Upper Saddle River, NJ; p.86: A. J. R. Russell-Wood, From Colony to Nation: Essays on the Independence of Brazil. John Hopkins University Press, 1975, p. 216; E. Bradford Burns, A History of Brazil. New York: Columbia University Press, 1993, pp 121-122.

Chapter 4 P.91: Colegio de Mexico, A Compact History of Mexico. 1995. p. 93; p.92: Visconde de Albuquerque, quoted in Alvaro Vargas Llosa, Liberty for Latin America: How to Undo Five Hundred Years of State Oppression. Farrar, Straus and Giroux, 2005; Simón Bolívar, quoted in Charles Gibson, The Spanish tradition in America. University of South Carolina Press, 1968; p.94: Sarah C. Chambers. From subjects to citizens: honor, gender, and politics in Arequipa, Peru, 1780-1854. Pennsylvania State University Press, 1999, p. 224; p.97: Adapted from SPODEK, HOWARD, THE WORLD'S HISTORY: COMBINED VOLUME, 5th Ed., ©2015, p. 552. Reprinted and Electronically reproduced by permission of Pearson Education, Inc., New York, NY; p.98: Frances Calderon de la Barca, William Hickling Prescott, Gino Doria, Life in Mexico During a Residence of Two Years in that Country by Madame C. de la B. Chapman and Hall, 1843; p.102: H.W.V. Temperley, "The Latin American Policy of George Canning," The American Historical Review 11, No. 4 (July 1906): 779-797, p. 796; p.103: Peter Blanchard, Under the Flags of Freedom: Slave Soldiers and the Wars of Independence in Spanish South America. University of Pittsburgh Press, 2008, p 112; Herbert S. Klein, Ben Vinson, African Slavery in Latin America and the Caribbean. Oxford University Press, 1986, 2007; p.105: El Rio de La Plata, September 1, 1869; p.106: El Observador, 1830; p.108: Rebecca Earle, The Return of the Native: Indians and Myth-Making in Spanish America, 1810–1930. Duke University Press, 2007, p xxxii; Rebecca Earle, The Return of the Native: Indians and Myth-Making in Spanish America, 1810–1930. Duke University Press, 2007, p. xxix; p.109: The Pamphleteer, Volume 14, Abraham John Valpy, 1819; Rebecca Earle, The Return of the Native: Indians and Myth-Making in Spanish America, 1810–1930. Duke University Press, 2007, p. xxix; p.112: E. Bradford Burns, The Poverty of Progress: Latin America in the Nineteenth Century. Berkeley: University of California Press, 1980; Rafael Carrera, quoted in E. Bradford Burns, The Poverty of Progress: Latin America in the Nineteenth Century. Berkeley: University of California Press, 1980; p.113: John Lynch, Argentine Caudillo: Juan Manuel de Rosas. Lanham, MD: SR Books, 2001, (reprint of Argentine Dictator: Juan Manuel de Rosas, 1829-1852, Oxford University Press, 1981), p 75; p.114: Rafael Carrera, quoted in E. Bradford Burns, The Poverty of Progress: Latin America in the Nineteenth Century. Berkeley: University of California Press, 1980; p.115: Rafael Carrera, quoted in E. Bradford Burns, The Poverty of Progress: Latin America in the Nineteenth Century. Berkeley: University of California Press, 1980; p.116: Belzú, quoted in E. Bradford Burns, The Poverty of Progress: Latin America in the Nineteenth Century. Berkeley: University of California Press, 1980; p.117: E. Bradford Burns, The Poverty of Progress: Latin America in the Nineteenth Century. Berkeley: University of California Press, 1980, 109.

Chapter 5 P.125: Octavio Paz. The labyrinth of solitude: life and thought in Mexico. Grove Press, 1962, p. 131; p.129: Excerpt from Ricardo Guiraldes, "Rosaura," p. 184, in Waldo Frank, Tales from the Argentine. NY: Farrar & Rinehart, 1930; p.130: The Capitals of Spanish America, by William Eleroy Curtis, Harper, 1888, 715 pages; p.132: Victor Bulmer-Thomas, The Economic History of Latin America Since Independence, 2nd ed., New York, NY: Cambridge University Press, 2003, 58. Reprinted with the permission of Cambridge University Press; p.137: Alcira Leiseron, Notes on the process of industrialization in Argentina, Chile, and Peru. Institute of International Studies, University of California, 1966, p 21; William Vincent Wells. Explorations and Adventures in Honduras. NY: Harper & Brothers Publishers, 1857. p. 47; p.139: James Bryce, South America: Observations and Impressions. NY: The Macmillan Co., 1912, pp 374-375; Aluísio Azevedo, The Slum. David H. Rosenthal, trans. Oxford University Press, 2000. pp 12 and 13. Originally published in the Portuguese in 1890; p.141: Silvia Marina Arrom, The Women of Mexico City, 1790-1857. Stanford University Press, 1985. p 20; p.142: Francesca Miller, Latin American Women and the Search for Social Justice. University Press of New England, 1991. p 44; Francesca Miller, Latin American Women and the Search for Social Justice. University Press of New England, 1991. p 69; p.144: Rafael Uribe; p.146: The Spectator (London), Vol 22, May 5, 1849, p. 419; E. Bradford Burns, A History of Brazil. New York: Columbia University Press, 1993, p. 219; Translated in A History of Brazil, by E. Bradford Burns, Columbia University Press, 1993, 544 pages; The Poverty of Progress: Latin America in the Nineteenth Century, E. Bradford Burns, University of California Press, 1980, 186 pages; p.149: Enrique López Albujar, quoted in E. Bradford Burns, The Poverty of Progress: Latin America in the Nineteenth Century. Berkeley: University of California Press, 1980.

Chapter 6 P.152: De Bow's Commercial Review, 1848; John O'Sullivan, "Annexation," United States Magazine and Democratic Review Volume17,

Number 1 (July-August 1845), pp 5-10; 1846-1850. Annexation of Texas-Compromise of 1850. 1881," in The Constitional and Political History of the United States, Dr. H. von Holst, Vol. III, Chicago: Callaghan and Company, 1881, p 270; p.153: Judith Ewell, Venezuela and the United States: From Monroe's Hemisphere to Petroleum's Empire. Athens: University of Georgia Press, 1996. p 46; Judith Ewell, Venezuela and the United States: From Monroe's Hemisphere to Petroleum's Empire. Athens: University of Georgia Press, 1996. p 87-88; p.154: Judith Ewell, Venezuela and the United States: From Monroe's Hemisphere to Petroleum's Empire. Athens: University of Georgia Press, 1996. p 86; Josiah Strong, Our Country. NY: Baker & Taylor Co., 1885. p. 175; José Martí.Inside the Monster: Writings on the United States and American Imperialism, Volume 1. NY: Monthly Review Press, 1975, p. 340; p.155: La RevistaIlustrada. Nueva York, January 1, 1891; José Martí, Letter to Manuel Mercado, Dos Rios Camp, May 18, 1895, quoted in José Martí, Philip Sheldon Foner, Our America: writings on Latin America and the struggle for Cuban independence. Monthly Review Press, 1977; American History told by Contemporaries, Vol. 1V, NY: The Macmillan Company, 1901. p 570; p.156: A. Pérez Jr., The War of 1898: The United States and Cuba in History and Historiography. University of North Carolina Press, 1998.p 12; Louis A. Pérez Jr., The War of 1898: The United States and Cuba in History and Historiography. University of North Carolina Press, 1998.p 14; p.157: Rubén Darío. Cantos de vida y esperanza: Los cisnes, y otrospoemas. F. Granada y Co., 1907. "A Roosevelt," 39; Drago doctrine, U.S. Department of State. Papers Relating to the Foreign Relations of the United States, Washington DC, General Printing Office, 1904; Theodore Roosevelt's Annual Message to Congress for 1904; House Records HR 58A- K2; Records of the U.S. House of Representatives; Record Group 233; Center for Legislative Archives; National Archives; p.158: Theodore Roosevelt, Autobiography, 1913, p. 198; George W. Crichfield, American Supremacy, Vol. II, NY: Brentano's, 1908. p 635, 638; William Howard Taft, 1912; p.160: Porfirio Díaz, quoted in Rodolfo Acuña, Anything But Mexican: Chicanos in Contemporary Los Angeles. Verso, 1996; p.161: James Bryce. South America: Observations and Impressions. NY: The Macmillan Company, 1914; p.166: El Proletario; p.170: José Batlle y Ordoñez, "Reivindicaciones Obreras" and "Asuntos Obreros" in Antonio M. Grompone, La ideología de Batlle, Seguido por Escritos de José Batlle y Ordoñez, 3rd edition (Montevideo, Uruguay: Editorial Arca, 1967), 118-122. Translated by Julie A. Charlip; p.171: José Batlle y Ordoñez, "Reivindicaciones Obreras" and "Asuntos Obreros" in Antonio M. Grompone, La ideología de Batlle, Seguido por Escritos de José Batlle y Ordoñez, 3rd edition (Montevideo, Uruguay: Editorial Arca, 1967), 118-122. Translated by Julie A. Charlip; The Law of Residence, 1902, Argentina Constitution and Citizenship Laws Handbook: Strategic Information and Basic Laws, International Business Publications, 2013.

Chapter 7 P.173: Octavio Paz. The labyrinth of solitude: life and thought in Mexico. Grove Press, 1962, p. 148; p.176: The Independent, Vol. LXVII, January-June, 1910, p 133; p.180: Porfirio Díaz, quoted in Peter Calvert, Revolution. Praeger, 1970; Based on http://users.erols.com/mwhite28/mexico.htm; p.181: John Womack Jr., Zapata and the Mexican Revolution. NY: Alfred A. Knopf, 1968. p 90; p.182: Michael J. Gonzales, The Mexican Revolution, 1910-1940, University of New Mexico Press, 2002, p. 82; p.183: Based on http://users.erols.com/mwhite28/mexico.htm; p.185: Based on http://users.erols.com/mwhite28/mexico2.htm; p.187: Based on http://users.erols.com/mwhite28/mexico2.htm; p.189: Based on http://users.erols.com/mwhite28/mexico2.htm; p.191: Article 27, quoted in Laura Randall, Reforming Mexico's Agrarian Reform. M.E. Sharpe, 1996; p.192: Michael J. Gonzales, The Mexican Revolution, 1910-1940, University of New Mexico Press, 2002, p. 192; p.193: Based on data in Adolfo Gilly, The Mexican Revolution (London: Verso Editions, 1983), 334; p.194: Gilbert Michael Joseph, Timothy J. Henderson, eds. The Mexico reader: history, culture, politics. Duke University Press, 2002. p 411; p.197: Michael J. Gonzales, The Mexican Revolution, 1910-1940, University of New Mexico Press, 2002, p. 250; Josephus Daniels, Shirt-Sleeve Diplomat, Volume 5, University of North Carolina Press, 1947, p 247; p.198: Michael J. Gonzales, The Mexican Revolution, 1910-1940, University of New Mexico Press, 2002, p.2.

Chapter 8 P.202: E. Bradford Burns, A History of Brazil. NY: Columbia University Press, 1993, p. 373; p.203: Franklin D. Roosevelt; p.204: Penny Lernoux. Cry of the people: United States involvement in the rise of fascism, torture, and murder and the persecution of the Catholic Church in Latin America. Doubleday, 1980. p 85; Brian Loveman, For la Patria: Politics and the Armed Forces in Latin America. Copyright 1999. Reprinted by permission of SR Books, now an imprint of Rowman & Littlefield Publishers,

Inc; p.205: Hernández Martínez, quoted in Ernesto Guevara, Brian Loveman, Thomas M. Davies, Guerrilla Warfare. Rowman & Littlefield, 1985; p.212: E. Bradford Burns, A History of Brazil. NY: Columbia University Press, 1993, p. 149; p.213: Stephen Schlesinger, Stephen Kinzer. Bitter Fruit: The Untold Story of the American Coup in Guatemala. Cambridge, Mass.: Harvard University, David Rockefeller Center for Latin American Studies, p 34; p.216: U.S. Department of State, Inter-American Series, Issue 48, 1954, p. 33; United States Congress. Congressional Record: Proceedings and Debates of the ... Congress, Volume 101, Part 3. U.S. Government Printing Office, 1955. p 2871; p.217: Los Angeles Times (July 5, 1984).

Chapter 9 P.220: George Jackson Eder, Inflation and Development in Latin America: A Case History of Inflation and Stabilization in Bolivia. Program in International Business, Graduate School of Business Administration, University of Michigan, 1968. p 87; p.223: Russell Humke Fitzgibbon, The Constitutions of the Americas: As of January 1, 1948. University of Chicago Press, 1948; p.225: Fidel Castro Speech at Colón Cemetery, Havana, April 16, 1961, quoted in Fidel Castro Reader. Ocean Press, 2007, p 193; Speech in Havana, June 30, 1961, http://lanic.utexas.edu/project/castro/db/1961/19610630.html; p.228: John F. Kennedy: "Address on the first Anniversary of the Alliance for Progress.," March 13, 1962. Online by Gerhard Peters and John T. Woolley, The American Presidency Project. http://www.presidency.ucsb.edu/ws/?pid=9100; p.228: President Juscelino Kubitschek (1956–1960); p.230: Carolina Maria de Jesus. Child of the dark: the diary of Carolina Maria de Jesus. New American Library, 1962. p 42; p.231: Republished with permission of SR Books, from Brian Loveman, For la Patria: Politics and the Armed Forces in Latin America. Copyright 1999; permission conveyed through Copyright Clearance Center, Inc; p.232: Che Guevara, "Socialism and Man in Cuba."First Published: March 12, 1965, under the title, "From Algiers, for Marcha, quoted in Ernesto Guevara, David Deutschmann, Che Guevara Reader: Writings on Politics & Revolution. Ocean Press, 2003; p.233: Compiled from information in Liza Gross, Handbook of Leftist Guerrilla Groups in Latin America and the Caribbean (Boulder: Westview Press, 1995); p.237: Salvadore Allende, quoted in Modern Chile: 1970 – 1989; a Critical History, by Mark Falcoff, Transaction Publishers, 1989, 340 pages; p.238: Secretary of State Henry Kissinger, Meeting of the "40 Committee" on covert action in Chile (27 June 1970); p.238: Richard Nixon; Author translation; p.239: United States. National Bipartisan Commission on Central America. Report of the National Bipartisan Commission on Central America, Volume 1, 1984. p 521; p.241: The Pope Speaks, Volumes 13-14, Our Sunday Visitor, Incorporated, 1968, p 231; p.242: "The Nicaraguan Peasantry Gives New Direction to Agrarian Reform." Revista Envio, No 51, September 1985. http://www.envio.org.ni/articulo/3409.

Chapter 10 P.246: Patricia Adams, Odious Debts: Loose Lending, Corruption and the Third World's Environmental Legacy, Ch. 10. Probe International, 1991. http://www.eprf.ca/probeint/odiousdebts/odiousdebts/chapter10.html; p.249: Republished with permission of SR Books, from Brian Loveman, For la Patria: Politics and the Armed Forces in Latin America. Copyright 1999; permission conveyed through Copyright Clearance Center, Inc.; p.250: Los Angeles Times, July 21, 1974; p.251: Luis Moreno Ocampo. "Beyond Punishment: Justice in the Wake of Massive Crimes in Argentina," Journal of International Affairs 52, no 2, (1999) https://www.questia.com/read/1G1-62686122/beyond-punishment-justice-in-the-wake-of-massive; p.253: The Los Angeles Times (February 22, 1983); NACLA Report on the Americas, Volume 16, p 42; Michael G. Mason, Mike Mason. Development and Disorder: A History of the Third World Since 1945. University Press of New England, 1997. p 80; Philip O'Brien, quoted in Michael G. Mason, Mike Mason. Development and Disorder: A History of the Third World Since 1945. University Press of New England, 1997. p 80; p.254: Richard A. Howard and Michael E. Kennedy, "Salvadoran Suffering Goes On—There and Here," Los Angeles Times, December 25, 1986 | http://articles.latimes.com/1986-12-25/local/me-351_1_pastoral-visit; p.256: William C. Thiesenhusen. Searching for Agrarian Reform in Latin America. Unwin Hyman, 1989, 478; p.260: Oscar Arnulfo Romero to U.S. President Jimmy Carter, quoted in Renny Golden, "Oscar Romero: Bishop of the Poor ", US Catholic; "p.261: Radio address, El Salvador, March 24, 1980, quoted in Renny Golden, "Oscar Romero: Bishop of the Poor ", US Catholic; p.262: Sects or new religious movements: pastoral challenge. Vatican Secretariat for Promoting Christ. Pub. and Promotion Services, United States Catholic Conference, 1986, p 17; p.264: Gilbert M. Joseph, Timothy J. Henderson, eds. The Mexico Reader: History, Culture, Politics. Duke University Press, 2002. p 580; p.267: James F. Petras,

Fernando Ignacio Leiva, Henry Veltmeyer. Democracy and Poverty in Chile: The Limits to Electoral Politics. Westview Press, 1994, p 46.

Chapter 11 P.271: Noticias Aliadas/Latinoamerica Press, Vol 28, no 31, 8/29/96; Senate Armed Services Committee hearing reports, 104th Congress, March 19, 1996. Washington, DC: US Government Printing Office, 1996; p.272: Bruce Arnold, U.S. Congressional Budget Office, The Pros and Cons of Pursuing Free-Trade Agreements; p.273: Report About Events of January 1, 1994, Documents of the New Mexican Revolution; Nicholas p. Higgins, Understanding the Chiapas Rebellion: Modernist Visions and the Invisible Indian. University of Texas Press, 2004. p 159; p.276: Jeff Faux, NAFTA at 10, Feb. 9. 2004, Economic Policy Institute http://www.epi.org/publication/webfeatures_viewpoints_nafta_legacy_at10/; p.280: Roberta Rice, The New Politics of Protest: Indigenous Mobilization in Latin America's Neoliberal Era. University of Arizona Press, 2012. p 55; p.284: Ted Galen Carpenter, Bad Neighbor Policy: Washington's Futile War on Drugs in Latin America. Palgrave Macmillan, 2003. p 46.

Chapter 12 P.287: Hugo Chávez (1999), quoted in Laura Mixon, Use of the Authorizing Figure, Authoritarian Charisma, and National Myth in the Discourse of Hugo Chavez: Toward a Critical Model of Rhetorical Analysis for Political Discourse. ProQuest, 2009; p.290: Naomi Klein, The Shock Doctrine: The Rise of Disaster Capitalism. NY: Metropolitan Books/Henry Holt, 2007. p. 457; p.294: Maxwell A. Cameron, "Latin America's Left Turns: Beyond Good and Bad." Presented in a panel on 'The Latin American Left: Social Actors, Political Parties, and Development Strategies,' the Canadian Association of Latin American and Caribbean Studies, 6 June 2008, Vancouver, BC. p 11 http://www.politics.ubc.ca/fileadmin/user_upload/poli_sci/Faculty/cameron/Cameron_TWQ.pdf; p.295: Evo Morales, quoted in Jean Friedman-Rudovsky, "Bolivia's Long March Against Evo Morales: An Indigenous Protest", Time, 10/17/11, http://content.time.com/time/world/article/0,8599,2097142,00.html; Tom Bawden, "The world has failed us: Ecuador ditches plan to save Amazon from oil drilling." The Independent, June 11, 2015. http://www.independent.co.uk/news/world/americas/the-world-has-failed-us-ecuador-ditches-plan-to-save-amazon-from-oil-drilling-8771506.html; p.299: Thelma Mejía, "HONDURAS: Joining ALBA 'A Step Towards the Centre-Left,' Says President," Inter Press Service News Agency, June 11, 2015. http://www.ipsnews.net/2008/08/honduras-joining-alba-lsquoa-step-towards-the-centre-leftrsquo-says-president/; p.300: Fernando Henrique Cardoso, César Gaviria and Ernesto Zedillo, "The War on Drugs Is a Failure," Commentary, The Wall Street Journal, Feb. 23, 2009. http://www.wsj.com/articles/SB123535114271444981; Latin American Commission on Drugs and and Democracy, Drugs and Democracy: Toward a Paradigm Shift. p 3 http://www.drogasedemocracia.org/Arquivos/declaracao_ingles_site.pdf; p.301: Fernando Henrique Cardoso, quoted in Somini Sengupta, "Coalition Urges Nations to Decriminalize Drugs and Drug Use". New York Times, September, 8, 2014; p.302: Evo Morales, quoted in "The Brics Bank Of Development", Global Learning, 2014; p.303: World Bank, Economic Mobility and the Rise of the Latin American Middle Class. Washington, DC: International Bank for Reconstruction and Development, 2013, 1-2; United Nations Development Programme (UNDP), Human Development Report 2013, The Rise of the South: Human Progress in a Diverse World, p 5; p.304: Data from www.indexmundi.com, accessed October 2014; p.305: Arturo Escobar, Encountering Development: The Making and Unmaking of the Third World. Princeton University Press, 1995; p.306: Martin Hopenhayn, No Apocalypse, No Integration: Modernism and Postmodernism in Latin America. Duke University Press, 2001. p 90; Nestor García Canclini, Hybrid Cultures: Strategies For Entering And Leaving Modernity. University of Minnesota Press, 1995. p. 264; Nestor García Canclini. Consumers and Citizens: Globalization and Multicultural Conflicts. University of Minnesota Press, 2001. p. 26; p.307: Consumers and Citizens: Globalization and Multicultural Conflicts, by García Canclini, University of Minnesota Press, 2001, 200 pages; Walter D. Mignolo, "Introduction: Coloniality of power and de-colonial thinking," in Walter D. Mignolo and Arturo Escobar, eds. Globalization and the Decolonial Option. Routledge, 2013. pp 17-18; p.308: Walter D. Mignolo and Arturo Escobar, eds. Globalization and the Decolonial Option. Routledge, 2013; Oswaldo de Rivero, The Myth of Development: The Non-Viable Economies of the 21st Century. London: Zed Books, 2001. p. 186; p.309: Thomas C. Fox, "Pope Francis speaks again on world poverty," National Catholic Reporter. June 7, 2013. http://ncronline.org/blogs/ncr-today/pope-francis-speaks-again-world-poverty; Evangelii Guadium, 202, http://w2.vatican.va/content/francesco/en/apost_exhortations/documents/papa-francesco_esortazione-ap_20131124_evangelii-gaudium.html.

Photo Credits:

Chapter 1 P.05: Source: Frederick J. Pohl. Amerigo Vespucci: Pilot Major. London: Frank Cass and Co. Ltd., 1944 (reprinted 1966, digital printing 2005), p. 132; p.09: Library of Congress Prints and Photographs Division[LC-USZ62-119651]; p.11: Julie A. Charlip p.13: Library of Congress Prints and Photographs Division[LC-USZ62-55139].

Chapter 2 P.48: Library of Congress Prints and Photographs Division[LC-DIG-det-4a03444]]; p.51: Album/Oronoz/Newscom; p.53: Ingram Publishing/Newscom.

Chapter 3 P.69: Library of Congress Prints and Photographs Division[LC-USZ62-51627]; p.75: Courtesy of the Library of Congress.

Chapter 4 P.91: Interfoto/Alamy; p.98: Library of Congress Prints and Photographs Division[LC-USZ62-53038]; p.113: Library of Congress Prints and Photographs Division[LC-USZ62-53041]; Library of Congress Prints and Photographs Division[LC-USZ62-90422].

Chapter 5 P.127: Library of Congress Prints and Photographs Division[LC-USZ62-83018]; p.136: Library of Congress Prints and Photographs Division[LC-USZ62-92714]; p.140: Library of Congress Prints and Photographs Division[LC-USZ62-115486].

Chapter 6 P.156: Library of Congress Prints and Photographs Division [PAN FOR GEOG - Cuba no. 2 (F size)]; p.159: Library of Congress Prints and Photographs Division[LC-USZ62-1733]; p.163: Library of Congress Prints and Photographs Division[LC-USZ62-26047].

Chapter 7 P.175: Library of Congress Prints and Photographs Division[LC-USZ62-100275]; p.179: Library of Congress Prints and Photographs Division[LC-DIG-ggbain-15609]; p.181: Library of Congress Prints and Photographs Division[LC-DIG-ggbain-14906]; p.184: Library of Congress Prints and Photographs Division[LC-USZ62-36672].

Chapter 8 P.203: Library of Congress Prints and Photographs Division[LC-USZ62-59103]; p.207: AP Images; p.216: Library of Congress Prints and Photographs Division[LC-USZ62-126866].

Chapter 9 P.222: Courtesy of the Library of Congress; p.232: Itar-Tass Photo Agency/Alamy; p.240: Julie Charlip; p.243: Julie Charlip.

Chapter 10 P.252: Library of Congress Prints and Photographs Division[LC-USZ62-130898]; p.258: Bettmann/Corbis; p.264: Don Rypka/Bettmann/Corbis.

Chapter 11 P.275: Scott Sady/AP Images.

Chapter 12 P.291: Diego Giudice/KRT/Newscom.

Color Inserts Granger, NYC — All rights reserved; Schalkwijk/Art Resource, NY; Schalkwijk/Art Resource, NY; SuperStock; Dea /G. Dagli Orti/ De Agostini/Getty Images; Christie's Images/SuperStock; Mechika/Alamy/© 2015 Artists Rights Society (ARS), New York / AUTVIS, Sao Paulo; The Artchives/Alamy/© 2015 Banco de México Diego Rivera Frida Kahlo Museums Trust, Mexico, D.F. / Artists Rights Society (ARS), New York; Mexico City, 1942 (tempera on masonite), O'Gorman, Juan (1905-82)/Museo de Arte Moderno, Mexico City, Mexico/Index/Bridgeman Images; Emmanuel Dunand/AFP/Getty Images; Wendy Connett/Alamy; MARTIN BERNETTI/Getty Images; Will & Deni McIntyre/The Image Bank/Getty Images; Fotos & Photos/Photolibrary/Getty Images.

Cover Juliasnegi/Fotolia